# Disasters and Accidents in Mann

Springer
London
Berlin
Heidelberg
New York
Barcelona
Hong Kong
Milan
Paris
Santa Clara
Singapore
Tokyo

David J. Shayler

# Disasters and Accidents in Manned Spaceflight

 Springer

Published in association with
 Praxis Publishing
Chichester, UK

David J. Shayler
Astronautical Historian
Astro Info Service
Halesowen
West Midlands
UK

SPRINGER–PRAXIS BOOKS IN ASTRONOMY AND SPACE SCIENCES
SUBJECT *ADVISORY EDITOR*: John Mason B.Sc., Ph.D.

ISBN 1-85233-225-5 Springer-Verlag Berlin Heidelberg New York

British Library Cataloguing-in-Publication Data
Shayler, David J.
  Disasters and accidents in manned spaceflight. –
  (Springer-Praxis series in astronomy and space sciences)
  1. Manned space flight – Accidents   2. Manned space flight –
  Case studies – Accidents
  I. Title
  363.1'24

  ISBN 1852332255

Library of Congress Cataloging-in-Publication Data
Shayler, David J.
  Disasters and accidents in manned spaceflight / David J. Shayler.
    p. cm. – (Springer-Praxis books in astronomy and space sciences)
  Includes bibliographical references and index.
  ISBN 1-85233-225-5 (alk. paper)
  1. Space vehicle accidents.   I. Title. II. Series.

  TL867.S53 2000
  363.12'4–dc21                                                          99-045244

Copy editing and graphics processing: R.A. Marriott
Cover design: Jim Wilkie
Typesetting: BookEns Ltd, Royston, Herts., UK

Printed on acid-free paper supplied by Precision Publishing Papers Ltd, UK

This book is dedicated to the men and women of the global human spaceflight programme, past, present and future, to the aeronautical research pilots, astronauts and cosmonauts who risk their lives in the interest of peaceful exploration of space, and especially to the memory of the crews of Apollo 1, Soyuz 1, Soyuz 11 and *Challenger* 51-L

*'If we die, we want people to accept it. We're in a risky business, and we hope that if anything happens to us it will not delay the programme. The conquest of space is worth the risk of life.'*

Gus Grissom, Commander of Apollo 1

*'Houston, we've had a problem.'*

Jim Lovell, Commander of Apollo 13, 13 April 1970

*'The crew of the Space Shuttle* Challenger *honoured us by the manner in which they lived their lives. We will never forget them, nor the last time we saw them, this morning, as they prepared for their journey and waved goodbye, and 'slipped the surly bonds of Earth to touch the face of God'.'*

President Ronald Reagan, 28 January 1986

# Table of contents

## LAUNCH TO SPACE

## SURVIVAL IN SPACE

# Foreword

In the fields of aeronautics and astronautics, when disasters occur they strike like lightning. There is seldom much warning, and the crew is instantly faced with very few options – and sometimes no options at all. The environment in which the pilot or the astronaut plies his or her trade is a hostile one, so when the calamity occurs there is rarely enough time for a considered response. One must train for the eventuality of disaster and be constantly on the alert – ready to react correctly and without hesitation. Training, then, is clearly the key to enhancing survivability.

There is an old adage in the world of flight: 'Flying consists of hours and hours of sheer boredom interspersed with moments of stark terror.' However, this is a somewhat misleading statement because it overlooks the magnificent vistas and sense of overview that addicts those of us who fly and keeps us coming back for more. A better word for 'boredom' might be 'routine'. One can easily depend upon routine to run the ship, and begin to lose touch with what is really going on. That is when that ever-present danger around the corner is most likely to have its way.

As I look back on my years as an aviator in the Marine Corps, as a NASA astronaut, and later as a private pilot, I am aware that a very large part of my training has involved readiness for the unexpected failure. They teach you to 'stay ahead of the aircraft'; that is, to be constantly considering your next move in the eventuality of something going wrong. Recently my pace has become more leisurely. Events moved more slowly in my private Mooney Ranger aircraft, and the number of calamities were far less than when I was flying the F-8 off the deck of an aircraft carrier, the T-38 in congested traffic areas, or the Apollo CSM on the Skylab 4 mission. But the drill remained the same – practise, practise and look ahead.

But as I implied earlier, there are some dangers for which no amount of preparation can save the crew. Certain kinds of failures leave no option because of the way in which equipment is designed or because the flight activity is just too close to the world of the unknown. Those of us who fly aeroplanes and spacecraft have had to come to grips with the high probability that our profession will be the death of us. Our families, too, must deal with it, and if the risk is too much to accept then we must find other work. That decision is rational and honourable, and the sensible professional cannot turn his or her back on it. It must be frequently reconsidered

David Shayler's treatment of *Disasters and Accidents in Manned Spaceflight* clearly brings to consciousness that flying in space is not easy and not without peril. If through his efforts he conveys some insight into the kinds of people who fly those machines and into the thousands of people who support their efforts, then his work has been well done.

Colonel Gerald P. Carr, USMC (Rtd.)
Commander, Skylab 4

Colonel Carr was a United States Marine Corps fighter pilot when he was selected to train for the Apollo lunar programme in April 1966. He was support crew and CapCom on Apollo 8 and 12, and worked on the development of the Lunar Module and Lunar Roving Vehicle before assignment as LM Pilot to the crew scheduled to fly Apollo 19 to the Moon. When the mission was cancelled due to budget cuts he transferred to Skylab and commanded the third and final manned mission flown between November 1973 and February 1974, setting a new world endurance record of 84 days. After working on Space Shuttle development, Colonel Carr retired from NASA in 1977 and entered the aerospace business. He later formed CAMUS to expand the resource of human potential, and became a consultant on human space engineering and habitation, often working with NASA and major aerospace companies in manned spacecraft development issues, including the International Space Station. He is now retired, and lives in rural Arkansas with his wife Pat Musick, who is a noted artist, therapist and consultant.

# Author's preface

Space has been called the New Frontier – a New Ocean for exploration, discovery and adventure, as well as scientific and engineering gain. 'The future of mankind', it has been said 'is written in the stars.' To explore and exploit outer space to the full, man himself must venture out there and witness firsthand the marvels and achievements to be gained from a yet untapped Universe. The modern day pioneering heroes of this new territory of mankind are the astronauts and cosmonauts who for almost 40 years have ventured beyond the fragile protection of our atmosphere, barely touching the shoreline at the very edge of the ocean of space.

To the general public, routine access to space is generally considered common-place and pretty routine after four decades of achieving it. Public understanding of what each mission is trying to accomplish normally suffers from the lack of adequate promotion of the benefits and gains from exploring space, some of which will not see immediate returns. However, rarely neglected are the setbacks of human spaceflight and the very few fatalities that have occurred. Presence of danger in space has always been known, if only because of the harsh unforgiving environment. Protection for spacecraft – and, of course, their human cargo – was always foremost in the minds of designers and mission planners, if not in those of the politicians and accountants.

Information on the developments and activities of each new flight by a human crew into space is normally available in one form or another. Therefore all setbacks and problems are logged along with the successes. With the real-time openness of the American programme this was dramatically witnessed during the Apollo 13 and STS-25 incidents. With the Soviet programme, however, the availability of real-time facts was not originally as forthcoming as they were in the West, and therefore the reporting of major incidents in the programme was not always as clear as with the Americans. The dramatic changes in the former Soviet Union over the past decade, and the move towards a more integrated international space programme, has seen a more open nature concerning the Russian space programme.

Failures do of course occur, and this book is intended to review the major spaceflight failures that have occurred in the human spaceflight programme over the past 40 years. Almost every mission has in one way or another recorded one or more

During the Apollo 13 mission, astronaut Jerry Carr explains the revised flight plan to Mary Haise – wife of Lunar Module Pilot Fred W. Haise – in the viewing room at Mission Control Center, Houston, Texas.

incidents of equipment failure, crew illness, or an inability to complete all of the pre-mission goals. These are not all recorded here. Instead there are presented nine major incidents that have resulted in either the loss of the crew or a mission being aborted due to life-threatening circumstances. In addition, each section is highlighted by several incidents to illustrate how close some missions have come to being a major accident or disaster

Space exploration does not start with a launch to Earth orbit but with preparations by the team of people selected to undertake the mission. They build upon the experiences of past missions and the pioneering developments of people who put their lives at risk exploring the upper reaches of the atmosphere in stratospheric balloons, rocket research aircraft, rocket-powered acceleration sledges and high altitude parachute descents. The dangers that those pioneers faced in the quest of the stratosphere to provide the answers to countless questions before attempting actual spaceflight are reviewed in the first section of the book. This first section also reviews the requirements to sustain human life away from the Earth and the added problems that arise from living and working in this new environment.

Every human space mission can be broken down into four main phases: training, launch, in-flight and recovery. Each brings its own unique dangers and risks and it is these phases that comprise the subsequent four sections of the book

Accidents in space have been the subject of space fiction for decades (an example being Martin Caidin's best-seller *Marooned*, published in 1964 and produced as a motion picture in 1969). When an accident in space actually happens, the skills of the astronaut crew and ground controllers in rescuing the situation are worthy of a 'best-seller'. Indeed, the Apollo 13 mission became a world-wide film hit in 1995, 25 years after making headlines itself.

Each main section of this book is concluded with a review of what was learnt from the past incidents discussed in that field. The sixth section looks forward to the immediate future of human spaceflight and reviews the dangers faced by space explorers of the next millennium, with the creation of the International Space Station, return to the Moon, and human exploration of Mars and beyond.

The text aims to present a balance between numerous incidents, serious accidents and fatal disasters, to highlight that every flight away from Earth has always been, and will remain, extremely dangerous. After almost 40 years of human exploration of space, all human life lost (at the time of writing) has been *within the confines of the atmosphere*. The chilling words 'Lost In Space' have yet to be added to a space explorers obituary or accident inquest report.

As manned spaceflight increases over the coming years into the next century, so the spectre of such mishaps will continue. 'If we die, we want people to accept it ... and not delay the programme', said Apollo 1 Commander Gus Grissom shortly before he lost his life in the 1967 pad fire. Riding on a rocket and flying in a spacecraft are, and always will be, dangerous modes of transport which will continue to claim the lives of those who dare to explore the depths of space. If the astronauts and cosmonauts are prepared to face the dangers of space 'for all mankind', are we prepared to witness and accept the setbacks as well as the triumphs, and let them do it again?

In compiling this book I have aimed at presenting a balanced account of the tragic accidents and near-miss incidents covering a period of 70 years. With almost 500 space explorers and more than 20 manned missions to Earth orbit in the record books, it is remarkable that there have been so few major accidents. The very nature of space is unforgiving and hazardous to human life, and is always treated as such in any preparation for venturing there.

These dangers are never ignored by the men and women who make the trip 'up there', and it is to those pioneering space voyagers of the past, present and future that this book is dedicated – in particular, to the memory of those who gave their lives in the pursuit of the peaceful exploration of space.

David J. Shayler
January 2000

# Acknowledgements

One of my earliest recollections of following human space exploration is of a tribute to the loss of the Apollo 1 astronauts during a school morning service on Monday, 30 January 1967. From this memory developed, many years later, the idea for this book. After following the developments of the successes and setbacks of humans in space for nearly 20 years, the loss of *Challenger* in 1986 resulted in the compilation of a detailed account of each major accident in the history of human space exploration.

In the decade it has taken to compile this book I have consulted and received helpful advice from numerous individuals and organisations, and I am indebted to all of them for assistance and direction in the compilation of the data and facts.

The help of the various NASA centres and staff in history, public affairs and flight operations over the years has been immense. Thanks are extended to: NASA Headquarters, Washington DC; History Office, Washington DC (Lee Saegesser and Roger Launius): NASA Johnson Space Center, Houston, Texas; Public and Media Affairs Office (Barbara Schwartz, James Hartsfield, Jeff Carr and Eileen Hawley); Still Photo Library (Mike Gentry, Lisa Vasquez, Debbie Dodds, Jody Russell and Mary Wilkinson); Audio Library (Diana Ormsbee and Pete Nubile); History Office (Janet Kovacevich, Joey Pellerin, Dave Portree and Glen Swanson): NASA Kennedy Space Center, Florida; Public Affairs (Kay Grinter), Still Photo Library (Margaret Persinger); History Office (Ken Nail and Elaine Liston): Dryden Flight Research Center, Edwards AFB, California; History Office (Dill Hunley): Rice University, Fondren Library (Joan Ferry and Lois Morris): and the history and public affairs office departments of the USAF, USN, USMC for details and illustrations of military programmes and incidents in the stratospheric, balloon and rocket research programmes.

Special thanks are extended to Colonel Jerry Carr for the Foreword, and to David Harland and Rex and Lynn Hall for their comments, suggestions and guidance through the manuscript stages, and Andrew Farrow in the early stages of compiling the text. Thanks are also due to my brother Mike Shayler, who devoted many hours using his computer wizardry in converting the original text to the finished draft.

Further appreciation and special thanks are due to Ms Lovisolo, Grumman History Office, for information on Grumman support of the Apollo 13 accident; to

Air Commodore Colin Foale, RAF (Retd.), for permission to use extracts from family e-mails from the 1997 Mir resident stay of his son, NASA astronaut Mike Foale. In addition, the two articles by Asif A. Siddiqi on Soyuz 1 (Soyuz 1 Revisited: From Myth to Reality, *Quest*, Vol. 6, No. 3) and Soyuz 11 (Triumph and Tragedy of Salyut 1, *Quest*, Vol. 5, No. 3) were of great assistance.

I am also indebted to fellow space researchers around the world. USA: Mike Cassutt, Curtis Peebles, Nick Johnson, Jim Oberg and Iva 'Scotty' Scott formerly of NASA JSC Public Affairs Office; Bill Thornton (former NASA astronaut); Staff of the Department of Space History, National Air and Space Museum, Washington DC. UK: Neville Kidger, Tim Furniss, Gordon Hooper, Phil Clark, Andy Salmon, Mark Wade and Anders Hansson. Europe: Bert Vis, the late Anne van den Berg (Holland), Bart Hendrickx (Belgium) and Brian Harvey (Ireland). Russia: Colonel V. Tolkov, Head of the Russian Air Force Museum, Moscow; the late Vladimir Molchanov, the Novosti/Tass Press Agencies (through 1991) for Soviet information and illustrations, and the Videokosmos Press Agency (since 1991)

Unless otherwise stated, all photographs are courtesy NASA for American subjects, or Novosti from the files of Astro Info Service for Russian subjects, and are used by permission. Extracts from *Waystation to the Stars* (Headline, 1999) by Colin Foale, are used by permission of the author and the publisher.

I am also grateful to general former space explorers for detailing aspects of their experiences. Interviews were conducted with Story Musgrave, August 1988, Houston, Texas; Gene Cernan, August 1988 and August 1989, Houston, Texas; Karl Henize, August 1988, August 1989 and July 1991, Houston, Texas; Yuri Romanenko, July 1989, London; Vance Brand, August 1989, Houston, Texas; Deke Slayton, July 1991, Houston, Texas; Gregori Grechko, April 1994, Manchester, UK; and Mike Coats, September 1994, Houston, Texas.

Finally, sincere thanks are due to Bob Marriott, Project Editor, and to Clive Horwood, Chairman of Praxis, whose enthusiasm and encouragement for the project spilled over to the author to create a much more comprehensive account, than first envisaged, of accidents and disasters in human space exploration.

# List of illustrations and tables

## TRAINING FOR SPACE

## LAUNCH TO SPACE

## Soyuz 11 decompression, 1971

## THE FUTURE IN SPACE

### Overview

### Tables

# Prologue

*Challenger* Space Shuttle flight deck
Launch Pad 39B, Kennedy Space Center, Florida, USA
28 January 1986

It was certainly a cold morning outside the window of *Challenger* as Commander Dick Scobee completed his routine checks of the controls in front of him with his pilot Mike Smith seated to his right. Behind and between them mission specialist Judy Resnik acted as Flight Engineer and helped in their preparations for ascent, planned for 11.30 am that morning.

Scobee and his crew of six were to spend a week in Earth orbit deploying a NASA communications satellite, tracking Halley's comet and performing a programme of experiments in the microgravity fun of space. *Challenger*'s tenth mission was also the 25th of the programme, but was the first flight into space for Smith, payload specialist Greg Jarvis – an engineer from Hughes Aircraft Corporation – and schoolteacher Christa McAuliffe. For Scobee, Resnik and fellow mission specialists Ellison Onizuka and Ron McNair this would be their second flight into space. For Scobee and McNair it would also be their second flight on *Challenger*.

They had all noted just how cold it was from the ice that hung from the gantry as they clambered into the orbiter. Any spaceflight crew always tries to feel confident that, despite previous launch delays, each time they climb into their spacecraft, this is the day they will launch. Scobee's crew had suffered their own delays for this mission for a variety of reasons. Weather was always a factor in any launch into space, and perhaps of even more importance in the Shuttle programme than in the Apollo days, as the orbiter also had to have good weather for landing at any of several emergency landing strips around the world. NASA had never launched in really bad weather since November 1969 when Apollo 12 had been struck by a bolt of lightning during its launch. That launch had occurred partly because President Nixon was on hand to witness in person the second manned lunar landing mission departure from Earth.

Several Shuttle flights had ignited engines before premature shut-down due to technical problems which ultimately scrubbed the launch. The previous mission, STS

61-C, was dubbed 'Mission Impossible' after it took seven attempts to get off the ground and almost as many to get back on it again. Every Shuttle is programmed for several possible abort modes, though none had been flown until the 19th flight in the summer of 1985, when an Abort To Orbit (ATO) mode was used after a main engine failed five minutes after launch. Despite this, the mission was flown successfully, and great confidence in the system and contingency plans were generated.

Scobee and his crew could reflect that at least they were at the end of the long and complex training program, and as he, Smith and Resnik completed the last procedures prior to launch, this was at last for real, and they were ready for the mission. Training had gone well, despite the usual small problems and several postponements of the launch date. Scobee knew that for any other future space explorers around the world, training for space would never be easy, and for some, just training for a flight would be the closest they would get towards an actual flight into space. The previous day had brought back memories of this. On 27 January 1967, nineteen years and a day before Shuttle 51L would be launched, three American astronauts perished in a pad fire during their final stages of training for the first manned Apollo mission, then planned for a February 1967 launch.

In front of Scobee and his crew was perhaps the most difficult and potentially dangerous phase of any spaceflight – the launch to orbit. They had some throttle control on the three Shuttle main engines on the tail of the orbiter, but the solid rocket boosters were uncontrollable. Once they ignited they burnt for the duration of their fuel, and the astronauts rode with them. Launch was never easy. The Soviets had suffered two serious launch aborts in 1975 and 1983, with the cosmonauts surviving because of in-built safety procedures. The Americans had come close in 1966 with the non-launch of a Gemini mission, the lightning strike on Apollo 12 and several Shuttle pad aborts. The 19th Shuttle mission suffered a lost engine in the ascent, and reverted to an Abort To Orbit (ATO) scenario to save the mission. This was the closest 'near-miss' on an American flight to date. Scobee's flight, however, was the 25th flight of the programme, and the system had been 'operational' since mission 5 – so what could go wrong?

Once in space the crew could relax and complete their programme of activities. But space itself was potentially dangerous. In March 1966, Gemini 8 suffered a stuck thruster and the crew had to wrestle with the controls to affect an emergency landing in the Pacific Ocean. April 1970 saw perhaps the most publicised space accident with the aborted flight of the Apollo 13 Moon mission. The skill of the astronauts, flight controllers and contractors, and the fact that the LM was still attached, added to the luck that returned the three astronauts.

Indeed, coming home also presented its own unique problems. Two other Soviet crews had lost their lives during entry and landing accidents in 1967 and 1971, and the Americans had their own near-misses. Scobee had trained for landing the Shuttle on a runway, and flying the orbiter back like an aircraft rather than splashing down as in the Apollo days. Although more desirable than wet feet, even a runway landing from orbital altitudes was never that routine. Several rocket research aircraft and lifting bodies – forerunners of Scobee's *Challenger* – had experienced emergency landings in their development years. They were certainly capable of performing a

Space Shuttle *Challenger* launches from Pad 39B at the Kennedy Space Center, Florida, 11.38 am EST, 28 January 1986. From this camera angle all appears normal as the crew begin their fateful journey.

smooth glide landing, but with no engines they had only one chance, first time, with no opportunity to perform an airline-type 'missed approach' to try again.

As career NASA astronauts they were well aware that although their mission was billed as 'routine' – with only one satellite deployment, some small mid-deck experiments and no planned spacewalks – it was far from being free of risk. For the payload specialists, chosen for just one flight, the thrill of actually being selected to fly in space, and the intense training programme, could easily act as a diversion from the risky adventure that they were about to undertake.

As *Challenger* finally lifted off, perhaps one or more of the veterans onboard sensed something was different about this flight; it could be due to the age of *Challenger*, or the conditions outside, or perhaps their imagination. But there was no time to think of it – they were off and had cleared the tower, performing the roll programme and heading through the period of maximum dynamic pressure on the vehicle. They were about to throttle up their engines, climbing ever higher towards the point where blue skies turn black.

The first seconds of the 25th Shuttle mission were ticking by as they all experienced the thrill of lift-off and accelerated over the Atlantic Ocean towards space. The thrill of launch, however, would not last long.

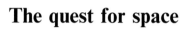

# The quest for space

# Overview

'Man must explore, and this is exploration at its greatest'

These were the first words uttered by Apollo 15 Commander Dave Scott as he set foot on the surface of the Moon in July 1971. At the height of the initial era of human exploration of another celestial body, it was an apt statement, coming as it did just ten years after humans first ventured into space. Across centuries of human history, exploration and discovery have been two key elements that have driven the species to progress from living in caves across the Earth during the Stone Age, to the exploration of the surface of the Moon during the Space Age.

## LAUNCH PAD TO KNOWLEDGE

Developments in science and technology and adventurous journeys of discovery have fuelled this desire to learn and acquire knowledge. But such knowledge has not been gained without cost. To earn this new knowledge and progress requires us to venture into the unknown, opening up new areas of risk and danger. Such dangers can serve as warnings and new learning in itself. At worst, the danger can lead to human deaths.

As the twentieth century dawned, so too began the desire to explore a new, challenging environment. The golden eras of land and sea exploration had all but exhausted the unknowns left of Earth. Human eyes turned, not for the first time, up to the skies above them and to the stars beyond. Science fiction writers had built upon tales of myth, magic and legend, creating new heroes who ventured from the Earth and on to strange new worlds. But it was not until the 1930s that the first steps in science fact would be taken, to begin human exploration of the upper reaches of our atmosphere – the stratosphere.

The skills and technology necessary to take human balloon flights to the very edge of space evolved from the exploits, and the failures, of early balloonists in the eighteenth and nineteenth centuries. Technological advancement led to the development of sealed, pressurised crew compartments, life support systems and the first pressure garments.

Piloting skills also developed. The early aviators that followed the Wright brothers evolved from the simple need to survive during World War I, to becoming the barnstorming and acrobatic stunt pilots of the 1920s and 1930s. This began the golden age of flight testing and cutting edge aviation technology. The men, and the very few women, who risked their lives in developing this new branch of technology, were the original test pilots.

With the evolution of the piston, jet and rocket-propelled aircraft, a new breed of test pilot emerged and gave birth to the legendary heroic, almost mystic, aura of the 'Right Stuff'. The holders of that title were those who risked their lives, pushing new aircraft to (and beyond) the limits of their operational envelope. Those who survived to tell the tale and report their findings would go and do the same again the next day and the day after that. But there is a long list of pilots who did not survive the day.

By the 1960s a new stage was available for these pilots to display their skills – space. From the ranks of the top jet and test pilots of two of the largest nations on Earth, a new breed of flyer and explorer emerged – the astronaut, or cosmonaut. With the assistance of the media, publicity and public adulation, these stellar voyagers became almost super-heroes. They were national and global pioneers in a new, dangerous environment beyond the very confines of Earth.

To many, it seemed that these new adventurers were a breed apart. They challenged adversity, and proved that pure skill and training could overcome almost any hurdle. They were calm, professional and determined in the pursuit of their personal Holy Grail – 'The Mission'.

But even these space explorers – highly skilled and motivated though they may have been – were still only human. They were trained by other humans to fly machines built by more humans. The whole process was built on the unknown, and would stand or fall on the simple fact that theory and practice often prove to be very different beasts.

Every time a human undertakes a journey that relies on technology – whether into space or down the road in a car – danger and risk accompanies them every step of the way. No matter how reliable, safe, tried and tested, there is no guarantee that every factor will behave exactly as theory suggests it should. But it is not just the technology that can fail; humans are also fallible. Adding the human element to a spaceflight increases the range of potential problems and dangers.

Without pushing the limits of both humans and machines, and testing ourselves against the unknown, no progress would ever be made. As with most affairs, preparation is paramount. For spaceflight, such preparation means endless hours in training, simulations and practice, and the evaluation of every possible contingency.

Any spaceflight begins, not in orbit around the Earth, or even on the launch pad, but with a mission training programme, building on the skills and knowledge gained from previous spaceflights and programmes, and developing a healthy awareness of the living and working conditions of a hazardous environment.

Through weeks, months or even years of practice, a crew learns every aspect of a mission, from the thrill of launch and the magic of spaceflight, to the relief of re-entry at the end of the flight. They also learn to understand and become familiar with the technology upon which their lives will depend. All this training provides them with a better chance of coping with problems and surviving potential disasters.

There is always the element of risk and the possibility of a simple mistake to consider; but without taking such risks we would never have evolved to take these first steps off our home planet. Crew training and experience eliminates, as far as possible, the potential for making errors, and increases the chance of completing a successful mission.

## THE RISK FACTOR

The machines built to explore the depths of space are one of the pinnacles of modern engineering. To develop vehicles capable of successfully supporting a human cargo, the designers and constructors have had to draw upon past experience, learning from both the triumphs and the tragedies of Earth-based programmes, in particular, the problems of structural failure. As described by Henry Petroski in 1985,[1] structures fail due to the limits of the component parts. Such failure is caused by one or more of the following: overloading, or under-strength materials; excessive motion, or component deterioration; or random hazard, such as fire, explosion or impact. Structural failure can also result from human error in design, construction and maintenance.

In 1962 Thomas McKaig reviewed building failures, and listed categories into which the reason for any failure could be classified:

- Ignorance in skills, experience and/or knowledge.
- Lack of precedent.
- Economy in budgets and/or maintenance.
- Lapses of concentration, carelessness and/or neglect.

A further study, by Scott Sagan[2] in 1993, reviewed safety in high technology fields. Sagan concluded that accidents with hazardous technologies (such as nuclear power or rocket fuels) could be prevented with careful preparation. He suggested the following:

- A good system of management and organisational structures should be in place, with the primary objective of safety.
- There should be duplication or overlapping of systems to provide a redundancy, or margin of safety.
- Important, decision making processes should be decentralised to increase the flexibility to respond to any surprises.
- A system of reliability, built upon previous experience, could help to enhance levels of safety.

---

1   *To Engineer is Human: The Role of Failure in Successful Design.* Henry Petroski, Macmillan, 1985.
2   *The Limits of Safety: Organisations, Accidents and Nuclear Weapons.* Scott D. Sagan, Princeton University Press, 1993.

- Development of training, simulations and continuous operations would create and maintain higher reliability levels in actual operations.
- Lessons learned from accidents – so-called 'trial and error' learning – could be supplemented by both anticipation of future mishaps as well as by the simulation of likely accidents.

Although these studies related to Earth-bound environments and technologies, they could also be applied to the preparation for, and execution of, human space exploration efforts. These procedures were designed to minimise the element of risk, defined in the dictionary as: 'A chance of danger or loss, a venture put into jeopardy.'

For spaceflight, this risk falls into five main areas:

- *Training* The preparation of a crew and their capacity to learn and understand, based upon their previous military or civilian, academic, scientific, or engineering careers, plus family and personal experiences. This includes medical and physiological evaluations of personal health, and the capabilities and responses of the crew-member to both normal and abnormal situations.
- *Launch* The acceptance of the skills and training of others, in the design, construction and flight operations of a high-technology machine. Riding a volatile launch system into a hostile environment, with the expectation that programmed sequences work as designed.
- *In-flight* Confidence in the structural integrity and systems support of a high technology machine operating in a hostile environment. The vehicle's capability to withstand extremes of heat and cold, pressure and vacuum, acceleration and deceleration, and radiation. Overcoming the additional hurdles of isolation, distance and stress.
- *Landing* Belief in the structural strength of hardware and supporting systems to withstand extreme variations of heat and structural loads. The trust in an adequate recovery system, support forces and contingency procedures, in the event of a variation from the normal flight plan.
- *Human error* The one area of risk, failure and accident that cannot be totally predicted. In the above studies, as well as in many others, it is suggested that a combination of adequate training and preparation can reduce the risk of human error, but it can never be totally ruled out.

In the early days of the programme – when the pioneering astronauts and cosmonauts first ventured beyond the protective cocoon of our atmosphere – experience had already been gained from high-altitude balloon flights and the rocket aircraft programmes. Sounding rockets and the first satellites had indicated that space was a hostile, unforgiving environment, but it was one that had been described and studied from Earth for decades.

Risk was one of the glamour elements of pioneering space flights, and a generally accepted price to pay to push the boundaries of knowledge; or, as one famous sci-fi TV series expresses it, 'To boldly go where no one has gone before.'

As missions became more frequent, and as more success was achieved, public appreciation of the risk element waned. Spaceflight became 'commonplace' very quickly, so that when a major accident happened it hit world headlines and once again reminded everyone of the risk of such hazardous journeys, designed to benefit all mankind.

For NASA, nothing failed like success. The apparent ease in reaching the Moon in just eight years, and then repeating the feat just four months later, followed by the 'routine' flights of the Space Shuttle, saw priorities shift from safety and pioneering engineering work to delivering immediate results under severe budget restrictions. In both the American and Soviet/Russian programmes, this pressure to succeed and deliver repeatedly led to serious consequences.

As with any venture, the more frequent and successful the operation, the more experience is gained, but with increasing complacency. This success and complacency shortens the odds against failure and a subsequent accident, and in any area of endeavour there is always a fine balance of total success and utter failure. This is certainly true of spaceflight. The hardware required, the unique challenges of the environment, and the dangers during return to Earth, are all vital areas to be considered when planning a trip to the heavens. One last, indefinable element features most prominently in all areas of human existence, not just in space exploration – luck.

## A HOSTILE ENVIRONMENT

If you believe that space explorers are vastly different from yourself, you could not be more wrong. In a way, we are all space explorers, travelling through the cosmos on spaceship Earth. But to explore outside the protective shield of our world we have to travel through the thinning layers of our atmosphere, taking a replica of Earth's environment with us. This is the main hurdle against the human exploration of space.

In the seventeenth century, travel in 'space' was thought to be not only possible, but also quite achievable. It was believed that our life-preserving atmosphere reached all the way to the surface of the Moon and, if this was the case, all that was needed to reach the Moon was the right mode of transport. Despite conflicting evidence from early mountaineers – that the higher you climb, the thinner the air becomes – it was long believed that access to space would come from the development of balloons. In part this became true, but not in the way it was at first thought.

During the seventeenth and eighteenth centuries, studies of the effects of gravity – using the newly developed mercury-filled barometer – revealed that the higher up a mountain the barometer was taken, the lower the mercury fell in the instrument. It was soon determined that the pressure of the air was much higher at sea level than at elevation. With this important information, early atmospheric scientists analysed the composition of the air at ever-higher ground elevations. It was discovered that the air that we all breathe is a mixture of gases that comprising nitrogen (approximately 78%) and oxygen (approximately 21%). A small percentage of argon (0.9%), a

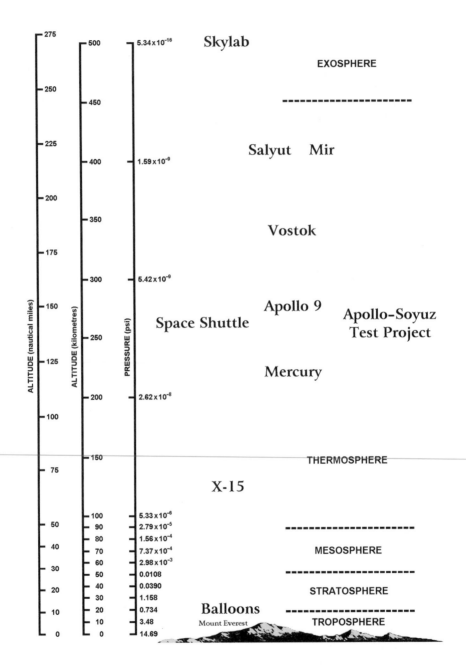

The layers of the atmosphere, showing pressure levels and heights at which vehicles usually operate.

fraction of carbon dioxide and traces of half a dozen other gases constitute the remaining percentage. It was many years before further research determined that these levels of gases remained the same up to around 50,000 feet. Pressure readings taken at sea level recorded the weight of air at 14.69 psi. The higher the altitude, the lower the psi measurement and amount of air molecules, making the air 'thinner'.

As a result of this research, atmospheric scientists gradually defined the layers of our atmosphere. From the surface of the Earth upward, these are: the troposphere (0–10 nautical miles – the life-sustaining region); the stratosphere (10–30 nautical miles – first layer to be conquered); the mesosphere (30–50 nautical miles); the thermosphere (50–250 nautical miles – where most orbiting spacecraft operate); and the exosphere (above 250 nautical miles).

### Air to breathe

The human body needs oxygen to survive. It cannot operate long without it, and the survival and evolution of the species has relied upon the correct amount of oxygen in the Earth's atmosphere at sea level, or just above it. It is only in the twentieth century that humans have begun to personally explore regions of our atmosphere above the troposphere, but it was soon discovered that to achieve this, and survive to tell the tale, an artificial means of supplying oxygen to our lungs would be one of the first important requirements. Essential to this was an understanding of how the human lung system delivers oxygen around the body. This observation led to the creation of suitable breathing apparatus to help divers penetrate the depths of the oceans and climbers to scale the highest mountains. From this, came the creation of life support systems to support high-altitude flights in balloons and aircraft, and this in turn led to the creation of fully pressurised cabins and personal pressure garments to help withstand the stresses and strains of exploration at altitude.

### Under pressure

Directly linked to the amount of life sustaining air supplied is the percentage of air pressure against the body at different altitudes. The effect of a gas mixture with 30% oxygen at an air pressure of 10 psi equates to a partial pressure level of 3.0 psi. It was soon discovered that at sea level and a little above, the human body is extremely comfortable living and working at a level of 3.08 psi.

It was also discovered that, as long as the psi of oxygen is greater than the psi of carbon dioxide inside the lungs, they will continue to absorb the oxygen and distribute it throughout the body via the blood system. Changes to this ratio can result in hypoxia (oxygen shortage), abaria (lack of pressure), hyperventilation (too much oxygen), nitrogen in the blood (more commonly called 'the bends') or carbon monoxide poisoning.

Clearly, therefore, it is vital to provide an air supply and an appropriate pressure level inside a spacecraft to allow the human crew to function in safety. Additionally, equipment would be needed to remove harmful carbon dioxide.

*Russian approach* From the start of their human spaceflight programme through to the Mir space station, Soviet/Russian spacecraft have operated with a cabin pressurised to sea level equivalent – a mixed-gas environment. This has brought additional weight penalties at launch, and the cosmonauts also need to pre-breathe pure oxygen, prior to EVA, to cleanse the nitrogen from their blood and reduce the risk of the bends developing during the space walk.

*American Approach* When NASA developed its first human-crewed spacecraft, it opted for the simpler design of a single-gas atmosphere (100% oxygen). This provided a safety margin against decompression sickness and removed the need to pre-breathe prior to an EVA. But this approach produced the added risk of fire. This system worked well for Mercury and Gemini, but the consequence of adopting such an atmosphere was tragically demonstrated in the Apollo 1 pad fire of January 1967 and the loss of three astronauts. The system was subsequently changed to a 64% oxygen and 36% nitrogen mixture, to prevent such a fire reoccurring.

*Compatibility* The different atmospheres and pressure levels used by the Soviets and the Americans brought an additional complication. If spacecraft from the two nations were required to dock and transfer crew, perhaps in a rescue situation, the incompatibility of the two systems would preclude this, making such a rescue unlikely, given the length of time it would take to equalise the atmospheres.

In the early 1970s this difficulty was the principal stumbling block for the Apollo–Soyuz Test Project. To obviate this, an intermediate docking module was proposed which would allow crews to transfer between both spacecraft, using the docking module as a kind of airlock. With the advent of the Space Shuttle, the United States had a spacecraft capable of operating with a sea level equivalent atmosphere, and compatibility with Soviet/Russian spacecraft was easier, as demonstrated by the successful series of Shuttle–Mir dockings in the late 1990s.

**The right temperature**

Temperature control is another consideration to be incorporated for the human body to perform normally. Providing the correct temperature, balanced with humidity control (the amount of water vapour in the air) is one of the most challenging aspects of design for life support systems. The human body attempts to maintain an internal temperature of 98.6° C and for the most important organs, such as the brain, there is little allowance for variation. The humidity, located in 'air pockets', is very much dependent on the levels of surrounding air pressure and temperature. The warmer the pocket, the higher the pressure surrounding it, holding in the water vapour. As the temperature drops, the pressure eases and the water vapour turns into clouds, converting these droplets into rain, sleet or snow.

To sustain a comfortable environment a balance of the temperature and humidity levels has to be created and maintained inside the spacecraft for the duration of the mission. This allows the human body to perform normally and maintain its vital organs, avoiding the risks of heat exhaustion, fatigue, hypothermia, or coma. These parameters have been breached in several missions over the years – most notably during the Apollo 13 mission (1970), the recovery of Soyuz 23 (1976) and the reactivation of Salyut 7 (1985) – while Shuttle humidity control has broken down several times, raising temperatures to 90–100° F. The environment must also allow other onboard systems to operate safely.

For a human being to survive and work in space – regardless of the duration of the mission – a very careful balance of factors must be considered. There must be a guaranteed and constantly monitored supply of atmosphere, pressure, temperature, humidity, food and liquid refreshment, clothing, hygiene and sanitation. Automated life support systems must operate totally independently from the Earth, faultlessly, for 24 hours a day, for the duration of the mission, and there must be sufficient supplies for each member of the crew and all their activities, as well as a margin for contingency.

## THE FINAL FRONTIER

### Microgravity

One of the main reasons for sending humans into space is to study the effects of reduced gravity on the human organism. This new environment has been called weightlessness, zero gravity and, more accurately, microgravity. A total absence of gravity is not possible, as even in deep space there will always be the gravitational influences of the Sun and planets, however minor. But in orbit, most of these effects are cancelled out, leaving the experience of 'weightlessness'.

On Earth, it is almost impossible to experience zero g, due, of course, to the Earth's own gravity (1 g). This is the basis from which other gravitation levels are measured. In order to overcome this influence and to provide the crews with some idea of what they would face in orbit, two primary methods of simulation were developed. One of these involves working underwater in a huge pool, wearing a weighted EVA suit to simulate floating around during a spacewalk, although the viscosity of the water renders this a far from perfect method. The second is to fly parabolic flight profiles inside a suitably padded aircraft, for about 20 seconds of weightlessness at the peak of the climb. A programme of 40 such rollercoaster cycles in one flight provides less than 15 minutes experience of weightlessness, in short bursts. It also earned the aircraft its more common name – the 'Vomit Comet'.

By varying the angle of the Vomit Comet climb, or using harness devices to carry a percentage of the body weight, simulations of different degrees of reduced gravity – such as the ⅙ g of the Moon – can be simulated.

The 'pull' of the Earth's gravity extends far out into space, until the influence of another celestial body takes over. This was clearly demonstrated during the Apollo flights to the Moon. Approximately two-thirds of the way there, the lunar sphere of influence began to have an effect on the approaching spacecraft and started to pull it faster towards its centre. A combination of orbital mechanics, celestial navigation and a good braking engine allows the trajectory of the spacecraft to be controlled. The crew then has the option to travel past the planet or moon (a fly-by), loop around it and head back in the direction they came from (circumnavigate), slow down enough to be captured by its influence, but not land (enter orbit), or attempt to rendezvous with its surface (entry and landing).

Gravitational forces – or the lack of them – can have significant effects on both the vehicle and the human crew. Upon entering space, the natural balancing system of the body, located in the inner ear, can take some time to adjust to its new environment – one for which it was not designed. This period of adaptability varies considerably with each individual, but generally the human system can adapt to 'zero g' within 2–3 days. Predicting who will suffer from 'space adaptation syndrome' is not easy. The early astronaut and cosmonaut selections targeted former military test pilots, accustomed to high g levels and fluid shifts, flying high-performance jets for a living. When some of these 'iron-belly' pilots flew into space, they became sick!

There is also a period of readjustment opposite reaction upon return to Earth. In a sustained zero g flight, the lower limbs are not used very much, and the heart muscles can deteriorate if an exercise programme is not maintained. The heart is not resisting the pull of gravity to feed blood to the brain. On long-duration missions, a regular exercise programme is required to sustain a level of fitness that allows the crew-member to readjust to 1 g at the end of the mission, beginning with re-entry.

On early, short flights, the occupants were cushioned against increased g loads in form-fitting couches, lying on their backs as the spacecraft descended through the atmosphere. When the Shuttle was developed, there was concern that the pilot's ability to control the aerodynamic surfaces (the rudder and flaps) during the descent would be hampered by the sudden exposure to g forces during re-entry. They had to achieve a one-shot, unpowered landing of the Shuttle, which was, in effect, a 100-ton flying brick! However, evidence from the X-series of rocket planes, the lifting bodies, and atmospheric tests of the Shuttle *Enterprise*, indicated that this would not be a problem.

More than 100 successful Shuttle landings, after flights of up to 17 days, on different types of runway and in varying winds, speeds and directions, have silenced these concerns. However, a quick return from a very long spaceflight of several months, piloting a winged vehicle (as envisaged for the rescue capability of the International Space Station), is a new challenge. The effects on performing a landing on Mars, after a flight in reduced gravity of several months (as opposed to the Apollo landing on the Moon after three days), is another problem that will need to be studied.

**Acceleration and deceleration**

To gain access to space you must (at least for the moment) ride on a rocket-propelled vehicle that ascends rapidly to attain an orbital velocity of 17,500 mph. In the early days of the programme, the astronaut or cosmonaut was pushed back into his couch by the g forces, as the rocket tried to break free of the gravity that kept it on the launch pad. Such high g forces on the body were first experienced in the middle of the twentieth century, by WWII combat pilots making rapid, sweeping turns, in dogfights. The unprotected and unprepared pilots often found themselves unable to move, and therefore unable to control their aircraft, as their eyes pressed back into their heads (later termed 'eyeballs in'), or almost pulled out of their sockets ('eyeballs out'). This often led to passing out (blackout), the structural failure of the aircraft, or inability to deploy the parachute safely.

Studies into the effects of rapid acceleration and deceleration continued between the 1940s and the 1960s, in the rocket research aircraft programmes and a series of high-altitude parachute descents and rocket sledge runs. All of these studies led to the development of the partial pressure garment, which placed pressure loads on the lower limbs to help pool blood during tight banking manoeuvres in combat aircraft, and increased the chances of surviving in the event of an ejection at great height or speed.

## Radiation

One of the most hazardous phenomena a crew will encounter, once free of Earth, is radiation. Radiation in space had been recorded for decades, and in some forms – such as light and heat from the Sun – is not harmful to the human organism, providing exposure is limited. The solar wind – the distribution of millions of gas emissions from the atmosphere of the Sun – carries solar radiation through space and to the farthest reaches of our Solar System. In 1958, America's first satellite, Explorer 1, mapped belts of a radiation 'shield' which, in conjunction with the Earth's magnetosphere, protects the planet from the constant solar radiation. Named after the inventor of the instrument that discovered them, these Van Allen belts were traversed twice by each Apollo crew – once on the way out to the Moon, and once on the way home.

NASA developed a scale of radiation exposure limits for space explorers that records levels for 30-day missions, 3- and 6-month missions, annual exposure and a career maximum. For Earth orbital and lunar missions these levels are manageable. By setting an annual, or career, limit (or maximum number of spaceflights – six, at present), radiation exposure can be kept to safe levels. However, the effects that would be felt on a two- or three-year flight to Mars, or the long-term effects on pioneering space explorers, decades after they last flew in space, are not known. Former astronauts and cosmonauts (and in some cases, their offspring) are constantly monitored with regular health checks. Flying older generation crew members into space (such as John Glenn at the age of 77, after a gap of 36 years between flights) helps provide medical baseline data to understand how radiation affects a human, both at the time of their spaceflight and in succeeding years. Future long-duration flights to Mars may mean that only one trip can be made in a lifetime. It could even be a one-way ticket, to begin the adaptation of generations of humans who will never have lived on Earth!

Solar flares were constantly monitored during periods of maximum activity in the 11-year solar cycle. Onboard the Salyut and Mir space stations, they needed adequate warning should the crew need to quickly return to Earth. Such a solar weather watch will become even more important during the expected 30-year lifetime of the much larger International Space Station complex.

## Space debris

Despite many suggestions that space is empty, it is actually quite cluttered. Although it is a vacuum, 'space' is filled with microscopic particles of solar wind, micrometeorites, chunks of rock (ranging in size from specks of dust to very large asteroids) and the evidence of the human exploration of space – space junk. With all this debris moving around, it is almost inevitable that, sooner or later, a human-crewed space vehicle will meet and collide with some it. During the Apollo 13 mission it was originally thought that the vehicle had been struck by something, before the true cause of the accident was determined.

After more than 40 years of space exploration, there is a lot of junk in Earth orbit. The problem of space debris has surfaced on several Shuttle missions, where the orbits of the vehicle and an old satellite or spent rocket stage have almost intersected. Avoiding elements of space debris does not only concern large, visible parts of exploded rockets, discarded equipment or dead satellites. A minute flake of chipped paint travelling at 17,500 mph can inflict serious damage on another vehicle travelling towards it at the same speed.

In venturing outside the relative protection of their spacecraft, EVA crews face new dangers from debris penetration. The design of the current 'soft-suit' allows the fabrics to 'bend' under pressure, and minute holes are fabricated into the layers of the suit. The design is such that no two holes can ever be lined up, preventing 'tunnel' access into the inner layers of the suit for a particle of dust.

The eventual size of the International Space Station offers a much larger 'target', and the odds of a serious debris impact will increase substantially. It has also been demonstrated – by the Mir collision in 1997 – that it is not only rogue items of space junk that can cause serious disruption to orbital life!

**Propellents**

A further danger to any human spaceflight crew is presented by the fuels used in launch vehicles and spacecraft. Liquid fuels are more difficult to handle than solid fuels because of their storage requirements. They require handling at super-low temperatures, and must be kept under pressure so that they are liquid instead of gaseous. For hydrogen this is $-423°$ F, and for oxygen it is $-297°$ F. For the Saturn V rocket, liquid oxygen (LOX) was used as an oxidiser, mixed with kerosene or liquid hydrogen.

Hypergolic fuels, such as aerozine and nitrogen tetroxide used in the Apollo missions, burn spontaneously when mixed, thus making ignition systems unnecessary. Solid fuels, in which both the fuel and oxidiser are mixed together in a solid sludge, or grain, are more easily handled and stored, and are simpler to use, although once ignited they burn for the duration of the fuel supply available, with no opportunity to throttle or shut them off apart from safety destruction from the ground.

The explosive nature of cryogenics was demonstrated in the 1986 *Challenger* accident when a solid rocket booster (SRB) punctured the external tank and resulted in a huge explosion of the mixing fuels. The twin SRBs continued to burn until destroyed by the Range Safety Officer.

The dangers of handling such fuels were tragically demonstrated as early as 1960, with the explosion on the pad of an intended Mars probe launch vehicle. The incident, on 24 October 1960, was the result of technical failure and the ignoring of basic safety standards. Launch commands had failed to ignite the launch vehicle, and safety commands were automatically issued to the vehicle. After two previous launch failures this was the third and final vehicle available to reach Mars before the planets lined up again two years later. Field Marshal Mitrofan Nedelin – then Commander-in-Chief of the Strategic Rocket Forces – ordered a full inspection of the rocket, and stood at the base of the pad while they tried to find the fault. While scores of workers were around the vehicle on the pad, the upper stage – which should have sent the Mars probe on its way from Earth parking orbit – had continued to count-down to its planned ignition time from its own independent timer. At the planned moment of its ignition the stage suddenly ignited just 120 feet above the pad, triggering explosions in the other rocket stages and creating a huge fireball of kerosene and liquid oxygen. Scores of workers, including Nedelin, were killed, and many more were seriously injured. The incident became known in the West as the 'Nedelin catastrophe'.

**Crew selection and compatibility**
Having determined what is needed to sustain a human crew in space, decided how to get them there (and hopefully back again), and addressed what they might encounter in the environment, the final major concern to ensure safety and success is to assemble a compatible crew.

Originally, flight crew selection was made from the very best test pilots who were accustomed to the rigours of flight tests and high-performance machines, and who were highly trained to respond to life-threatening situations in a split second. As the programmes developed, so did the composition of the flight crew, to include engineers, scientists and specialists who had more of an academic background than flying experience. With the advent of the International Space Station and eventual exploration and colonisation of the Moon and Mars, the role of space explorer in the new millennium will again shift in focus, to new astro-skills in construction, maintenance, systems management, farming, mining and resource extraction.

Training for a flight into space will, however, remain essentially the same, but the crew will more than ever need to be multi-skilled and cross-trained, in addition to simply being compatible with each other. Long flights out to the far reaches of the Solar System will rely increasingly on self-sufficiency and teamwork, and less on support from the Earth.

**Conclusion**
A spaceflight is not just the journey from launch pad to landing; it is an awareness of what faces the human cargo and the machines that carry them. Before the first attempts were made to lift astronauts and cosmonauts into space, an enormous amount of pioneering work was accomplished in exploring the upper atmosphere by balloon, parachute and rocket plane, during the period from the 1930s to the 1950s.

The knowledge and experience in understanding how the human body works and what is required to keep it working, combined with the early knowledge of the space environment, allowed the first steps to orbit to be taken, as a prelude to actual human exploration of space.

Without expanding and building upon centuries of earlier medical and theoretical data, human spaceflight would not have been possible.

# Pioneers of the stratosphere

## EARLY ENDEAVOURS, 1930s–1940s

By 1927 no aircraft had ascended to 40,000 feet above sea level, and no balloon with a human crew aboard had reached over 36,000 feet. No human venture into the atmosphere had returned useful information about the composition of the stratosphere; indeed, most of those who had tried to explore this new frontier had died in the attempt. There were some developments in flying automatic instruments aboard small balloons, but no real scientific data were available. However, in 1931 Auguste Picard, a Belgian, reached 51,775 feet (9.81 miles), and in the following year reached 55,152 ft (10.07 miles) with his first ascents into the stratosphere and successful recoveries. The success of Picard flying in his pressurised capsules – named FNRS in honour of his sponsor, the Fonds National de la Récherche Scientifique – triggered a sudden expansion of several programmes to claim ever higher altitude records in Europe, the United States and Russia. The 1930s were the pioneering years in high-altitude research by humans, and marked the beginning of the exploration of the upper limits of our atmosphere at the frontier of space.

In a subsequent tour of the United States, Picard promoted his research into cosmic rays from his pressurised capsule, and suggested that even a flight to the Moon may be possible in such vehicles. In the United States, the Army Air Service (later the Army Air Corps) had investigated the development of pressurised cabins for aircraft but this had been abandoned in 1921 when difficulties in regulating air pressure were encountered. The record altitude attempt of 42,240 feet by Capt Hawthorne C. Grey in May 1927, was disqualified when he had to bail out of his open basket. A Federation Aeronautique Internationale (FAI) requirement was that a pilot must land in the craft to claim a flight record (which would later cause the first Vostok cosmonauts to state that they landed in their spacecraft and not, in truth, that they ejected and parachuted to the ground). Grey's later attempt in November of that year resulted in his death. It was clear that further progression into the stratosphere would have to be supported by the development of special pressurised cabins or pressurised flight suits. The most famous of early pressure suits

Soviet stratospheric balloon USSR-1 during preparations for its ascent to the stratosphere in 1933. (Astro Info Service Collection via Exclusive News Agency (1936).)

were those developed by American pioneer aviator Wiley Post in the early 1930s, but it was the pressurised cabins that would finally explore the stratosphere.

When Picard travelled to America in January 1933, the Century of Progress Exposition had recently opened in Chicago. Interest in manned ballooning was growing in the United States, and support was raised to fund an American balloon to be flown from the Exposition. This soon became a programme of national importance, as the exposition was run on the theme of national pride and service to science and, in return, science in the service of American national needs.

What evolved was the Century of Progress flights supported by the US National Academy of Sciences National Research Council, with sponsorship from several corporate ventures in Chicago. Subsequently the National Geographic Society/US Army supported flights of Explorer to achieve scientific objectives which also pointed towards several military advantages. Following the Second World War, renewed interest in manned stratospheric balloon exploration continued with programmes sponsored by the USAF and the USN during the 1950s and early 1960s.

In the Soviet Union of the early 1930s, manned stratospheric ballooning was a combined effort of scientific goals and military objectives with co-operation between the Soviet Army and the Soviet Academy of Sciences, who jointly developed not only stratospheric balloons but also high-altitude aircraft and pressure garments.

Plans to fly the first Soviet balloon crew were aimed at a December 1932 launch, but a series of organisational and financial hurdles resulted in a delay of a further 10 months before this was achieved with the September 1933 flight of the USSR-1 carrying a crew of three to a height of 12.6 miles on a flight lasting 8 hrs 15 min. Severe weather cancelled seven previous launch attempts, but success was finally achieved on the eighth attempt for this gondola, and it is interesting to note that advanced news of the pending ascents were issued to a group of official guests who were on hand to witness the failed launch attempts. With this embarrassment, Soviet officials decided that all future attempts would be made in private, to be announced only after a successful launch. However, not only the successes but also the failures were hidden from the West for many years, and were not revealed to the world until 30 years later.

Throughout the 1930s, until the onset of the Second World War, the Soviets continued to strive for aerial supremacy in the stratosphere. Following the War the Soviets, like the Americans, resumed high-altitude balloon exploration as a stepping stone to orbital spaceflight with the Volga programme in the 1950s and 1960s.

The progress of human stratospheric balloon exploration during the 1930s–1960s is presented in the table on p. 20. It was during this time that there arose the argument of supporting either a programme that risked a human crew or a programme utilising automated equipment. This argument occurred as frequently as it did during the development of the human and automated space programme several decades later, and continues to the present day.

There is no dispute that stratospheric flight was dangerous, as the environment is not exactly healthy for the human body without adequate protection. But Picard argued that although automated balloons had their advantages, only a human crew could obtain more accurate and detailed recordings, and after gathering data they could also recharge instruments which would not be left to run down after a few minutes. Manual photography was becoming more frequently used in data recording, and Picard would not promote the study of cosmic ray investigation outside manned ballooning. Indeed, it was clear that his enthusiasm for, and promotion of, manned balloon flights was without any planned scientific agenda. This was very similar to the operation of the Apollo programme in the early 1960s, in that the desire was simply to reach the Moon and return safely to Earth, without detailing what was to be done when man reached there. Scientific objectives, as well as biomedical objectives, became more important in later programmes once the achievement of sending a human crew into the higher reaches of the atmosphere, and retrieving them safely, had been demonstrated. Again, this was reflected in the Apollo lunar landings where, once the skill of achieving a lunar mission had been achieved, more complex scientific objectives were added to the later missions.

As human stratospheric ballooning progressed ever upwards, so too did the risks and accidents associated with such a bold adventure.

## Stratospheric manned balloon flights, 1927–1966

| Date | Balloon | Country | Altitude (feet) | Crew |
|------|---------|---------|-----------------|------|
| **Pre-Second World War, 1927–1939** | | | | |
| 1927 Nov 4 | | USA | 42,470 | Gray |
| 1931 May 27 | FNRS | Belgium | 51,775 | A. Picard |
| 1932 Aug 18 | FNRS II | Belgium | 55,152 | A. Picard |
| 1933 Sep 30 | USSR-1 | USSR | 60,698 | Prokofiev, Birnbaum, Godunov |
| 1933 Nov 20 | Century of Progress | USA | 61,237 | Settle, Fordney |
| 1934 Jan 30 | Osoaviakhim-1 | USSR | 72,182 | Fedeseenko, Vasenko, Usykin |
| 1934 Oct 23 | Century of Progress II | Belgium/ USA | 57,579 | A. and J. Picard |
| 1934 Jul 28 | Explorer 1 | USA | 60,613 | Kepner, Stevens, Anderson |
| 1935 Jun 26 | USSR-1 Bis | USSR | 52,496 | Christopzille, Prulutski, Varigo |
| 1935 Nov 11 | Explorer 2 | USA | 72,395 | Stevens, Anderson |
| 1939 Oct 12 | SP-2 Komsomol (VR60) | USSR | 55,154 | Fomin, Krikun, Volkov |
| **Post-Second World War, 1956–1966** | | | | |
| 1956 Nov 8 | Strato-Lab I | USA | 76,000 | Ross, Lewis |
| 1957 Jun 2 | Manhigh I | USA | 96,000 | Kittinger |
| 1957 Aug 19 | Manhigh II | USA | 101,500 | Simons |
| *(1957 Oct* | *Sputnik 1* | *USSR* | | *First satellite in Earth orbit)* |
| 1957 Oct 18 | Strato-Lab II | USA | 86,000 | Ross, Lewis |
| 1958 Jul 26 | Strato-Lab III | USA | 82,000 | Ross, Lewis |
| 1958 Oct 8 | Manhigh III | USA | 99,700 | McClure |
| 1959 Nov 16 | Excelsior I | USA | 76,400 | Kittinger |
| 1959 Nov 28 | Strato-Lab IV | USA | 81,000 | Ross, Moore |
| 1959 Dec 11 | Excelsior II | USA | 74,700 | Kittinger |
| 1960 Aug 16 | Excelsior III | USA | 102,800 | Kittinger |
| *(1961 Apr 12* | *Vostok 1* | *USSR* | | *First human to orbit Earth)* |
| 1961 May 4 | Strato-Lab V | USA | 113,740 | Ross, Prather |
| 1962 Nov | Volga | USSR | 93,970 | Andreyev, Dolgov |
| 1962 Dec 13 | Stargazer | USA | 81,500 | Kittinger, White |
| 1966 Feb 2 | Strato Jump II | USA | 123,500 | Piantanida |
| 1966 May 1 | Strato Jump III | USA | 57,600 | Piantanida |

### 1934: Osoaviakhim-1 (USSR)

On 30 January 1934, an early morning launch by the Soviets saw a three-man crew of Commander Pavel Fedeseenko, physicist Ilya Usykin and engineer Andrei Vasenko onboard the Osoaviakhim-1 balloon. Following the success of the earlier USSR flight, the aeronauts were to continue scientific observations of the atmosphere. At 3 hrs 18 min into the flight they achieved a new altitude record of 72,182 feet, and a few minutes later they began their descent to reap the rewards of their achievement as the new Heroes of the Soviet Union. However, this was not to be.

In attempting to achieve the record height the crew had dropped over 793 lbs of ballast, and as a result of expansion by heating they had also lost a large quantity of the hydrogen stored in the balloon envelope. To achieve a safe controlled descent they needed 1,500 lbs of ballast, but had only 925 lbs left onboard. Solar heating helped keep them aloft, but as the Sun set the hydrogen cooled, and there began a rapid descent. As the speed increased, the thickening layers of the atmosphere tore at the envelope and resulted in violent buffeting of the crew module. Inside, the three men were thrown around like rag dolls, making illegible entries in their flight log as they tried unsuccessfully to dump the remaining ballast and heavy equipment overboard in an attempt to slow the rate of descent. At 57,000 feet the support cables snapped, and the crew – still inside the doomed gondola – plummeted to Earth.

At a mission elapsed time of 7 hrs 8 min, Osoaviakhim-1 smashed into the ground near the village of Potish-Ostrog, in the Insar Raion, 10 miles east of Kadoshkino railway station located on the Moscow–Kazan line. All three men were instantly killed. Three days later they were honoured with a full military funeral, and were interned in the Kremlin Wall.

From the recovered wreckage the tragic sequence of events was investigated. The crew's last radio contact occurred at 15 hrs 28 min Moscow time (MT), and until that time the flight seemed to be going well. At 16.05 MT, however, they recorded in the log book that they were 'descending very rapidly, we have jettisoned a lot of ballast', (to lighten the load and maintain height). They scribbled their last entry into the log book at 16.10 MT before the crash, which occurred at 16.23 MT according to the time recorded on the broken watch of Vasenko. Post-flight examination of the data in the log-book, together with the last recorded comments over the air-to-ground communications link, indicated that all three men were in a cheerful mood and that all aspects of the flight was progressing normally up to the last radio communication.

The investigation committee determined that the accident was a combination of the lack of available ballast and the instability of the gondola, which contributed to the tearing of the balloon envelope and the subsequent severing of the support cables. The crew had been unable to exit the capsule to make individual parachute descents as a result of the buffeting, and had managed to remove only seven of the 24 hatch nuts before the flight ended.

Soviet development of the use of automated balloons for future research was a direct result of the tragic loss of the three aeronauts, and the programme of unmanned flights continued for several decades. Even the volunteering of the USSR-1 crew to ascend into the stratosphere in honour of their fallen comrades soon after the accident was not pursued. It would be 18 months before the three-man crew of USSR-1 Bis, in June 1935, took the latest Soviet balloon to almost 10 miles in a flight of 2 hrs 30 min to continue studies begun by the USSR and Osoaviakhim crews.

Two months later it was revealed that the USSR-1 Bis crew nearly became a second Soviet crew to be killed. A large unidentified rip was reported in the envelope, and both the flight engineer, M. Prilutski, and the scientific observer, Professor

Alexandr Varigo, had bailed out. The Commander, M. Christopzille, remained in the now lighter gondola, and was able to steer towards a safe landing.

### 1934: Explorer (USA)

Just seven months after the loss of the Soviet aeronauts on Osoaviakhim, the Americans had their own experience of the dangers of manned balloon ascents during the NGS/Army Air Corps Explorer balloon flight of 28 July 1934. Launched from the Stratobowl in the Black Hills of South Dakota, with a three-man crew consisting of Capt Albert W. Stevens, William E. Kepner and Capt Orvil A. Anderson, all aspects of the launch and ascent went remarkably well, and set a new altitude record of 60,613 ft (11.48 miles).

By mission elapsed time of 7 hrs 15 minutes the crew noticed a series of rips in the bottom of the bag, and although they were growing larger these tears had fortunately not reached the expanding gas within the balloon. It was not a difficult decision for the crew to leave the stratosphere and attempt to land Explorer as soon as possible. Their good fortune was not to continue with Explorer as the bag ruptured at 5 km and began a rapid decent at 656 feet per minute, the remnants of the envelope acting as a large, flimsy parachute.

The aeronauts decided to wait until they had descended to a sufficient altitude that would allow them to open the hatch and not cause a rapid decompression. The three aeronauts then prepared to exit the gondola and descend by personal parachute; but as they did so, one of them became caught in the exit hatch, adding to the tension. Kepner climbed out on the support ring to provide more room inside and thus help his colleagues. Stevens had previously walked on his parachute pack during the rushed preparation to exit, and it had become bunched up. There was no time to straighten the pack, so the crew quickly decided to let him jump first – backwards, holding his parachute in his arms as he went. There would have been little the other two could have done had it not deployed. All three had only just managed to leave the gondola when the hydrogen-filled balloon, contaminated with the oxygen in the atmosphere, exploded and sent the Explorer gondola on a collision course with the ground on the plains near Holdrege, Nebraska, destroying most of the photographic records and two or three spectrographs as it did so. Fortunately, all three aeronauts landed safely.

The review of the accident, held between July and September 1934, revealed that during subsequent tests and examination of recovered fabric material, the balloon envelope had not opened evenly during the ascent, and that the initial tears which appeared in the fabric were a result of stress on the folds as they failed to expand correctly. The launch had been delayed for a month because of unfavourable weather in the launch area, and some adhesion of the rubberised cotton – of three different weights – had occurred in the folds, so that they stuck together when they expanded. The explosion was a result of the mixing of the hydrogen in the ripped bag with the oxygen in the atmosphere. At the time there was no mention of a second flight to try to regain some of the lost scientific results. As a result of the Explorer accident and those experienced by the Soviets, the second Explorer was eventually supported, but with a much lighter scientific payload and a two-person crew. It

The Soviet balloon SP-2 Komsomol (VR60) is launched on 12 October 1939 for the final pre-war stratospheric ascent.

launched successfully in November 1935, and reached a record 72,395 feet (13.71 miles) before landing safely. A third Explorer was planned, but not supported, despite the previous success. Indeed, parts recovered from the second Explorer balloon envelope were cut up into a million strips and offered as commemorative bookmarks to members of the NGS who supported the programme, noting that this was the end of American manned balloon research for some time.

### 1939: SP-2 Komsomol (VR60) (USSR)
The Soviets attempted at least one more ascent prior to the start of the Second World War. During 1937–38, a young student engineer, working on a graduation thesis, proposed a balloon–parachute project to be flown under the civil management of the civil air fleet Aeroflot. The concept of a parachute–balloon was that during a normal landing, or in the event of a rip in the balloon, the canopy would open into a parachute and allow the crew to achieve a safe landing while still inside the gondola.

After months of preparation and delay the three-man crew of Commander A. Fomin, pilot A. Krikun and scientific crew member M. Volkov took off in the Komsomol balloon on 12 October 1939. They reached a maximum height of 383,254 feet. The only problem during the ascent was the 30 minutes spent in trying to release the ballast sacks of small shot. After an hour at their maximum height, having completed their scientific observations, the crew began the descent. At 29,500 feet above the ground the balloon changed shape into a parachute, as planned. At 23,600 feet Volkov was looking up into the envelope and noticed a fire burning the remaining hydrogen, probably caused by a spark as the hardware changed configuration. Fomin immediately separated the gondola from the burning canopy, and as the craft dropped Krikun hand-operated the back-up parachute system. As the vehicle descended they decided to rescue scientific results and equipment before attempting to exit the balloon by personal parachute. They did so approximately 5,000 feet above the ground, and all three achieved a safe landing.

They walked towards to the unmanned gondola, which was sitting in the snow-covered peat bogs of the landing area; but they were unable to approach for some time, as the outer surface of the gondola hissed as it cooled, and smoke and flame – probably from short circuits in the damp environment – spat out of the open hatch as the capsule cooled.

## POST-WAR MANNED BALLOON FLIGHTS, 1950s–1960s

Following the end of the Second World War, interest in the military aspects of manned stratospheric research by balloons resumed in the interests of the Cold War and in allowing various government contractors and manufactures to test new technology and hardware planned for rocket and missile programmes. Developed from the cancelled Helios programme, the USN Strato-Lab programme took the first Americans into the upper reaches of the atmosphere for the first time since the Second World War. In August 1956, USN Lt Cdrs Malcolm Ross and Lee Lewis rode in an open gondola to 40,000 ft to test pressure suits and balloon control. This was the first time Americans had entered the stratosphere since 1935, and it was the start of a new push into space.

Research and development continued in support of these programmes by the USN Strato-Lab and USAF Manhigh series of balloon ascents, and the Excelsior and Strato Jump high-altitude parachute jumps, which also provided biomedical data for later work on the rocket research programmes and the Mercury Man In Space programme that were to follow. (See the table on p. 25). Concurrently, programmes of high–altitude parachute descents were accomplished under the Excelsior and Strato Jump programmes.

The Soviet Union continued research in the area of manned stratospheric balloons with the Volga balloon gondola, which logged several flights, and by supporting altitude parachute descents and engineering studies related to the Vostok space programme.

Key events in X-plane and lifting body programmes, 1946–1975

| Date | Vehicle | Pilot | Remarks |
|---|---|---|---|
| 1946 Jan 19 | Bell XS-1-1 | Woolams | First glide flight of an X-1 aircraft |
| 1946 Aug 29 | XS-1-1 | Yeager | First powered flight, Mach 0.85 |
| 1947 Aug 20 | D-558-1-1 | Caldwell | New world air-speed record 640.663 mph in Skystreak aircraft |
| 1947 Oct 14 | XS-1-1 | Yeager | First supersonic flight by manned aircraft: Mach 1.06 at 43,000 feet, flying at 700 mph |
| 1948 Mar 26 | XS-1-1 | Yeager | Fastest flight of original X-1 (957 mph) during dive at Mach 1.45 |
| 1949 Aug 8 | XS-1-1 | Everest | Reached 71,902 feet |
| 1951 Aug 15 | D-558-2-2 | Bridgeman | Altitude record: 79,494 feet in Skyrocket |
| 1953 Nov 20 | D-558-2-2 | Crossfield | Mach 2.005 flight in slight dive in Skyrocket |
| 1954 May 28 | X-1A | Murray | Manned aircraft altitude record: 87,094 feet |
| 1954 Jun 4 | X-1A | Murray | Reached 89,750 feet |
| 1954 Aug 26 | X-1A | Murray | Reached 90,440 feet |
| 1956 Sep 27 | X-2-1 | Apt | Mach 3.2 at 65,500 feet; crashed; pilot killed |
| 1959 Jun 8 | X-15-1 | Crossfield | First glide flight of X-15 |
| 1959 Sep 17 | X-15-2 | Crossfield | First powered flight |
| 1961 Mar 7 | X-15-2 | White | Achieved Mach 4.43 |
| 1961 Jun 23 | X-15-2 | White | Achieved Mach 5.27 |
| 1962 Jul 17 | X-15-3 | White | FAI world altitude record: 314,750 feet |
| 1963 Aug 22 | X-15-3 | Walker | Highest X-15 flight: 354,200 feet; unofficial world altitude record |
| 1966 Jul 12 | M2-F2 | Thompson | First lifting body flight |
| 1966 Nov 18 | X-15-2 | Knight | Achieved Mach 6.33 |
| 1967 May 10 | M2-F2 | Peterson | Landing accident |
| 1967 Oct 3 | X-15-2 | Knight | Speed record: Mach 6.70 |
| 1967 Nov 15 | X-15-3 | Adams | 261,000 feet; fatal flight |
| 1968 Oct 23 | HL-10 | Gentry | First powered flight; engine malfunction |
| 1969 May 9 | HL-10 | Manke | First supersonic lifting body flight: Mach 1.13 |
| 1970 Feb 18 | HL-10 | Hoag | Fastest lifting body flight: Mach 1.86 |
| 1970 Feb 27 | HL-10 | Dana | Max altitude for lifting body: 90,303 feet |
| 1975 Nov 26 | X-24B | McMurty | Last lifting body flight |
| *(1977 Aug 12* | *OV-101* | *Haise/Fullerton* | *First Shuttle ALT free-flight from Boeing 747)* |
| *(1981 Apr 12* | *OV-102* | *Young/Crippen* | *First Shuttle orbital spaceflight)* |

## 1956: Strato-Lab I

The American programme continued remarkably well, with only a few minor setbacks. On 8 November 1956, Ross and Lewis crewed Strato-Lab I to a peak altitude record of 76,000 feet, when, just as they recorded their success, the balloon began to drop rapidly. At first the two aeronauts thought the balloon had ripped, and immediately radioed the ground to report their emergency situation. They were dropping rapidly, strapped into their seats but not sealed into their suits systems. They reported a rotation and a 'decided elevator feeling'.

At a descent rate of 4,000 feet per minute their attempts to drop steel-shot ballast had no effect. They considered detaching the balloon and dropping under the recovery parachute canopy, but realised that they were over terrain which was far too rough for a comfortable landing, and decided to stay with their craft as long as possible. As they finally entered the lower atmosphere they opened the porthole and threw everything they could lay their hands on overboard to help slow their sink rate. The weight thrown out totalled more than 200 lbs, and included the radio once they had informed the ground that they would be out of communications for a while!

Both men landed in the gondola, safely, on an isolated patch of sandy soil on a ranch near Brownlee, Nebraska. Lewis later informed the press that they were more concerned than scared, and Ross also contributed in playing down what was in effect an essentially uncontrolled descent by giving the impression of a smooth and uneventful landing on what was a relatively level stretch of land. This nonchalant attitude was the first seedling of what was to be later termed the 'Right Stuff' – a display of calmness and serenity under extremes of pressure, without letting human frailty cracks appear in the professional, military façade. The post-flight investigation indicated that a helium relief valve began leaking gas prematurely.

### 1957: Manhigh I

On 2 June 1957 USAF Captain Joe Kittinger departed Earth for the maiden flight of the Manhigh programme and climbed steadily into the atmosphere keeping in contact with the ground by a long-winded telegraph key system. At 45,000 feet the Manhigh I gondola was hit with 100 mph blasts of wind as it entered the area where most balloon flights had failed. Kittinger sat and watched as the effects of the wind made strange and weird shapes of his balloon, and he later wrote that the experience 'frightened the hell' out of him.

As he climbed above the jet stream he noted that the contents of the precious oxygen supply was apparently reading only half full, when it should have been almost full. In reporting this to the ground, it was deduced that the technicians, in installing the equipment, must have installed the oxygen supply valves backwards, and the main supply was being vented overboard while the bleed-off was what Kittinger was breathing. The reality of the situation suddenly struck home. Kittinger had to descend immediately – and fast – before the oxygen supply ran out. Realising that it might already be too late, Kittinger shut the cabin system down and reverted to his pressure suit supply to service the gondola instead. This stopped the leak. At 96,000 feet his ascent stopped and he took time to look out the window. Only X-2 rocket plane pilot AF Captain Iven C. Kincheloe – who had spent a fleeting moment at 126,000 feet – had ever been higher.

There was not much time to admire the silent view, however, as the onboard gauge indicated only 1 litre of oxygen left – probably not even enough for the descent. As Kittinger began to release the helium from the balloon to begin his descent a message came over the one-way radio, instructing him to return. Kittinger replied in his halting Morse code, 'C-O-M-E A-N-D G-E-T M-E'. On the ground the support team though Kittinger was suffering from 'breakaway phenomenon', in which high-altitude pilots suffered a surreal dreamy state due to extreme isolation

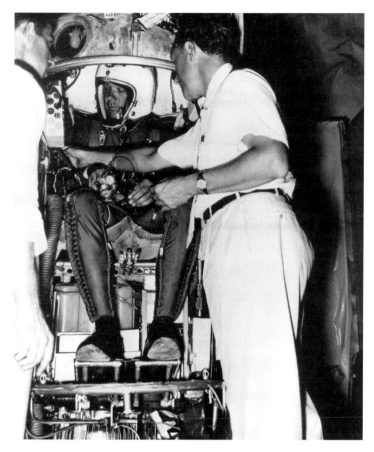

Preparations for one of the record-breaking Manhigh balloon ascents to the upper reaches of the atmosphere during the 1950s. (Photograph courtesy USAF.)

and separation from Earth. Kittinger was fine, and was just euphoric and joking as he faced the dangers of his mission.

But his luck held. The oxygen lasted as he vented the gas for a faster descent. Manhigh I gondola splashed down in Indian Creek – a small muddy tributary of the great Mississippi River. He had reached an altitude of 19 miles, and had survived.

### 1957: Manhigh II

In extreme environments or situations, the reaction of a human to a sudden unexpected incident can have a significant effect on subsequent actions. Interpretation of a sudden event can sometimes make it appear much worse that it is in reality. This was experienced by the pilot of the second Manhigh ascent in 1957.

After being sealed in the Manhigh II cabin for 11 hours, Air Force Major David G. Simons finally lifted off aboard the gondola for what turned out to be an unexpectedly long flight. His peak altitude was 101,500 feet, and from this vantage point was able to appreciate the curvature of the Earth far below. Only the X-2 rocket plane had flown higher. Shortly afterwards the first indications of problems were recorded. He was informed over a crackling VHF radio link that not only had high-frequency radio contact been lost, but that also the tracking vehicle could no longer track the progress of the capsule, and that all radio contact may be lost. The choice was an immediate descent in the dark, or remain aloft without radio contact throughout the night.

Simons weighed his options. Being alone was not really a concern, and the loss of ground tracking did not concern him; and he was sure he would be found after he landed – somewhere. It was the loss of telemetry that concerned him the most. After almost 24 hours with no sleep he was close to exhaustion, but he knew that the ground was monitoring his vital signs. He ached all over, but in the front of his mind was the overriding desire to complete the mission for which he had trained for two years. He told the ground that he would stay at altitude to complete his research objective and study the effects of a whole day in the stratosphere on the primary research subject – his own body.

After a few more hours of observations, he finally slumped forward to try to sleep, with his head resting on his parachute pack. Suddenly he was awoken by a bright flash of white light in the cabin ... another flash ... and another. He realised he was flying inside a huge storm. He later wrote that for the first time in the flight he was afraid. Even a slight spark could ignite the oxygen-rich air in the gondola, resulting in a very bad day. What if he was hit by a billion volts of lightning and fried to a crisp? But it was not lightning; it was just the automated camera programmed to take a picture of the instrument panel every five minutes. Condensation had shorted the system early in the day, but it had now dried out and come to life – scaring Simons half to death. He wanted to laugh, but was too tired. He tried to nap again.

It was not long before he was abruptly awakened again. He had dropped rapidly and felt the gondola spinning. He thought the envelope had split. Looking outside, he realised what had caused him to suddenly drop 500 feet. This time he could see the thunderclouds all around him, and this time he actually had dropped into the middle of a storm. He wanted to drop the 100 lbs of reserve batteries mounted on the landing ring beneath him, but ground control advised him via VHF that this could send a severe shock up the cables to the balloon envelope, which would probably not survive such an event.

Simons felt he had no choice. With the storm affecting any airborne tracking of the gondola this would only add to the time it took to find him. Neither was he looking forward to being drawn deeper into the storm. He triggered the release and waited for the shock wave to hit the balloon – which it did, hard, eight seconds later. It worked, and the balloon rose rapidly out of the storm. He survived the event, and the next evening safely guided the balloon to a slightly rough landing in a freshly ploughed field of flax near the South Dakota town of Elm Lake.

**1958: Manhigh III**

When First Lt Clifton McClure, USAF, finally lifted off in the third Manhigh ascent on 8 October 1958, he had already saved the mission by repacking his emergency parachute. Sealed in his capsule for launch, he was completing his pre-flight preparations when he inadvertently knocked his hand against the rescue parachute pack which was hanging on the capsule wall. As he realised what he had done, he looked down between his feet and was horrified to see layers of white nylon neatly folded on the floor of the cabin. By hitting the tightly packed parachute its restraint pins shot out and spilled the contents onto the gondola floor. Terrific! If he informed ground control the flight would be cancelled.

Everything in the developing space programme was a balance of size, weight and safety. There had been arguments for and against personal parachutes, and the decision was to go with a personal parachute; but the problem was where to put it in such a confined space. The pilot could not wear it, so it was hung on the wall in the only place possible. On this particular day, this was not a good place for McClure to put a parachute pack.

No one outside the capsule had seen the incident, and the parachute was still folded. Not wishing to risk a long delay due to cancellation of the flight, McClure decided to keep quiet and try to repack the parachute. He had never actually packed one before, but he had seen someone else do it, and had watched with curiosity and interest. His questions and observations that day in training would pay off. However, he was not in the spacious hanger; he was in the close confines of the capsule, and was wearing a restricting pressure suit. He may not have room, mobility or experience, but he had plenty of time awaiting the lift-off.

It took McClure two hours, and he documented every step on the onboard tape recorder. After a great struggle to lift the pack and re-hang it on the wall with the pins in place, he relaxed and rested. As he did so he casually looked at the parachute that had given him so much trouble, and noticed the restraining pins were secured backwards. He had been so concerned in replacing them that he forgot to line them up properly, but after another 90 minutes he finally corrected the error and prepared for launch, which by then was less than two hours away.

Launched without incident, Manhigh III climbed to a peak altitude of 99,700 feet at just over three hours into the ascent. McClure reported feeling a little tired at this point, but no one on the ground seemed concerned. He later reported a cabin temperature of 89° F, which raised eyebrows on the ground, and it was decided that the positioning of the temperature gauge above the air generation system could be the cause. McClure thought that his extra efforts in packing his parachute did not help keep his own temperature down. When he subsequently reported a possible error, with the capsule temperature gauge reading over 90° F, his voice appeared a little sluggish. When he responded to a request for a body temperature reading from the gauge on his suit, which was connected to a rectal thermometer, he reported 101.4°. The response of the ground team was afterwards likened to a double take from a Hollywood movie. This could mean serious trouble. With such a high temperature while wearing the tight partial pressure suit, McClure could suffer due to a fever from the heat and sweat.

McClure was told to take on drinking water, but he found that he could not extract any water from the drink dispenser. The cabin temperature was recorded as 97° F. Manhigh III was turning into a flying oven. It was decided to bring McClure down, but he argued to stay up, and that apart from feeling a little tired he was fine. He had not been able to unhook the hose from the air regeneration fan to blow cool air into the cabin. The hose felt hot, but he had repaired the drink dispenser and had had a good drink of cool water.

The option to remain in the air until sunset and fly through the night – which would lower the temperature – was not considered. It was six hours until sunset, and McClure could not survive those temperatures for that length of time. Manhigh III had to come down – and quickly. Careful not to raise alarm in the pilot, the ground decided to bring him down but not to tell him that they were terminating the ascent. They would advise him to reduce the altitude to help lower the temperature, but in reality begin gradual descent to the ground. McClure guessed the real motive for the descent and voiced his objections. He was overruled, and ordered to come down. He remained silent in his cabin.

Landing would be tough at night, and probably in a remote mountainous region. The option of trying for a forced landing by a late separation of the envelope to prevent dragging the capsule was not a favoured option, as the pilot could be crushed or burned in the leak of hazardous chemicals aboard the gondola. With the sustained temperature of 104.1° F he may well also pass out. He could not touch the skin of the capsule, as it was hot enough to burn, and as he struggled to release the gas to begin his descent his vision burred. As he descended, his body temperature was recorded at 105.2° F and as high as 107° F. Ground control had lost radio communication, and therefore did not know whether or not McClure was conscious; but they were monitoring steady heart and breathing rates, so at least he seemed to be alive.

Onboard the gondola, McClure had gradually been able to slow his rate of descent. He had also considered using his parachute to leave the capsule, but doubted if his strength would be sufficient to achieve this. He therefore decided to stay onboard and ride out the descent and expected hard landing. He was also faced with a more pressing personal problem: he needed to urinate. Trying to repack a parachute was nothing compared with trying to urinate inside the gondola. He knew the danger of experiencing a hard landing with a full bladder. He had learned that in an impact accident a full bladder could burst and kill him – if the impact, of course, did not. It was no use – he had to go. He spent several long uncomfortable minutes trying to undo the partial-pressure suit, and being extra careful not to again dislodge his parachute, he finally peed into a bottle and then fastened himself up again. All this exhausted him even more, although he was much more comfortable.

As he landed with a thud, a helicopter landed as close as possible, expecting to find the occupant close to passing out inside the capsule. As the rescue team ran to the capsule they were amazed to see the hatch open and McClure stand up, take off his helmet and flash a wide grin. He wondered what all the fuss was about. He soon recovered, but had been extremely lucky.

Post-flight reports indicated that errors had been found in the calculation of insulation and temperature. It was also reasoned that the repacking of the parachute was a contributing factor to the overheating. In his post-flight report, McClure's observations of aurorae like lights at 95,000 feet was interpreted as virtual images under extreme stress. McClure argued that it was not an image and that stress was not a problem – in fact, he easily coped with it. It was the excess heat that was the problem. In true Right Stuff legend, he overcame his obstacles and survived to tell the tale – with a grin.

## 1961: Strato-Lab V

The balloon programmes had clearly revealed the harsh environment that the flights were trying to explore, but what was sometimes forgotten was that the events after the flight could be as dangerous as those during the flight itself. By May 1961, preparations for the fifth and final Strato-Lab ascent were almost complete. Manned ballooning to the higher regions of the atmosphere was not the pioneering adventure it had once been. Indeed, the month before, a young Soviet Air Force Pilot named Yuri Gagarin had flown higher (maximum of 203 miles), faster (17,400 mph) and further (25,000 miles) than anyone before – and all in 108 minutes. But in those few short minutes he became the first human to orbit the Earth – only once, but it moved mankind into the realm of a new mode of transport and exploration: human spaceflight.

With the news of Gagarin resounding around the world, early on the morning of 4 May 1961 Navy Commander Malcolm Ross and Lt Cdr Viktor Prather were sitting side by side in the Strato-Lab open frame gondola, wearing a type of full pressure suit designed by B.F. Goodrich for the Project Mercury astronauts, and fully pressurised to 5 psi. Their aim was a new record of 102,800 feet. This time they would be launched not from dry land but from the deck of the aircraft carrier *Antietam*, cruising 138 miles south-east of New Orleans in the Gulf of Mexico.

The flight was plagued by minor problems from the lift-off, when parachute lines snagged one of the carrier's deck cleats used for mooring ropes. Once cleared they were released to begin their rapid ascent. Unlike previous attempts in an open-framed gondola, the searing cold and winds made the temperatures bitter, despite the protection of layers of clothing and electrically heated gloves and socks. At 43,000 feet the temperature had dropped −73° F. As the aeronauts ascended and the pressures dropped, so their suits inflated around their limbs.

As they climbed, both pilots' helmet face-shields began to fog up. The helmets were fitted with heating elements to prevent this condensation, but these also had the capability of actually melting the plastic face-plates if the temperature rose too high. A warning light would alert the pilot if the temperature of his heated shield reached 135° F. Aware of this, but still bitterly cold, both men cautiously increased their helmet heater levels.

At the same time, Ross was aware of a hissing sound. Fearing that his own suit had developed a fatal leak he saw a vapour cloud emerging from beneath the two men and dispersing overboard, but due to the pressure restrictions of his suit he was unable to bend to determine where the leak might be coming from. After trying to contact the ground, all radio contact on both the primary and emergency channels was lost. Luckily, it was not a leak in the suit, and its source remained a mystery.

Ross also noted a decrease in the ascent, which he thought could be the result of a leak in the envelope of the balloon; but in fact they were ascending through a pocket of warmer air which had slowed them down. Then, with no warning, the hissing stopped and suddenly the radio came back on line. The first communication from the ground informed them that their face-plates were overheating – which was particularly good timing, as their own warning lights had failed only minutes before the face-plates would have blistered and cracked, releasing the internal pressure and probably killing both men.

They reached their peak altitude of 113,740 feet (21.5 miles) above the Earth – the highest altitude ever reached by a manned balloon, and the highest anyone had attempted an altitude climb without a pressurised compartment. The suits were working fine, and the view was spectacular, but it was time to descend. The two men began their descent slowly, at first releasing the gas valve and allowing the rising Sun to warm the helium. After an hour their descent was still only a modest 70 feet per minute.

The onboard oxygen limited the time that they could remain at such a high altitude, and so Ross decided to leave the valve open for 15 minutes, which resulted in an increased descent of 750 feet per minute. Their feet were feeling the cold as they dropped through the lower atmosphere at a rate of 1,140 feet per second. This was fast – a little too fast – and so they decided to lighten the load by offloading their ballast. After they had offloaded all 125 lbs of steel shot, they started throwing out unwanted equipment like the battery power supply unit and the radio. The lower they became, the slower the descent, and so the less urgent was the need to offload unwanted equipment. They even had time to enjoy a cigarette and soak in the experience of the last few minutes of the descent back into the Gulf.

It was hoped that Strato-Lab could be recovered back on the deck of the carrier in a 'running catch' but when the gondola splashed down in the waters of the Gulf the carrier was more than two miles way. The crew released the balloon and parachute lines, the balloon canopy separated, but the parachute did not, although luckily it did not inflate enough to drag the two men across the water. Ironically, in waiting for the carrier the two men noticed items of debris, tossed out by them during their descent, floating around like flotsam from a shipwreck. When the first helicopter reached them, Prather deferred the ride to Ross who put on the sling and stepped onto the rescue hook. As he was pulled up he slipped and half submerged into the ocean, but managed to hang on, although he received a good soaking in the attempt. A second helicopter dropped a line for Prather, who also slipped as the helicopter pulled him upwards; but unlike Ross he did not have a good grip, and he fell into the sea. With an open face-plate, water immediately began to flood into his helmet and pressure suit. Having survived one of the most rewarding flights in aviation history, and despite the quick efforts of the navy support divers, he drowned.

Ross was devastated. Despite his insistence that the flight was a success and that Prather had not died in vain, and the award of the Distinguished Flying Cross to

Colonel John Stapp MD and his team of aeromedical researchers demonstrate the effects of high-speed flight using 2,000-foot rocket sledge at Edwards Air Force Base, California. They ultimately proved that humans could survive 48 g deceleration when properly restrained. (Photograph courtesy USAF.)

Prather's widow by President Kennedy at a White House ceremony, this was the end of the Strato-Lab programme.

The very next day – 5 May 1961 – NASA astronaut Alan B. Shepard became the first American to fly in space in a 15-minute sub-orbital flight down range into the Atlantic Ocean.

## PARACHUTES AND ROCKET SLEDGES

By the early 1960s the frontier of space exploration had moved from the upper reaches of the atmosphere to orbital flight itself. While the first astronauts and cosmonauts orbited the Earth in vehicles launched by rocket and recovered by parachute on either the land or ocean, work continued in developing the next generation of spacecraft – hopefully with wings and a rocket in the tail.

The X-series of rocket research aircraft brought their own unique brand of danger and risk, and a new breed of explorer – the test pilot. While the flights of the X-series of rocket planes were constantly pushing the limits of aerodynamic research which would, it was hoped, lead to winged hypersonic aircraft and space vehicles, there were other experiments in rocket sledge acceleration runs and high altitude parachute descents being completed.

Pioneering the rocket sledge experiments was Colonel John Stapp, an Air Force doctor who experimented on himself by riding a rocket-powered sledge. The purpose of the runs was to investigate the effect of high-speed acceleration and deceleration forces on the human body, in not only the aircraft programmes but the developing space programme.

The first runs in the programme were held at Muroc Air Force Base (later renamed Edwards Air Force Base) in California, where a 2,000 foot track was constructed. A mock-up aircraft crew compartment was fabricated, and was mounted on the rail-track sledge. Behind the sledge were three rocket engines which combined to produce a speed of 200 mph. On this track, 255 runs were completed on the sledge (94 empty or with mannequins, 88 with chimpanzees and 73 with human test subjects). During the early runs, the harness broke, and one of the mannequins slammed through the 1-inch thick wooden windscreen and continued for a further 700 feet into the desert when the sledge was stopped. Stapp was the test subject in the first run with a human crewman, and participated in several of the human runs. The tests proved that it was possible to survive up to 18 g, but several injuries could be incurred by the abrupt stop. Stapp himself suffered broken ribs and wrists, severe headaches that lasted for days, haemorrhages of the retina of both eyes, small cuts, abrasions and lots of bruises.

Due to Stapp's experiments, the theory that a human body could not withstand forces up to 18 g was no longer considered acceptable. But what were the limits? To answer these questions, Stapp designed a 3,500 foot high-speed track with a gauge of 7 feet. A Northrup-designed sledge would be propelled by nine rockets delivering 40,000 lbs of thrust. The test runs were held at Holloman Air Force Base, New Mexico. The final run on the Holloman sledge, on 10 December 1954, would see Stapp sustain stresses that were in excess of 600 mph, with a 20 g force on his body at the beginning and 46.2 g at deceleration. He had proved that the human form, adequately protected, could withstand a higher g force than most aircraft could survive, and probably a higher force than most space flight crews would encounter.

The research into high-altitude parachute descents was a logical extension of both the balloon programmes and the research with rocket sledges and X-planes. With safety a prime consideration, a method of escape for a rocket plane pilot would require a combination of ejector seat and parachute recovery from high speed, high-altitude and high-g-force ejection. The altitude parachute descents also provided valuable medical data on the effects of long-duration descents and engineering data for the design of proposed crew personal rescue systems for spacecraft returning from orbit. The Russian experiments were directly connected with the development of the cosmonaut ejection and parachute descent system incorporated in the Vostok spacecraft.

**1959: Excelsior I**

The Excelsior programme was in many ways much simpler than the Manhigh or Strato-Lab programmes. As the objective was to have the occupant leap out of the gondola and parachute to a landing, there was no requirement for pressurised capsules or complicated systems to heat the cabin or provide observation and experiments. It was, however, a challenging design

The first descent began with an ascent by Captain Joe Kittinger on 16 November 1959. A press release indicated an heroic and daring parachute jump from 76,400 feet, but the truth was that the ascent had almost been a disaster and the descent had almost killed Kittinger. Achieving a survivable parachute jump from an aircraft flying at stratospheric heights was one of the most difficult challenges for aviation medicine at the end of the Second World War. The natural tendency for the body to tumble and roll creates a spin that can increase to hundreds of revolutions a minute and the death of the parachutist long before reaching the ground. If a pilot could survive the breakup of his aircraft but die during the descent, there was not much point in wearing a parachute.

The Excelsior programme was devised to evaluate a body-controlled descent called 'skydiving' and the use of stabilising drogues before deployment of the full parachute. This application would also have use in plans for returning spacecraft which encountered difficulties high in the atmosphere, allowing the crew to escape and land by personal parachute.

As he climbed in Excelsior I, Kittinger was troubled by the known phenomenon of the spinning gondola's moving in and out of direct sunlight blocking his view of the gondola's instrument panel, including the important helmet pressure gauge. This was also hampered by the face-plate fogging up, which, added to the sunlight, meant that he was flying blind. At 58,000 feet another problem surfaced in the pressure garment helmet rising off his shoulders. Fearing the helmet might rip off the neck seal, he tried a trick that had worked for another pilot in an altitude chamber who experienced the same phenomenon. Kittinger tied lengths of parachute cord to clasps on the helmet and then onto the parachute harness. It worked. At the same time the condensation in the helmet disappeared, and as he flew through 76,000 feet he prepared to leave the capsule and take a step towards the ground – a long step!

As Kittinger stepped off the gondola he automatically took a gulp of oxygen and held his breath – a natural reaction. He was really alone in the stratosphere, but strangely could not feel himself moving. He seemed to be hanging in mid air. He feared he was too high for the pull of gravity to take effect and that he would remain in free-fall until his oxygen ran out! He waited for the drogue parachute to deploy, but nothing happened. As he tried to reach behind him, he twisted around so that he no longer looked down, but up towards the black sky; and for the first time could sense his movement – very fast, and downwards. The temperature was –104° F, and he was travelling at 400 mph; but he was protected by his pressure garment and he became history's fastest man without an engine.

As the parachute eventually deployed and he hung beneath, his vision greyed and he blacked out, only to awake lying on his back on the desert floor. He had survived. Before a second Excelsior balloon could be launched, the team had to address problems in five areas:

- Difficulties in arming the parachute opening devices.
- Condensation on the helmet face-plate.
- Sun-blindness resulting from intermittent Sun-glare.
- The rise of the pressure helmet from the neck due to internal pressure.
- The activities required by the jumper prior to leaving the gondola.

The lessons learned from the ascent were incorporated into Excelsior II, launched with Kittinger aboard on 11 December 1959, from where he would again jump at 74,700 feet. This time he encountered no problems, and landed safely in a descent of just 12 min 32 sec.

### 1960: Excelsior III

On 16 August 1960 the third flight of Excelsior with Kittinger onboard left the launch site in Turarose, New Mexico. Meteorologists had been keeping an eye on the weather front, and with an approaching storm a call for cancelling the mission was given as the restraining lines on Excelsior III were released. Kittinger rose at 1,200 feet per minute, and also noticed the clouds gathering. He was confident that if there was a weather problem the go-ahead for launch would not have been given.

As he rose he found that the pressure glove of his right hand had no resistance to his movements, as the oxygen feed line to the glove had split. This was dangerous. Without adequate protection on his hand as he rose, pressure would expand the air trapped beneath the skin, pooling the blood and swelling the hand, which would become very painful and very cold.

Kittinger had decided not to tell the ground immediately. There had been too much effort put into the project to terminate it, and his goal was an altitude jump at 100,000 feet. Anything less than that would be a failure, to abort would mean a termination of the programme, and the Air Force was already threatening to cut the budget. He knew the pain would be unbearable and his hand damaged beyond repair, but he had already decided that only a threat of death would force him to abandon this ascent. The pain he would suffer was a price he was willing to pay for the success of the mission. It was not critical to the mission that he would have restricted use of his right hand.

He continued to ascend to over 103,000 feet, and with 99% of the Earth's atmosphere below him he looked out into the void of space. His right hand was twice the normal size, and his eyes burned, but he was distracted by what he could see. At 90 seconds before the planned jump, Kittinger informed ground control in a casual way that his right-hand glove was not pressurised. He received the third degree from ground control on the state of the rest of the suit, and confidently reported that he was fine and that it was 'no sweat'. He then turned off the radio as planned, and with a pulse rate of 136 beats per minute he placed his feet over the sign on the edge of the gondola that proclaimed he was at the 'Highest Step In The World'; and with a quick request to a higher power to protect him, he stepped off – with a record pulse rate of 156 beats a minute (twice normal) – and then dropped towards the ground 103,000 feet (20 miles) below him. At an altitude of 90,000 feet his recorded air speed was 614 mph.

He descended through the layers of the atmosphere, and at 90,000 ft recorded 614 mph – close to the speed of sound at that altitude. His one problem was a brief but choking sensation as the pressures against his body increased and then suddenly subsided as quickly as they appeared. Down and down he fell, until the parachutes opened to slow his descent. It was then that he finally thought that he may survive after all. He tried to release some of his equipment prior to landing, but was not able to as his right hand was still in great pain. He hit the ground 13 min 45 sec after leaving Excelsior. Three hours after landing his swollen hand and his circulation were back to normal, and he was sitting up in the programme office at Holloman Air Force Base, retelling his story with a big grin on his face. He openly admitted that he was afraid most of the way down. He entered the pages of test-pilot legend and helped mould the myth of the Right Stuff. When NASA astronaut Alan Shepard – the first American to ride a rocket into space – was asked if he would have attempted the jump, he gave a typically brief reply: 'That's easy. Hell, no. Absolutely not.'

**1962: Volga**
High-altitude parachute descents by the Soviets developed from the military use of parachutes and the creation of the parachute sporting clubs across the Soviet Union in the 1930s, which were mainly associated with breaking world records a variety of fields, including that of altitude. Part of this programme supported the human spaceflight programme which was being developed in the late 1950s. In the Vostok programme, cosmonauts were to eject from their descending spacecraft at the end of their mission and descend by personal parachute. Tests of these ejection and parachute descent techniques would first be conducted by air force test pilots from the rear cockpits of high-flying jets, and from high-altitude balloon ascents under the Volga programme.

On 1 November 1962 Major Yevgenny Andreyev and Colonel Pyotr Dolgov took off for such a test from Saratov, in a Volga capsule. Dolgov would test the Vostok pressure suit from a high-altitude parachute descent, while Andreyev would test a non-explosive simulation of the Vostok ejection system. Dolgov was wearing a full Vostok pressure suit, an independent oxygen supply, two parachutes and a barograph. Andreyev wore a high-altitude pressure suit overlaid with winter flight clothing, and a pressure helmet completed the suit. Amongst the support equipment he carried was a belt containing batteries for heating, as well as one prime and one back-up parachute.

At 2 hrs 20 min into the flight, Andreyev, seated in a simulated Vostok ejection seat, was mechanically catapulted from the Volga at 83,500 feet. After 270 seconds he separated from the ejection seat and began a free-fall descent for 79,560 feet before opening his parachute at 3,940 feet. He landed after a descent of 7 min 30 sec.

Dolgov remained in the balloon until it had attained an altitude of 93,970 ft whereupon he jumped out 78 seconds after Andreyev had left the gondola. This was Dolgov's 1,409th parachute jump in a long and distinguished career. As planned, he activated his personal parachute during the descent (he was already holder of the world altitude record for parachute opening at 48,671 feet set in June 1960), but at an unspecified time during the descent his suit depressurised, and he died. For 37 minutes the body of Dolgov descended to Earth. Initially the Soviets announced only that he had lost his life 'while fulfilling his duties'. The cause of death was not immediately revealed, and it was speculated was that he either froze to death in the long descent, ran out of oxygen, or that the shock of the parachute opening at such a high altitude contributed to his death. It was many years before the true cause of his death became known in the West.

## 1966: Strato Jump III

In 1963 Nick Piantanida heard about Soviet parachutist Andreyev's parachute jump from a balloon at 83,000 feet to set a new world free-fall jump record. Piantanida wanted to recapture the record for the US. He was not an astronaut or a pilot, or even a member of the military; he was a pet dealer with an interest in parachuting and rock climbing. Over the next two years he worked every moment he could in preparing to beat the Soviet record. He gained support and sponsorship, but not much help from NASA. However, he gained some assistance from the USAF and established the Survival Programs Above a Common Environment (SPACE) Inc. He also had a project name – Strato Jump.

On 1 October 1965 Piantanida finally flew Strato Jump I to 22,700 feet and jumped successfully. On 2 February 1966 he set a world record of 123,500 feet in Strato Jump II, but could not disconnect the oxygen hose, as it was frozen. He struggled against his pressure suit as his balloon climbed two miles higher than anyone else had travelled in a balloon. The record of the longest free-fall descent was not to be his, as he could not release the oxygen hose attaching his suit to the gondola system. He eventually realised that he would not be able to release the hose and that the automatic separation from the balloon in the gondola was the only option. In a wave of disappointment he braced himself for the release and a 32-minute ride to the ground under the recovery parachute system. Severe buffeting caused him bouts of sickness, but he survived the landing. He immediately used a pen-knife to free the jammed hose connection.

On 1 May 1966 Piantanida tried again in Strato Jump III. An hour into the flight, at 57,600 feet, ground control recorded a rush of escaping air and a painful one-word scream – 'Emergency!' It seemed that the pressure helmet or suit had failed, and they had to retrieve Piantanida very quickly. They activated the gondola release mechanism automatically at 57,600 feet but in doing so activated the recovery parachute to slow the descent into a controlled fall. This added precious minutes to the descent, and it would be 25 minutes before Piantanida was safely on the ground near the town of Lakefield, Minnesota.

A USAF Air Rescue team was at the landing site, and dragged him from the gondola as he moaned and sucked in air to breathe. They forced a tube down his throat and rushed him to nearby Worthington Municipal Hospital, where surgeons performed emergency surgery on him for an hour. He had suffered perhaps the most severe case of explosive decompression recorded. He appeared to stabilise, but brain damage was expected if he ever regained consciousness. The truth of what happened would never be learned. On 29 August 1966 he died, never waking from the coma. He was not a member of the official programme to explore the stratosphere, but contributed to the knowledge gained and set a new balloon altitude record that is unlikely to be surpassed. By the time of his death the dangers of stratospheric exploration by humans had been replaced by those of astronauts and cosmonauts risking their lives for the exploration of space. In just five months, three Apollo astronauts would be killed in a tragic pad fire, and three months later a cosmonaut would lose his life in trying to return to Earth.

## ROCKET RESEARCH PILOT ACCIDENTS: LEGENDS OF THE RIGHT STUFF

By the mid-1940s, sufficient information and experience had been gained to provide a clear understanding of the upper atmosphere and the requirements to support humans at high altitudes. Stratospheric balloons had clearly demonstrated that pressurised compartments were the answer, and further development was underway around the world in refining the technology to incorporate sealed cabins into high-flying aircraft. In addition the development of a personal pressure garment also demonstrated that this would provide each crew-member with added protection in the harsh environment. As we have seen, these quite soon led directly to the series of record altitude balloon ascents and delayed parachute descents in the 1950s and 1960s.

But there was another hurdle in aviation that challenged technology and human limits. It had been found, during aerial dog-fights in the Second World War, that during high-speed manoeuvres, steep dives or higher speeds, a wave of air resistance would build up in front of the aircraft, creating a resistance through which the aircraft could not penetrate and which often resulted in mid-air break-up.

During the nineteenth century an Austrian physicist, Dr Ernst Mach, discovered a connection between the rifle bullet shock wave through the air and the velocity of the sound it produced. His name was given to the method of assigning a numeric scale (Mach) to the travel of an object through the 'barrier' of sound. This 'speed of sound' is also dependent on temperature and pressure of the surrounding air, and the speed of the sound generated is in direct relationship to the speed of the air particles. Therefore the speed of sound is not constant, and the changes in the air temperature at different altitudes result in a reduction of the speed of sound due to the changes in temperature, and are not due the altitude itself or a decrease in pressure levels.

At sea level the speed of sound is 760 mph, while at 36,000 feet and above, where the air is much thinner, it drops to 660 mph. To attempt travel beyond this barrier, and to move faster than the speed of sound, an aircraft would have to fly at more than 660 mph at 36,000 feet.

This was the challenge in the late 1940s: to design aircraft that could not only safely meet and exceed this barrier and survive, but also be able to operate and push the limits of aviation further still – even faster, and eventually even higher, towards Mach 2 or 3 (twice or three times the speed of sound) and beyond.

It was the penetration of this invisible barrier that caught the imagination of the public, who saw it as a physical barrier against which a daring pilot would risk life and limb smashing through a wall of sound in his fragile aircraft. The Western pilots assigned to do this became heroes and legends – secretly at first, in the cause of national security, although they eventually became well-known media stars – while the Soviet test-pilots were secret anonymous figures hidden behind the Iron Curtain.

This myth of a physical barrier was quite false. The 'barrier' is a combination of air particles flowing over the surface of the aircraft, which, at speeds lower than that of sound, contract. With speeds in excess of that of sound, the air expands. The areas are two distinct regions where airflow behaves differently. Between them is a sort of 'no-man's land' wherein lies the 'barrier' to breaking the speed of sound. By flying through the three regions of subsonic, transsonic and supersonic, a pilot has to control his aircraft through two conflicting regions of contraction and expansion of the air, and at the level of sonic speed these rise and fall like ripples on water in decreasing levels of pitch, creating a sound wave. On penetrating this wave the forces are equally balanced, and by cancelling themselves out all resistance falls away and the passage of the aircraft beyond Mach 1 is streamlined until it approaches the next Mach barrier, when the process is repeated.

The first person to successfully break the sound barrier was Chuck Yeager, flying the Bell X-1 on 14 October 1947, despite his having two broken ribs produced by a fall from a horse just a few days before the flight. He had a civilian doctor strap him up, and with the help of a sawn-off broom handle to lever the door latch shut, was able to manually lock the X-1 hatch in place. Only a few close friends and family knew of the broken ribs, and the crew of the B-29 drop aircraft knew only of the fall. Almost immediately upon landing, the achievement was placed under a veil of national security, but eventually Yeager became the first of many heroes of the X-series of research aircraft programmes who pushed the limits of man and machine to the very edge of the operational border (known as the 'envelope'). These pioneering steps to space were achieved over the deserts of Muroc Air Force Base on the dry lake beds of California. This later became known as Edwards Air Force Base – the home of the USAF Test Pilot School and the landing site of early Space Shuttle missions.

The successes of the X-planes were also balanced with the tragedies, and a regular feature of test-pilot and family life at Edwards AFB in the 1940s and 1950s was in attendance at the funeral of a colleague who paid the price of flight-test error, and 'bought the farm' earlier than retirement.

### X-1 incidents

There were 237 glide and powered flights of the X-1 and its variants (X-1A, X-1B, X1-D and X-1E) flown between 1946 and 1958. Being a research, development and test programme, small incidents were a feature of this and all research programmes of this type over the years (see the table on p. 25).

During the X-1 programme, Major Frank Everest, USAF, encountered perhaps the most life threatening of these incidents. On 2 May 1949, on his fourth flight in the X-1-1, shortly after dropping from the belly of the B-29 aircraft he ignited the XLR-11 rocket chambers and began his climb, when he heard a loud explosion behind him as all the rocket chambers promptly shut down. Instinctively he tried to move the rudder to gain control, but it was jammed. He calmly reported the incident to the ground and began his glide back to the air base, jettisoning fuel as he descended. As he passed to 30,000 feet the F-80 chase plane pilot flew behind the X-1 to inspect the damage, and reported that the No. 1 engine chamber had blown up and the damage had affected the rudder. Everest landed safely, and post-flight examination revealed that a faulty rocket igniter on No. 1 engine chamber had failed to ignite the propellant gases, whereupon exhaust from two other rocket chambers had mixed and ignited gases from the No. 1 chamber and caused the fire to track back to the propellant feed, resulting in the explosion. It was to be six weeks before the X-1-1 aircraft flew again.

On his seventh flight, on 25 August 1949, Everest became the first person to use a partial pressure suit in an in-flight emergency. As he waited for the drop from the B-29 he noted a small 1-inch crack in the inner canopy of the X-1. It did not increase in size, however, and he continued with the drop and the powered flight. After passing the 65,000-foot level he heard a loud 'poof' and felt immediate inflation of his pressure suit. The 1-inch crack had extended to 6 or 8 inches and had broken through to the outside, his cockpit air rapidly escaping into the void. He immediately cut the engines and nose-dived back to the 20,000-foot level, where he would not have to rely on the functioning of his pressure suit

On 22 August 1951, during preparations for the drop of the X-1D, pilot Everest noted the gauges were relining and tank pressures were dropping. After consulting on what to do next it was agreed that the launch attempt should be cancelled. Everest returned to the X-1D cockpit to jettison the fuel and as he reached to the fuel, control handle a huge explosion rocked the X-1D and sent a tongue of flame into the B-29 bomb bay. Instinctively Everest leaped from the X-1, and the aircraft crew jettisoned the rocket plane in a shower of debris as it plummeted to the ground and was destroyed on impact. The cause of the incidents was the failure of an Ulmer leather gasket in the LOX system used to seal the access doors of inner bulkheads.

On 8 August 1955 the X-1A aircraft suffered a similar explosion and rupture of a liquid oxygen tank while still attached to the B-29 carrier aircraft. Once again it was fortunate that none of the crew was injured, but the damage to the X-1A meant that the fuel onboard could not be offloaded, and so the B-29 crew headed for one of Edwards AFB's bombing ranges, where at 6,000 feet the X-1A was jettisoned and struck the ground with a large explosion. The Ulmer leather gasket was again the cause of the explosion.

Further investigation and experiments revealed that the organic gaskets were perfectly acceptable for use at room temperature. However, when heated to $200°$ F during the engines tests, a quantity of phosphate and wax leaked from the gaskets. When collected and cooled, these solidified and produced a chemical reaction of explosive violence when in contact with liquid oxygen. Such ignition would normally require a triggering impact, and tests with a small hammer revealed that any force on a few drops of the phosphate would detonate it when hit. There were $1\frac{3}{4}$ lbs of leather used in sealing the tanks of the X-1, and this contained 0.45 lbs (approximately a cup-full) of phosphate. Any small impact in the tanks – such as vibrating vent struts – could have triggered the explosions. As a result, all Ulmer leather gaskets were removed from the current and subsequent X-series research aircraft.

Other emergency situations with the X-1 highlighted the humorous side to what was a very dangerous occupation. During one flight, Jack Ridley reported that he had a fire in the cockpit just one minute after leaving the B-29. The chase pilot reassured him that it could only be an electrical fire and that he was not in any danger as there was nothing else to burn in the cockpit Ridley replied rather rapidly that there was: he was still sitting in the cockpit!

X-2 pilots with one of the aircraft. Milton Apt is seated inside the aircraft, while Iven C. Kincheloe stands to the side. (Photograph courtesy USAF.)

## 1956: X-2

The X-2 programme suffered from a series of setbacks, but it also provided several advances in higher Mach levels of aircraft stability and control, and was a pioneer in the development of a jettisonable pressurised cockpit section and personal parachute for pilot rescue, eliminating the need for an ejector seat. The X-2 programme achieved both stunning success and tragic setbacks.

On 12 May 1953 – during liquid oxygen top-off tests, with the carrier aircraft flying at 30,000 feet over Lake Ontario – the X-2 (46-675) in the bomb bay suddenly exploded in a ball of flame. At the time, pilot Jean 'Skip' Ziegler was not in the X-2 but was recording instrument levels in the bomb bay of the carrier aircraft. The force of the explosion killed Ziegler as he was thrown out of the bomb bay. A Bell employee, Frank Wolko, who was in the rear of the aircraft, was thought to have bailed out, but his parachute failed to open. Neither body was ever found. The blast pushed the aircraft 100 feet in the air, and despite the explosion there was no fire and the flight crew were able to land the seriously damaged aircraft. Again, investigation pointed to the explosive properties of the Ulmer leather gaskets and contributed to the growing evidence to omit them from all X-series aircraft.

On 7 September 1956, Captain Iven C. Kincheloe set an altitude record of 126,200 feet (23.9 miles) in the second X-2 (46-674), flying at the very edge of space and earning the media title 'First Man in Space'.

The tragedy of the twentieth and final X-2 flight demonstrated the problem of controlling a high-speed aircraft tumbling in three axes. On 27 September 1956, Captain Milton Apt, USAF, dropped from the carrier aircraft and ignited both rocket chambers, climbing at an angle of 33°. When he reached 72,000 feet at Mach 2.2, he pushed over the top of his flight profile and began a long steep dive with the engine still burning. During the descent at a 6° dive he passed Mach 3 and set a new

An artist's impression of the X-15 in flight.

speed record of Mach 3.196 (2,000 mph – almost 35 miles a minute). As the engine automatically shut down after a burn of 145 seconds, Apt attempted to turn the vehicle to begin his long and very fast descent to the ground.

The X-2, now out of control, dropped 40,000 feet in just 16 seconds, and as Apt fought to regain control he was tossed against the sides of the cabin as the X-2 rolled over and over, pitched up and down and swung side to side, all at the same time, like a wild but deadly fairground ride.

Apt reached for the ejection handle between his legs, and pulled, and immediately the escape capsule rocketed away from the falling aircraft and put further g loads on the ailing pilot. Next, he ejected the canopy and released his seat harness, but was probably too dazed to get himself out of the cockpit, which landed heavily, nose down on the desert floor, killing him on impact.

The subsequent inquiry pointed to the inability of onboard instruments to record real-time data, so Apt thought he was flying much slower than he actually was. As a result, when he tried to turn at high speed it created a sequence of events that led to a supersonic and inverted spin, from which he had difficult recovering. There were also suggestions that Apt had insufficient rocket-powered aircraft experience to attempt such a demanding mission on his first rocket-plane ride.

## 1959: X-15
On 5 November 1959, X-15 pilot Scott Crossfield experienced an in-flight explosion and fire in one of his rocket engine chambers. It caused him to abort his scheduled

The X-15 landing accident in 1959, showing the fuselage sustaining a broken back.

third powered flight of the research vehicle, attempt an emergency landing as soon as possible, dump the remaining fuel, and land. However, due to the steep descent required by the emergency landing, not all propellants were offloaded and so the landing weight was heavier than planned. Aiming for Rosemund Dry Lake, he jettisoned the lower ventricle fin in order to provide ground clearance upon landing, and was guided down by chase planes. At 200 mph, the tail skids of the X-15-2 hit the lake-bed, and because of extra weight the aircraft dropped heavily. Crossfield heard the sound of his plane buckling and twisting behind him, and observers saw the vehicle break its back just behind the cockpit before sliding to an agonised halt. The rescue team sped towards the craft, with Crossfield still inside, expecting the aircraft to burst into flames at any moment. Miraculously the explosion never came, and rescue teams soon made the vehicle safe and evacuated Crossfield. The vehicle had buckled between the cockpit and the fuel tanks!

The design of the nose wheel was such that to conserve space the landing gear was stowed compressed and held in place by oil. When deployed for landing, nitrogen gas, held in place in flight by the oil, would deploy the strut by pressure. On this occasion the gas mixed with the oil and rendered useless about one-third of the compression volume. The excessive landing loads this flight had imparted on the structure highlighted a design oversight. The accident could have happened on any of the previous landings. The aircraft was taken to the Los Angeles facility for repairs that would take several months.

The X-15 landing accident in 1962, during which the aircraft turned upside down. Pilot McKay escaped, but with serious injuries.

## 1962: X-15

On 9 November 1962, Pilot John McKay was making his seventh X-15 flight – the programme's 74th free flight. Flying the 31st flight of No. 2 aircraft, all went well until the pilot attempted final approach. Shortly after ignition, the huge XLR-99 rocket engine had stuck open at 33% thrust, and the pilot had no alternative but to abandon the flight. Abort occurred at 1,000 mph at 53,950 feet altitude. Jettisoning remaining fuel, McKay headed for landing, but as he attempted to deploy his flaps he found he had no response from his controls. At over 290 mph, McKay and X-15-2 streaked towards the ground. Onlookers could only watch as disaster loomed before their eyes.

Acting with instinctive pilot reactions, McKay delayed deploying the landing gear – which would produce drag and decrease his landing speed and altitude – until the very last moment to preserve his all-important lifting capability. Almost immediately after the landing gear was down and locked, the vehicle slammed into the dry lake. The tail skids and nose wheels struck the dirt, and immediately the nose wheels rebounded and collapsed as the nose of the vehicle burrowed into the ground. Inside the vehicle, McKay could do nothing but sit and ride out the crash. The left-hand tail landing skid collapsed under the strain of the forced landing, and the vehicle suddenly lurched sideways and rolled, breaking up as it tumbled along the desert floor. The X-15 eventually stopped upside down, with the badly injured pilot still strapped inside and trapped in the cockpit. For more than four hours, McKay lay inside the wreckage of the vehicle until rescuers freed him from the mess of tangled live wires, leaking fuel, smoke and dust. Again the X-15 returned to the factory for extensive rebuilding, eventually to be flown several more times in the programme. The accident left McKay with three crushed vertebrae in his spine, constant pain and ¾ of an inch shorter. However, always the true pilot, this Second World War veteran flew a further 22 X-15 missions, including one flight of the No. 3 aircraft, on 20 September 1965, which earned him the title of 'astronaut' by flying a ballistic mission reaching more than 50 miles altitude. Shortly after his 29th and last flight on 8 September 1966, he retired. This was forced upon him by the results of the 1962 crash. On 27 April 1975 McKay finally lost his battle with the injuries sustained in that crash, and died aged 52.

### 1967: M2-F2 lifting body

During the 30 years from 1946 to 1975, America conducted an evolving series of research aircraft programmes that were ultimately the forerunners of the Space Shuttle. From the famous X-1 series through the X-2, Skyrockets, X-3, the X-20 studies and the series of manned lifting body designs, most of the aerodynamic work for the Space Shuttle was completed many years before its design was finalised. The aim of most of the work in that period was to obtain data on high performance vehicles flying at very high speeds on the fringes of space, and detailed research into landing such vehicles on runways as conventional aircraft. The success of the research conducted by these vehicles has led to the highly successful Shuttle landings of recent years. However, despite these successes, actually landing these vehicles was not easy, and in another case in 1967, almost fatal.

One of these vehicles, the M2-F2 (M, Manned; F, Flight) lifting body was completing its glide flight programme on 10 May 1967, and was scheduled as the last planned glide flight of the vehicle. At the controls was NASA pilot Bruce Peterson, who already had three lifting body flights in his logbook. Separation from the B-52 parent aircraft and initial manoeuvres in the atmosphere proceeded smoothly, and Peterson pulled out of the final turn of a series of 'S' turns to line up for landing approach. Almost immediately, the vehicle began to 'Dutch roll', imparting a vibrating roll and a repetitive yaw at the same time. This usually occurs at low angles of attack as wind gusts change the vehicle's attack angle, or if a pilot inputs pitch down control on the nose to increase speed, in order to achieve a pre-landing flare manoeuvre. Drawing on his experience, Peterson knew that neither aileron control or rudder application would overcome this problem. The vehicle was now imparting 200° per second rolls, and over the following 11 seconds the pilot compensated for the sudden manoeuvre by increasing the angle of attack of the M2-F2. This was sufficient to bring control of the vehicle back to the pilot, but at the end of this manoeuvre Peterson found himself with a new problem. The lifting body was out of sight of all runway markers, which were a third of a mile off to his right.

In order to avoid generating turbulence during recovery of his vehicle, Peterson's chase planes had departed seconds before, and were unable to help guide him down to the unmarked desert floor that he was now blindly rushing towards. Without their

M2-F2 following the crash in 1967 which injured pilot Peterson.

observations, his only source of aerial assistance was a rescue helicopter directly in his flight path angle.

Peterson lowered the undercarriage just before it impacted with the desert floor, after stalling the vehicle and imparting a larger sink rate than during descent. Travelling at 217 mph, the vehicle suddenly bounced back up and twisted into the air before crashing down into the desert in another cloud of dust some 80 feet away. The vehicle skidded to one side and began a six-roll phase across the desert floor, shearing the undercarriage, the cockpit canopy, the tail fins and other parts. With the dust settling, the vehicle came to rest upside down on the pilot's headrest and the buckled tail fin.

Rescue teams were racing to the scene before the vehicle had stopped rolling. At the site, all hands were mobilised to help free the trapped and unconscious Peterson, who was still strapped in his seat. He was taken immediately to the hospital on Edwards Air Force Base, where he was diagnosed seriously injured with severe facial lacerations, partial blindness in his right eye, several broken bones, a fractured skull and cuts and bruises. He was lucky to be alive.

Over the next two years, surgeons almost rebuilt Peterson, including an 18-month programme of operations to work on his face. He returned to test-pilot duties (though not flying a lifting body) in 1970 before taking up a desk job at NASA in 1971. The film of his accident was used in the 1970s TV series *The Six Million Dollar Man* – the story of a former astronaut, cybernetically rebuilt with artificial limbs after a near-fatal crash. In real life, Peterson was the world's first bionic man.

# The loss of the X-15, November 1967

In the 1950s and 1960s, while the space race gathered momentum, a joint NASA, USAF and US Navy research aircraft programme was looking towards the development of a reusable vehicle. It would be capable of flying at high Mach numbers routinely, to the fringes of space, using a conventional runway landing and an aircraft-style control system. This largely forgotten programme of hypersonic research was called X-15. It was the platform for developments that would culminate in today's Space Shuttle.

The X-15 programme consisted of three sleek, black, rocket-propelled research vehicles. Between 1959 and 1968 they completed nearly 200 missions to the edge of space, 13 of which broke the 50-mile altitude level – the recognised line between air and space. The eight pilots who flew these 'space' missions qualified for the title 'astronaut'. Despite some dispute to this claim, the USAF awarded Pilot Astronaut Wings to its pilots who attained this goal. One such Air Force pilot, Captain Joe Engle, flew three X-15 missions above 50 miles before becoming a NASA astronaut in 1966. He had held USAF Astronaut Wings for almost 20 years before making his first NASA spaceflight on the second Shuttle mission in 1981.

By the autumn of 1967 the three X-15 vehicles had flown regularly from beneath the B-52 launch aircraft without the loss of a pilot. This was a remarkable achievement, considering the complexity of the hardware and the demands of the programme. Budget restrictions were also already in the pipeline, which would terminate the programme towards the end of the decade, by which time all the programme's major research objectives had been met. The X-15 missions had become so 'routine' that most media sources ignored them as the Apollo programme began to recover from the tragic loss of the Apollo 1 crew at the beginning of the year.

The day after the tenth anniversary of Sputnik, 5 October 1967, pilot Mike Adams was selected to fly X-15 mission 191. He would fly the No. 3 aircraft on its 65th mission – his third flight in the vehicle. Adams completed more than 23 hours of simulations in preparation for the flight.

No. 3 aircraft was configured for a programme of high-altitude research experiments, similar to those it had completed on its two previous missions. The

X-15 3-65 (191st programme) flight chronology

Pacific Standard Time (PST) am, 15 November 1967

| | |
|---|---|
| 09.12.00 | B-52 carrier aircraft takes off from Edwards Air Force Base, California. |
| 10.30.07.4 | X-15 launches at 45,000 feet and Mach 0.78 near Delamar Dry Lake, Nevada. |
| 10.30.08 | Engine ignited. |
| 10.30.42 | Velocity 2,000 fps; altitude 60,000 feet. |
| 10.31.07 | First indication of electrical disturbance that lasts for 2 min 46 sec. |
| 10.31.28 | Inertial system computer and instrument failure lights are indicated as a result of electrical noise. |
| 10.31.29 | Engine shut-down by Adams, four seconds late. Velocity at shut-down is 5,236 fps, 136 fps higher than planned. Altitude at shut-down is 150,000 feet instead of the planned 140,000 feet. |
| 10.31.36 | Adams attempts to reset computer malfunction lights. |
| 10.31.45 | Wing rocking experiment initiated. |
| 10.31.54 | Altitude 200,000 feet. |
| 10.32.10 | X-15 begins a slow deviation of the flight plan. |
| 10.33.00 | Peak altitude 261,000 feet, velocity 4,600 fps. Heading misalignment is 15° to the right of the flight path. |
| 10.33.05 | Adams makes three yaw reaction control inputs with his manual controller, in the wrong direction, to the right. |
| 10.33.25 | Adams completes two further right yaw inputs which increases drift to right. |
| 10.33.28 | Adams inputs right, left, then right yaw control inputs, again increasing drift to the right. He is now misaligned 50° to the right of the flight path. |
| 10.33.39 | X-15 is now flying sideways 90° from the heading alignment. The pilot is apparently unaware of gross misalignment. |
| 10.33.49 | X-15 now 180° from its heading alignment, and flying backwards. |
| 10.34.02 | X-15 has turned 360°, and for a while heads in the correct alignment before continuing the spin. Adams informs ground control for the first time that he is in a spin manoeuvre. |
| 10.34.36 | The spin manoeuvre stops, and the aircraft begins to oscillate. At this point it is falling at 3,000 fps. |
| 10.34.54 | Aircraft fuselage buckled due to excessive side loads; altitude 80,000 feet. |
| 10.34.59 | Telemetry data indicates forces in excess of ±13 g; altitude 62,000 feet; speed 2,500 fps. The aircraft breaks up into many pieces and telemetry is lost. The pilot, probably incapacitated by high g forces, does not eject and is killed by ground impact. The wreckage covers a ground footprint of 10 × 1.5 miles. |

vehicle was being used as a test bed, having attained most of its flight objectives some years before. Some of the experiments supported high-speed, high-altitude flight. There were 28 experiments ranging from astronomy to micrometeorite collection, some of which were related to work on the Apollo–Saturn programme, such as development of navigational equipment and thermal protection and other areas of NASA research. This made full use of the unique opportunities of numerous short-duration, high-altitude, high-speed research flights which the X-15 programme afforded to researchers and engineers.

Adams stands beside X-15-1.

Flight 3-65 was originally attempted on 31 October 1967, but was aborted prior to release from the B-52 mother-craft when Adams noticed an engine malfunction. The replacement of the faulty XLR-99 engine, plus unfavourable weather, pushed the launch date back to 15 November.

On 15 November 1967, 09:52 Pacific Standard Time (PST), the huge B-52 took off, carrying X-15-3, with Adams onboard, under its starboard wing. The planned maximum flight altitude was 250,000 feet (qualifying Adams for USAF Astronaut Wings). The programme of scientific experiments included measurement of the solar spectrum, the collection of micrometeorites from the upper reaches of the atmosphere, a check on a boost-guidance system and the measurement of the stand-off distance of bow shockwaves at very high Mach numbers. The results of such experiments would prove valuable in the design of the Shuttle some years later.

The vehicle was released from the B-52, over Delamar Dry Lake, Nevada, at a height of 45,000 feet, at 10:30:07.4. Travelling at Mach 0.78, Adams began his seventh X-15 flight and was scheduled to land at the dry lake-bed at Edwards Air Force Base about ten minutes later. Seconds after release from the wing pylon, Adams ignited the huge XLR-99 engine, and the X-15 began its steep climb to space. The following sequence of events of Adams' last X-15 flight came from a combination of flight data and extracts from the post-flight report.

Less than one minute into the powered flight an aircraft electrical system

The X-15 in its launch position under the wing of the carrier aircraft.

disturbance – later attributed to a possible arc in the bow shock experiment – affected the quality of information transmitted to the ground. The disturbance also interrupted the information provided by the Inertial Flight Data System's (IFDS) altitude and velocity computer software, cockpit displays of the inertial guidance system and the automatic operation of the MH-96 Adaptive Flight Control System (AFCS), the vehicle's primary control system.

After less than five minutes in operation, the bow shock experiment was turned off as planned. Shortly afterwards, all electrical disturbances ceased. The 'noise' this had generated could not initially be distinguished from the radio frequency disturbance, which had been common on several previous X-15 flights, as a result of 'contamination' from engine noise. At the same time, direct communication between Adams and the ground also proved difficult and had to be diverted through the B-52 carrier aircraft.

For a while Adams was slightly distracted, and recorded flight data indicated a drop in the availability of the automatic control of the RCS (Reaction Control System) below 50%. It was a feature of the X-15 that, should any usage of the RCS drop below 60%, the automatic system would disarm and force the pilot to use manual input only. As the vehicle was still in the denser layers of the atmosphere, the aerodynamic surfaces were still being employed and the RCS subsystem was not required. Adams still had sufficient flight instrumentation and control capability within the vehicle to continue with the flight.

The X-15 is dropped from the carrier aircraft, and the engines are ignited to begin the research mission.

As the X-15 climbed, all guidance information was being provided by the boost-guidance experiment. Real-time data measurements were compared with pre-flight data, stored in the onboard computer. The differences between the two sets of data became the guidance command, to correct the flight path from the one actually being flown to the one planned. Additional data displayed included heading altitudes and movement about the three axes, in pitch (up and down), yaw (turn left or right) and roll (left or right). Of these, only the boost-guidance was affected by the electrical disturbance, but the available data still apparently failed to reveal any problem for the pilot.

Adams then either inadvertently applied a small amount of back pressure on the control stick as he reached forward to the engine throttle – a normal reaction seen in earlier X-15 flights – or responded to a command from the guidance display. This resulted in a small oscillation down the length of the vehicle, which dampened out after 20 seconds.

Immediately prior to the scheduled engine shut-down, the electrical disturbance became so severe that it caused the onboard guidance computers to malfunction. As a result, until the electrical disturbance subsided all flight data provided by the computer was erroneous. This included data on the total speed of the aircraft, as well as altitude, rate (angle) of climb, and whether or not the nose was pointed in the direction of flight, or if not, its angle to the left or to the right. With this malfunction, warning lights on the instrument panel were illuminated. Eight seconds later, Adams attempted to reset the computer, but the computer warning light remained illuminated for the duration of the flight, as the recovered cockpit film recorded.

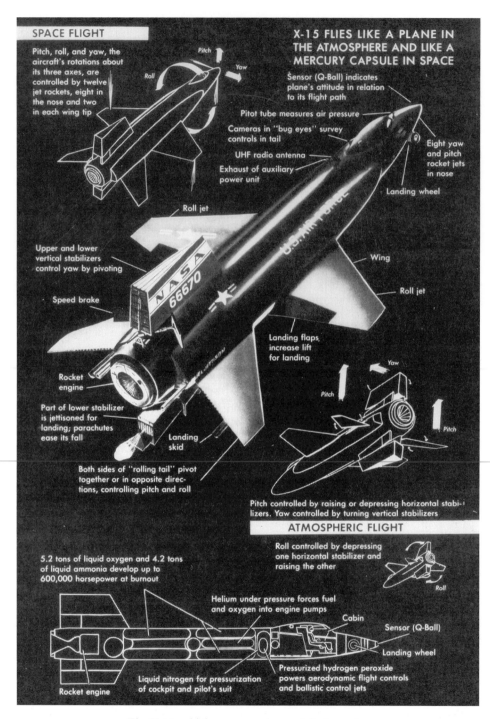

The X-15 vehicle manoeuvring and fuel supply.

With the warning light still on, and using data supplied by inertial total velocity and burn time of the engine, Adams signalled shut-down of the XLR-99 engine, four seconds later than the planned 78 seconds of burn time. His actual velocity was 5,236 feet per second (fps) at shut-down, instead of a planned 5,100 fps. However, the instruments on his panel displayed his velocity as only 5,020 fps – more than 200 fps less than he was actually flying. It was later determined that the electrical disturbance problem had caused a time delay in the real-time data.

Adams' action, in deciding to shut down the engine just a fraction of a second before the fuel ran out, was typical of the X-15 pilots. By ensuring as little fuel as possible was on board, maximum thrust could be attained when the vehicle was at its lightest. Adams' flight was 10,000 feet above the planned 150,000 feet at shut-down, but his instrument display recorded only 140,000 feet. Adams was heading for a maximum altitude that would far exceed the planned 250,000 feet.

At this point, the flight data system should have recovered and reset automatically within a couple of seconds. It did not. The warning light remained on, indicating that the problem was either a continuous one, or it was repeatedly going off-line, adding to the erroneous information supplied to the computer. The recovered flight data indicated that during the electrical disturbance the onboard computer had dumped at least 61 times in less than three minutes.

Immediately after engine shut-down, Adams switched on the precision altitude steering commands. Their data were displayed alongside that of the boost guidance indicator as the vehicle headed towards its peak altitude. The solar spectrum experiment and the micrometeorite experiment were then also activated.

Adams' next action was to 'push over' the control stick to 0° angle of attack, which initiated a wing-rocking experiment designed to obtain data for the UV exhaust plume measurement experiment, located in the tail cone box. A camera located in the box to record the equipment activity was also recovered and analysed, along with experiment records, to determine the attitude of the aircraft at this point in the mission. It was planned that the wing-rocking exercises would be no greater than 10° left or right, at no more than 5° per second. The data recovered indicated that the actual roll was much greater than planned.

From his experiences on past flights and simulator runs, Adams should have been aware that the vehicle's response was incorrect, and fired the RCS steering jets to compensate. By now, they would have been at an effective altitude to override the aerodynamic surfaces. However, the RCS did not fire, indicating either that Adams was not aware of the problem, or that he had a flight control problem that was more serious. The continued electrical disturbance and servo-actuator activity probably prevented automatic operation of the RCS at an altitude where it would normally be in operation.

Observers on the ground were aware of the problem, noting that the RCS was not activated during the wing-rocking experiment, as it should have been. But for some unexplained reason they did not try to identify the nature of the problem.

More than 30 seconds after the computer dump, Adams finally made his first call to the ground, as communications had improved enough to inform them of his problem. 'Okay, I'm reading you now. I've got a computer and instrument light.'

What he failed to tell them was that he had tried to cancel the light but had been unable to do so. This would have indicated the true seriousness of the problem to those at ground control. Ground control advised him to continue flying, using the data displayed in front of him (which was wrong). He was to fly to 230,000 feet and record the degree of sideslip and angle of attack to the flight path. This data was obtained from the 'Q'-ball sensor located in the nose of the aircraft, and was recorded by pushing an Alpha/Beta button on the console with one hand, while holding the attitude steady with the other hand, on the control stick.

By using the data obtained by the Q-ball, Adams entered the new information on sideslip and attack which was automatically cross-referenced by the flight control and then updated. This allowed the flight control system to recover its operational levels to 90% and to release the lock on the RCS, allowing Adams to manually fire the jets to stop a yaw and roll of the aircraft to the right. After two seconds he achieved this, and started to bring the nose back towards the left and zero. Almost immediately, the MH96 flight control system dropped off-line, deactivating the RCS, just as Adams was completing a manoeuvre to stop motion to the left. Despite depressing the trigger on the control stick which commanded the firing of the RCS jets for eight seconds, nothing happened. Recovered film from the cockpit indicated that the general attitude of the X-15 at this point was quite steady, despite the loss of the RCS system.

Ground control requested Adams to immediately check his angle of attack. The cockpit film shows Adams with his finger poised over the switch for an unexplained seven seconds before complying with the request. At this time the data showed that the nose of the X-15 was pointing 6–8° to the right, when it should have been at zero. Three seconds later, both the angle of sideslip and precision heading displays went off the scale high and to the right, but Adams apparently did nothing to stop the drift, or correct the heading alignment.

By now, warning lights from the flight control system dampers were indicating a problem in both pitch and yaw. Adams noticed this and immediately reset the flight control system to re-engage it. Normally this would not have been a problem, but there was a time delay of 20 seconds as a result of the electrical disturbance, effectively disengaging the pitch and yaw modes during this period. In the flight plan there was no call for automatic RCS input at this stage, so when the automatic system failed to activate, Adams did not attempt to input any manual firings.

As the X-15 approached its peak altitude, ground control informed Adams that he would actually peak at 261,000 feet and advised him to re-check his attitudes to begin the precision attitude tracking for the solar spectrum measurement experiment. Pitch and roll was almost zero, but the heading drift was 14° to the right of the flight plan at this point. The advice that he was approaching peak altitude, given by the ground, prompted Adams to activate the solar spectrum measurement experiment. As he manoeuvred the X-15 to the required position, flight control display informed him that he was sideslipping at an angle of 19.5°, a full 15° more than intended. His first action was to fire the RCS to control the roll, when he should have made a combined pitch and roll command. At this time the aircraft was travelling at 4,600 fps.

Adams had apparently mistaken his roll for a yaw. He tried to correct his misalignment, but in the wrong direction. His natural instinct to fly by instruments rather than by visual stimuli, combined with several distractions and the extra experiment workload, added to his confusion. As the vehicle reached the top of its flight path Adams was flying outside most of the atmosphere, but with the nose of the aircraft pointing in the wrong direction. He tried to use the automated reaction control system to regain control, but when he received an inadequate response he instead opted for the manual system. Adams input the commands into the control system, but incorrectly, increasing the error instead of correcting it. On the ground, the heading of the aircraft does not affect the ground track, so there was no indication from telemetry that the X-15 was pointing in the wrong direction as it began the descent back through the atmosphere.

Adams saw a deviation of 2° on the control panel roll indicator, and apparently mistook the reading. He fired the RCS to zero the reading, but actually compounded the problem, firing the jets too excessively and increasing the roll to the right. The ground was still unaware of the problem as he tried to bring the nose back to the direction in which he was flying. The X-15 was now 53° to the right of the heading when it left the B-52. 30 seconds after descending from peak altitude, as aerodynamic forces against the vehicle built up through the dense layers of the atmosphere, the heading alignment error had increased to 90° right of the flight path, and the nose of the X-15 was pitched down 20°. In effect, the X-15 was flying sideways, and Adams was apparently unaware of this major heading misalignment as he came back down through the atmosphere. For 109 seconds he had constantly fired the RCS jets to stop the deviations, but being so preoccupied with the task in hand, had failed to inform ground control of his predicament.

Almost immediately the vehicle entered into a spin at around 230,000 feet, travelling at Mach 5 (3,000 mph). Adams informed Pete Knight of his problem. Knight was flying as co-pilot on the B-52 (NASA 1), and was acting as a communications relay between the X-15 and ground control. Adams: 'I'm in a spin, Pete.' NASA 1: 'Say again.' Twice more, Adams repeated his statement. Knight apparently misheard or misunderstood the call and replied, 'Let's get it straightened out.' At ground control, Adams' wife and mother were monitoring the flight. They were quickly ushered out of the viewing area once it was realised how serious the situation had become.

Ground control found the report of his spinning to be incredible. There was no one monitoring the X-15's heading, which apparently had not been anticipated. No one had thought it possible to 'spin' an aircraft at 3,500 mph, and they could see only pronounced, slow rolling and pitching motions recorded on their consoles.

After 43 seconds of spin, as the aircraft passed the 120,000 feet level, the RCS was firing almost continuously to correct the spin. Incredibly, there was no recommendation in the Pilot Handbook on how to recover X-15 from a flat spin! In addition, there had been no previous experience of such a spin at this altitude and speed. A combination of Adams' piloting skills, the vehicle's own stability characteristics, aerodynamic forces and the action of the MH-96 onboard control system, recovered the spin at Mach 4.7. Immediately after recovery from the spin,

One of final stills taken by the in-cabin camera, showing the instrument panel and readings.

however, the X-15 developed a pitch oscillation (a repeated up-and-down motion of the nose). With this, the vehicle's automatic control system became saturated, causing the pitch damper, used to correct the movement, to remain at its maximum setting. In addition, the horizontal stabiliser was 'saw-toothing' in a $\pm 10°$ oscillation, resulting in a 26° per second vibration. Because the setting was so high, the automatic control system caused the oscillations to increase in severity and become self-sustaining.

The only effective way to stop these motions would have been to reduce the amount of pitch gain to minimum or turn the damper off. Adams did not achieve either of these. The heavy stresses and violent motions of the aircraft had probably incapacitated him. The g loads he had encountered became very violent, and buffeted him in the cockpit. Adams would have been unable to activate the ejector seat as the vehicle dropped at 160,000 feet per minute (2,600 fps), with drag forces increasing at almost 100 lbs per square foot each second.

The X-15's side load limits were the first to be exceeded, causing the loss of the rudder and directional stability. The repeated vibrations in pitch and yaw forced an overload of the flight control system. Adams made no voice contact with NASA 1 or the ground during this period, although Knight tried to keep contact with him: 'Keep pulling up. Do you read, Mike?' The structure of the X-15 repeatedly

The wreckage of the last flight of the X-15-3. The aircraft has come to rest on its left side with the nosewheel well open and the broken nosewheel at front right.

sustained loads of both plus and minus 15 g, exceeding the structural design limits of the airframe and causing the vehicle to break up at over 60,000 feet.

The fuselage buckled at 62,000 feet and 318,000 fps, with dynamic pressure loads of 1,300 lbs per square foot. As the X-15 broke up around him, Adams did not attempt to eject, as he was apparently still trying to recover the hopeless situation. As the vehicle broke apart, Adams was probably already unconscious from the violent g forces, the loss of the canopy, or from debris hitting him. A black mark was later found on the side of his helmet, probably caused either from hitting the cockpit in the buffeting or from debris hitting him.

One of the chase planes reported, 'I've got dust in the lake down there.' The recovery helicopter had scrambled on the first report of the spin being received by ground control. On arriving at the crash site, they reported that Adams' body was still in the cockpit seat.

The wreckage fell over a 10 × 1.5-mile ground area north-east of Johannesburg, California. The major elements recovered from the crash site were three sections of the fuselage, the two wings and the cockpit and, after a long search, the film from the cockpit camera.

A NASA Board of Review was set up to investigate the accident and pinpoint the cause of the break-up of the aircraft and the death of Adams. The report, published in July 1968, presented a detailed evaluation of the events of 15 November 1967. It analysed all available data and put forward several recommendations to improve safety on future flights.

The report stated that although Adams was well qualified for the flight, from previous experience and pre-flight medical tests, he did not challenge his instrument readings, either by radio call or by moving the aircraft to a new attitude to observe any changes. This led the review board to suggest that he was not aware of the serious control problem that was developing early in the flight. As the problems increased he still seemed not to recognise the seriousness of the situation until he entered the spin. This was evidenced by the lack of acceleration forces in the zero g phase of the flight, common on this type of X-15 mission. Normal procedure for this event was to abandon the research objective and recover the aircraft. Major Adams' actions could not, for certain, point to whether he attempted this or whether he continued with the experiment programme, misinterpreting the profile he was flying while occupied with his experiments.

Medical evaluations from the inquiry pointed out two physical abnormalities of the pilot. One was a minor heart condition, but this was dismissed by the flight surgeon as having no bearing on his flying ability or contributing to the accident. The other was an unusual susceptibility to vertigo, which could have played a significant part in the accident. Indeed, it was later revealed that Adams had experienced vertigo on a previous flight. In the debriefing he had indicated that he had no comprehension of what he was doing when the vertigo occurred. He did not mention other incidents of vertigo to the flight surgeons, as it was common knowledge that most X-15 pilots suffered from this during the climb to altitude as a result of the high g forces which the rocket-powered flight incurred.

No anomaly was discovered in the recovered oxygen equipment, and medically it seemed as though Adams was disorientated early in the mission. Evidence from recorded flight data and in cabin camera film suggested that he did try to control the vehicle in a logical manner up to break-up. There was no evidence of ejection failure, or even whether ejection was attempted. In conclusion the board stated:

1    The accident was precipitated during the ballistic portion of the flight, when the pilot allowed the aeroplane to deviate in heading. He subsequently flew the aeroplane to such extreme attitude, with respect to the flight path, that there was a complete loss of control during the re-entry portion of the flight. Destruction of the aircraft resulted from divergent aircraft oscillations, which exceeded the aircraft's structural limits.

2    The MH-96 control system, operating in the automatic mode, contributed to the accident by (a) allowing the intermittent loss of normal reaction control and stability during the ballistic portion of the flight; (b) sustaining a subsequent control-system oscillation that eventually resulted in the aircraft being forced beyond its structural design limits.

3    An electrical disturbance, probably emanating from the bow-shock stand-off measurement (traversing probe) experiment located in the forward section of the right wing-tip pod, adversely affected the normal operation of the integrated flight data system computer and the adaptive flight control system. These effects

were apparent to the pilot as a deterioration of the aircraft's response to control input, pitch and roll-damper trip-outs, inaccurate inertial velocity and attitude indications, and inertial computer system malfunction lights.

4   Prior to the loss of control, the pilot had essential subsystems, adequate display information and sufficient aircraft control capability.

5   The pilot's improper control of the aircraft was the result of some combination of display misinterpretation, distraction and possible vertigo.

In addition, the X-15 Accident Investigation Board offered the following recommendations as possible means of improving the levels of safety of future flights of the X-15 and comparable aircraft.

1   A telemetry indication of aeroplane heading should be placed in the X-15 control room where it would be visible to the flight controller.

2   The destruction of the only X-15 equipped with an adaptive control system made detailed recommendations in this area unnecessary. However, there appeared to be inadequate information available to aircraft designers and operators, calling attention to the inherent characteristics of self-adaptive control systems, which under certain conditions would be detrimental to the operations of an aircraft. Therefore, it was recommended that NASA and Air Force engineers publish a report within 90 days, summarising experience with this type of control system at the Air Force Flight Test Center.

3   All experiments and other equipment would be environmentally checked before being placed aboard X-15 and other high-altitude aircraft.

4   All X-15 pilot candidates would undergo astronaut-type physical examination, including specific testing for labyrinth sensitivity (vertigo).

5   The following actions would be taken to reduce the possibility of attitude deviations of the X-15 aircraft while under predominantly ballistic conditions:
   a)   The primary attitude indicators and associated vernier functions would be used only in the conventional manner and only for basic flight parameters.
   b)   Insofar as practicable, flights would be kept within the regions where the 'ball-nose' was usable and the 'ball-nose' information (angle of attack and angle of sideslip) would be continually displayed to the pilot.
   c)   If higher-altitude flights are necessary, a source in addition to the stable platform should be provided in order to maintain redundancy in regard to heading information.

Pilot error was not indicated in the report. A combination of severe vertigo, malfunctions and distractions all contributed to the tragic end result. The high g loads, to which the pilot's sensory system had been subjected, were still being evaluated. Adams probably did not want to admit that he had a problem which he could overcome. Unfortunately, the sequence of events moved too swiftly for him to recover.

After the tragic death of Mike Adams, only eight flights were left for the programme, with funds available until the end of 1968. The historic programme was to end with its 199th flight, on 24 October 1968, as the 200th flight was delayed and then cancelled. All flights after Adams' crash were made by the No. 1 aircraft, which had started the long series back in 1959. The tragic accident that ended Adams' last flight was a bitter blow to the programme and its remarkable safety record.

The lessons learned from the X-15 flights played a significant role in the development and evolution of the Space Shuttle in areas of operating flight data systems in a high dynamic pressure and space environment: turbulent heating rate levels, measurements of hypersonic aircraft skin friction; discovery of hot spots which were generated by irregularities on the airframe surface, and in positioning the vehicle for runway landings.

The death of Adams had also once again revealed that at all times the high-speed and high-altitude test pilot must keep his wits about him, and that hypersonic return from space by a winged vehicle was not an easy operation. The successes in returning the Shuttle from space began with the X-planes and the contribution of pilots like Adams in ensuring the effects of landing an unpowered winged spacecraft at high speeds and angles of approach were understood before the Shuttle left the launch pad. Once launch is secured, there can be only one attempt to land a Shuttle – and it has to be right first time.

# Summary

By the dawn of 1957, considerable experience had been gained from the human exploration of the stratosphere by balloon, aircraft and high-altitude rockets. The next logical step was to orbit a satellite around the Earth, which was achieved with Sputnik 1 on 4 October 1957. The following month the Soviets orbited the first living creature, a dog called Laika, onboard Sputnik 2. It would not be long before humans would follow. With the orbiting of Yuri Gagarin and the flight of Alan Shepard in 1961, the era of human space exploration had dawned. The operational phases of stratospheric balloon, rocket research aircraft and lifting body programmes continued for the next 14 years through 1975, and direct applications from those programmes have continued both in space and aeronautical research to the present day.

By drawing upon the experiences gained by the high-altitude balloon and rocket plane experiments and from the military ballistic missile programmes, the first spacecraft to carry a human cargo was developed in the United States and in the Soviet Union. At a time of strained relations between the two superpowers, the first to place an object and finally a human in orbit would clearly demonstrate technological superiority. The Soviets had won the first leg of the race with the orbiting of Sputnik and Laika; and with the contest to put the first man in space, the Cold War had also spawned the Space Race.

Using the experiences of crew-members from the balloon programmes and early rocket research aircraft allowed spacecraft designers to refine the layout of pressurised cabins, pilot controls and displays, and to provide pressure garments with extra facilities. The requirements for the support of human life were learned from experiences during the early pioneering exploration of the stratosphere, and were incorporated into the first space vehicles to carry man.

The rocket research plane development in supersonic flight up to and including Mach 6 helped not only important studies that were later used on the Space Shuttle programmes but was also directly applied to both operational military and commercial air transport. NASA – founded in 1958 – began a programme of aircraft safety with the X-planes, and continues to use knowledge from its space and aircraft programmes to provide improvements to aeronautical research worldwide.

The X (eXperimental) series of aircraft also provided valuable research baseline data for aircraft performance in flying to, operating in, and returning from the very edge of the atmosphere at a variety of speeds and altitudes. This had direct applications in the development of the Space Shuttle, and with the data provided from the series of lifting bodies allowed Shuttle designers to confidently plan safe landings of the 100-ton unpowered orbiter from space, travelling at Mach 25 down to below Mach 1, and a first-time landing in about 45 minutes.

From the safety aspect, experience in using life support and crew rescue systems in balloon gondolas, high-altitude parachute descents and rocket research aircraft provided valuable information that was directly applied in the evaluation of proposed designs for the early spacecraft planned to carry a crew. The development of a collapsed balloon becoming a parachute for gondola descent, in both the US and USSR programmes, also became a redundant system after the provision of personal parachutes – if the crew were able to use them. The USSR Volga balloon descents featured tests of the Vostok ejection system and qualification of the Vostok pressure garment used by cosmonauts on Vostok orbital missions.

Training and simulation developed during these pioneering programmes, along with the development of the aircraft flight simulator. Ground simulations were identified as a vital preparation for a test flight, and the prompt reaction of crew members in difficult situations – instinctively following procedures used on countless simulations – saved their lives on several occasions.

The development of these vehicles during the 1930s–1950s also saw the development of relatively new technologies such as rocket engines and supercold fuels, which produced higher performance but also higher risks, as witnessed by the Ulmer gasket incidents on the X-planes and the explosive nature of the fuels used. Personal escape systems also evolved during this period, and the parachute and the ejector seat were developed, as was the already complex ground support in mission preparation, tracking and recovery.

These recovery systems also included, for Strato-Lab, the first ocean recoveries that the Americans later adapted for the Mercury–Apollo programmes. The Soviets, on the other hand, learned that with their large landmass and smaller ocean support fleet, land recovery would be the best option for their returning spacecraft.

As early as the 1930s the Soviet practise of not announcing an event until it had been successfully accomplished had begun to emerge. They had been caught out with the announcement of pending balloon launches, only to have them abandoned in full view of invited witnesses. They soon opted for a secret launch followed by the release of a news bulletin of the event after it had occurred, and this practise was carried on during the Soviet space programme well into the 1980s.

The experience of obtaining data from the upper levels of the atmosphere also saw the first indication of rivalry over what was then termed 'manned' or 'unmanned' vehicles. The scientists wanted more experiments flown, but by using a human crew this was curtailed by the need of providing life support systems. This argument between human versus automated exploration of the stratosphere has continued into the space programme to the present day.

Sending men into the stratosphere also allowed for the collection of preliminary medical information from adequately protected human bodies at extreme altitudes. This in turn provided further references for baseline data on the changes that occurred in high-altitude, high-speed flight, the response to normal and stress conditions, and the ability of the crew-member to perform simple or more complex tasks again under normal or stressed situations. This knowledge was generally used to refine the design and layout of subsequent vehicles (such as the X-15, X-20 and lifting bodies), which in turn influenced the design of spacecraft, such as using pilot input in designing the Mercury instrument panel and allowing for pilot control rather than fully automated systems.

The rivalry between the test pilots at Edwards Air Force Base and the astronauts at NASA over piloted aircraft (such as the X-15) against sitting on a rocket or under a parachute (as in Mercury) became part of the lore of the early space programme.

Overall, the experiences of the stratospheric programmes of the 1930s through the 1960s combined with the rocket research aircraft and lifting body programmes of the 1940s to 1970s, and were supplemented by the rocket sledge experiments, developments in pressure garments and ballistic missiles, and other related areas in science and technology, all of which provided a database from which to build on in order to send humans into space. However, nothing could be learned of the effects of putting a human into orbit until the first person actually accomplished the task. On 12 April 1961, Yuri Gagarin made that first step into the unknown.

# Training for space

# Overview

If you think that flying in space is all that astronauts do, then you could not be more mistaken. There is a well-known saying that 'practise makes perfect', and this is certainly the case for spaceflight. Years of training go into any flight – be it a few short days, or several weeks in space. If you are lucky, you are assigned to a flight crew, but this also brings its own problems, both in flight and upon return.

As each crew-member repeatedly practises phases of a mission in simulations on the ground, in the air, or underwater, so the ground crews and flight control teams conduct their own simulations and training. They team up with the astronauts for integrated simulations, so their roles in the flight are well rehearsed in the event of an emergency. As the teams undergo their complex training, the trainers throw in sudden problems or malfunctions both for the crew and for the controllers to deal with. Sometimes multiple failures are thrown at the crews, who frequently think the trainers are not as friendly as they seem. The simulator training is so realistic that when problems have developed on space missions the crew often questions why the trainers have put the problem into their real mission!

Before any astronaut or cosmonaut can be assigned to a mission he has to undergo intensive training from the very first days of entering the programme. Irrespective of the years of study, experience, and academic work accumulated prior to selection as an astronaut or cosmonaut, each candidate starts afresh in astronaut basic training school (called 'Grubby School' in the US). Following the period of basic academic and orientation training, all candidates support other missions on the ground, or work on future flights and programmes, performing engineering assignments. Added to this is a period of personal tours, visiting and training at contractors and support facilities. The life of an astronaut or cosmonaut is mostly far from glamorous. Indeed, the mundane and repetitive engineering simulations are, to some, downright boring.

For the lucky few, the chance of making even one flight in space comes after long years of preparation and disappointment. In the case of American astronaut Don Lind, the wait from selection to lift-off was 19 years. For some, the big day never comes and, sooner or later, they leave the programme without a flight. A few unfortunate ones give everything they have, training for a flight they will never make, and lose their lives in a training accident.

In the early days of the programme several Russian cosmonaut candidates were rejected before the end of their training programme, having for one reason or another been deemed to be unsuitable to continue. Unlike the NASA astronauts (who are announced to the world's media upon selection), the Soviet cosmonauts were selected in secrecy, and were only identified upon being launched into space. Some of the early space accident stories feature mystery cosmonaut trainees who supposedly perished on secret space missions. Others were seen on still photographs or on training films, but no details were released for many years. The skill of 'space sleuthing' came into its own in trying to unravel the story of the Soviet cosmonaut team between 1960 and the early 1990s. But it was a tricky business, because many of those caught on film and suspected of being cosmonaut trainees were actually technicians and other ground personnel.

At the beginning of human exploration of space, since any spacecraft returning to Earth might be required to land virtually anywhere in an emergency situation, crews were trained for a variety of survival courses in arctic and desert conditions, in mountainous areas, jungles and oceans. Accidents were remarkably few and far between during this training, given the discomfort and dangers inherent in the courses.

One cosmonaut is known to have suffered serious injury in the early phase of training. Alexandr Viktorenko joined the cosmonaut team in 1978, and following a year at test pilot school he commenced basic training for flights on the Soyuz–Salyut programme. While undergoing isolation chamber tests in 1979 he was badly burned in a needless, avoidable accident. He was forced to withdraw from the team while he recovered, but was eventually restored to 'flight' status, completed training, and flew three successful space station missions. Official accounts of the incident refer to his 'very long and difficult course of training.' This incident occurred 18 years after the loss of cosmonaut Valentin Bondarenko in a similar isolation chamber fire.

It seems remarkable that the lessons learned from that tragic incident, and from the loss of the Apollo 1 astronauts in the 1967 pad fire, had *not* been learned. On 11 July 1993, cosmonaut Sergei Vozovikov was participating in water survival training exercises in the Black Sea, with cosmonauts Alexandr Lazutkin and Sergei Treshchev, when he became caught in a discarded fishing net and drowned. He had been selected as one of three military cosmonaut candidates in May 1990, and had completed the candidate training programme to qualify for assignment to Soyuz TM missions to Mir in March 1992.

Medical problems certainly affected a large number of astronauts and cosmonauts in this area of their flight preparation. A review of the most serious incidents of medical disqualification is presented here, followed by an overview of perhaps the most frequent area of danger to training crews – flying accidents. For both American and Soviet/Russian spaceflight preparations, training for elements of their missions involves the use of high-performance aircraft. As the development of the International Space Station renders crew training more global, so the risks increase for the individual.

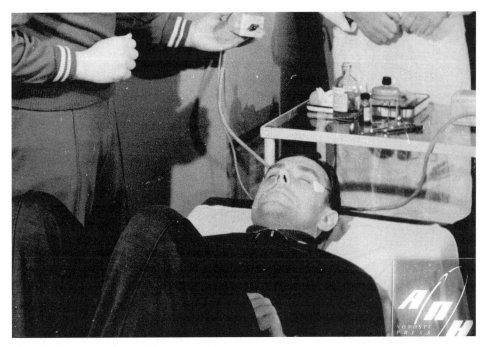

Cosmonaut Vladimir Komarov undergoes medical tests during preparations for Voskhod 1 in 1964. He had been grounded from Vostok 4 due to a heart murmur, but secured the command of Voskhod 1 and later the ill-fated Soyuz 1 in 1967. (Courtesy Novosti Press Agency via Astro Info Service Collection.)

## MEDICAL SETBACKS

To fly in space a person has to be reasonably fit and healthy. In the early days of the programme, because of the unknowns and the complexity of the vehicles flown, health and fitness was as important as mission training. Today there is still a requirement to attain certain physical standards and pass fitness tests, although some criteria for astronaut and cosmonaut selection have been relaxed. Frequent medical check-ups allow you to keep your 'space boots' on for some years, though the chance of flying may be pretty slim.

A prime example is John Young who, after almost 38 years, is still a flight status astronaut (technically, at least) although he has not flown since 1983. However, pass the tests and you qualify. Story Musgrave took his sixth and last flight at the age of 61 in 1996. John Glenn left NASA in 1964, spent the next 25 years in the Senate, but kept fit and healthy, and flew on the Shuttle in 1998, aged 77. There are now several hundred people from around the world who have flown one or two missions The most missions flown by an American or Russian is normally six, but this can take many years to achieve and requires regular medical examinations to remain on the active flight list.

In the early days, all astronauts assigned to a flight crew were also 'backed up' with an alternative, who would step in should the 'prime' become ill or injured. Back-up and support crews were then recycled to later missions as prime crew, with their own set of back-up colleagues. This is still sometimes the case today. For American astronauts, the frequency of Shuttle flights provides a pool of flight-ready astronauts from which to draw replacements for medically disqualified astronauts, if necessary. For the Russians, back-up crews continue to provide an alternative flight crew if needed. With the advent of the ISS programme and the complexity of training which it involves, the role of back up crew-members is likely to be called upon more frequently for resident long-duration crews.

America's first astronaut, Alan Shepard, developed an inner ear ailment that grounded him for ten years before corrective surgery allowed him to select himself for a Moon flight. During that time he served as Chief of the Astronaut Office. Astronaut Deke Slayton and cosmonaut Vladimir Komarov developed heart murmurs that grounded them for years, although both eventually flew in space. In a 1991 interview, Slayton (who waited 16 years to make his first spaceflight) stated that one of the worst things he ever did in the programme was to 'allow some fool of a doctor to put a stethoscope on me.' It cost him his Mercury flight, and probably also a Gemini and an Apollo flight.

Broken limbs were not uncommon in training. Donn Eisele was removed from Gus Grissom's ill-fated Apollo 1 crew due to a shoulder injury; Michael Collins was removed, due to surgery on a bone spur, from the Frank Borman crew that eventually flew Apollo 8, but fate (and the rotation cycle) put him back in line for the Apollo 11 mission as a result; and cosmonaut Valentin Lebedev was replaced just three weeks before launch for a six-month mission, because he broke his leg in a trampoline accident.

In May 1993, Shuttle astronaut Rhea Seddon broke four metatarsal bones in her left foot, at the Orbiter Crew Training Facility, at Johnson Space Center (JSC), near Houston. Practising routine post-landing emergency egress from the full-size orbiter mock-up, she slid down an inflatable slide, but her left foot became pinned under her, causing the injury. But she was training for her third mission, and her past experience allowed her to miss out on much of the refresher course training that her crew was completing at that time. After two weeks rest she was able to resume training, catch up, and fly the mission later that year.

In June 1993 Story Musgrave was slightly injured while performing simulated EVA activities for his impending Shuttle mission to repair the Hubble Space Telescope. Wearing a full EVA spacesuit, he was working in freezing vacuum conditions inside one of the altitude chambers at JSC when he suffered from a mild case of frostbite of the fingers through his EVA gloves. The incident was minor and he soon recovered, with no impact on his flight assignments. However, if the suit had leaked or been ripped, then it would not have been such a lucky escape.

Illness also took its toll. Ken Mattingly was exposed to German measles from the children of a friend of back-up astronaut Charles Duke shortly before the launch of Apollo 13, and was replaced by Jack Swigert three days before launch. He never did contract the illness, and later flew with Duke on Apollo 16. Ironically, Duke almost missed that mission due to a bout of pneumonia in the weeks prior to launch. A bout of influenza also caused NASA to think of delaying the Apollo 9 launch when all three prime astronauts caught colds. However, they launched on time. The first mission to be postponed was STS-36 in 1990, which was delayed to allow mission Commander John Creighton to recover from the influenza, as there was no astronaut qualified to replace him at short notice. As the mission was not time-critical, a delay to wait for the prime crew was judged to be the safer option.

This was not the case in 1996, when the prime resident crew of Soyuz–Mir (EO-22) was grounded and replaced by their reserve crew a week before launch. This was not such a drastic measure as it at first seems, as the Russians train two crews and do not finally decide which will fly until the medical check-up a week before launch. In this case, Commander Gennadi Manakov was found to be suffering from undisclosed heart problems, and he was admitted to the Burdenko Hospital in Moscow as a precaution. As he could not fly the long-duration flight, his Flight Engineer Pavel Vinogradov was stood down and paired with a new Commander, Anatoly Solovyov, to fly to Mir at a later date. The French cosmonaut on the flight, Claudie André-Deshays, was unaffected, as it would be a short visiting mission, so she was able to fly in space with the new crew of Valeri Korzun and Alexandr Kaleri.

Perhaps the luckiest crew to be replaced was that of Alexei Leonov, Valeri Kubasov and Pytor Kolodin, the original Soyuz 11 crew trained to spend three weeks aboard Salyut 1 in 1971. On 4 June, just two days before launch, Kubasov was diagnosed with a lung ailment, and the whole crew was replaced by their back-ups. The back-up crew flew the mission, but perished on the deorbit phase.

## GROUND TESTING

As previously mentioned, one of the more mundane elements of spaceflight training is the hundreds of hours of engineering simulations and systems testing, mostly before assignment to a crew. The danger of dealing with state-of-the-art technology, and with hazardous materials such as gases and fuels, is illustrated by the series of events which befell X-15 pilot Scott Crossfield during the early days of the programme.

### 1960: X-15

When a new engine is test-fired, a crew is not normally required to sit inside the vehicle. However, for the X-15 programme the pilot had to man the vehicle during tests and simulations. In the early part of the programme the large XLR-99 rocket engine (which had about the same power as the Redstone missile used in the Mercury programme) was running behind schedule. Two X-1-type engines were used until the big engine was available for more powerful flights into the upper atmosphere. By 1960 the big engine was ready for testing in an X-15 in a ground test in which the vehicle was bolted into a test frame. The third vehicle had been kept in reserve for the arrival of the engine, and No.2 aircraft was also to be modified to use it.

In the late afternoon of 8 June 1960, Crossfield was ready to put the XLR-99

The result of the explosion on 8 June 1960, which split the X-15 in two. This view shows the XLR-99 engine damage still bolted to the static ground test support frame.

through its paces. The third aircraft had been specially clamped to the concrete apron at the ground test stand. Concrete observation bunkers were built underground, and equipment was positioned to record everything that happened to the X-15 and its engine. As this was 'just' a test, Crossfield arrived wearing street clothes, and donned an oxygen face-mask in the cockpit to provide breathable air via an external supply. As the ground crews retreated to their bunkers, fire and rescue crews stood by at a safe distance.

The plan was to conduct an engine ignition after a simulated drop from the B-52, following the sequence of events as they would occur on a real mission. Following the command 'drop', Crossfield waited a couple of seconds and then started the test firing. As the throttle was moved, the engine roared into life at half thrust. The whole ship vibrated in the steel mounts and, as Crossfield edged the throttle to full power, the noise was terrific. In a normal flight the power of this new engine would enable the X-15 to exceed Mach 2 and the 100,000 feet altitude restrictions and hurtle towards the fringes of space. For this demonstration, Crossfield would shut down the engine for a few seconds, then restart it again at 50% thrust, just as he was to do in the first in-flight demonstration a few weeks later. Onboard safety devices on the X-15 sensed abnormalities in the burn, and automatically shut off the engine. However, with no indication of trouble in the cockpit, Crossfield began the restart procedure, effectively overriding the system that had switched off the engine in the first place. As he hit the restart button, disaster struck.

In the blink of an eye, the X-15-3 blew up with tremendous force, as 900 gallons of ammonia and 60 gallons of hydrogen peroxide ignited. The airframe, from the trailing edge of the wing backward, was completely destroyed. The front section of the aircraft, including Crossfield in the cockpit, was hurled 20 feet across the apron. It was the shortest and fastest rocket ride in history! Having his head already reclined in the headrest probably saved Crossfield a broken neck due to the 50 g acceleration. In the wreckage of the cockpit, Crossfield shut down as much instrumentation as he could to prevent a second explosion shooting him across the airbase, and then waited for rescue. Fire engulfed the apron, but within a minute, on-site rescue teams were trying to get him out. He later wrote that he would have preferred to remain inside the cockpit of the X-15, which was one of the safest places in the world in the event of a fire. Fire and rescue crews were soon controlling the blaze, and he was quickly helped out of the wrecked plane. Luckily, he was uninjured.

With all the telemetry and film that had been recorded, this became the best-documented aircraft accident in history. Post-flight analysis determined that it was a malfunctioning relief valve and pressurising gas regulator that caused the explosion. According to Crossfield, they had been caught in a typical sequence of events, hidden in the mysteries of the fabled Murphy's Law (if a something *can* go wrong, it *will* go wrong). The X-15-3 was rebuilt, and the new engine flew on X-15-1 and X-15-2.

The forward fuselage of X-15-3 is hoisted from the static test stand. Pilot Scott Crossfield survived a massive explosion and fire while sitting in the cockpit.

When contacted by news media, Crossfield was ready in true test-pilot style with the answer that they expected. The only casualty was the crease in his trousers! The firemen had soaked them when they sprayed the aircraft with water. One reporter later innocently enquired, 'Are you sure it was the fireman?' 'Yeah, I'm sure,' replied Crossfield, silently fearing the possible headline: 'Space Ship Explodes, Pilot Wets Pants!'

## TRAINING ACCIDENTS

### 1961: Bondarenko

For more than thirty years the early part of the Soviet space programme was full of 'secret history', rumour and guesswork. It was not until 1986, and Glasnost, that new information began to trickle out. Since the fall of the Soviet regime in 1991, the true story of the origins of the Soviet programme has emerged. One of these revelations was the identification of the full team of first cosmonauts selected in 1960. In the spring of that year, twenty men had been selected. Twelve of them eventually flew in space; but eight did not, and their names were withheld for many years. Western observers had known from photographs of the Gagarin group, and from snippets of information, that these eight men had all been dropped from the programme for a variety of reasons. They had played vital support roles during several missions, without getting a chance to fly in space. In 1977 former cosmonaut Georgi Shonin finally identified their first names in his book, and details of their fates were released in 1986. One of them tragically emerged as probably the first person to die as a result of spaceflight training: 25-year old Senior Lieutenant Valentin Bondarenko, the youngest of the group.

On 23 March 1961, three weeks before Yuri Gagarin flew on Vostok 1, Bondarenko was completing a ten-day series of experiments in a pressure isolation chamber. As with the others he had endured long periods of complete isolation and silence, in a chamber in which the atmospheric pressure had been reduced and a higher oxygen content added, to prevent the bends at simulated high altitudes. The young cosmonaut trainee had just completed a series of medical tests when, as he removed the sensors attached to his body and cleaned himself with pieces of cotton dipped in alcohol, he carelessly tossed one of the swabs aside. It landed on a hot plate, and ignited immediately, turning the oxygen-rich chamber into an inferno and severely burning the unfortunate Bondarenko.

The chamber took several minutes to open, due to the intense heat and the procedures required in equalising the pressure, opening the door and rescuing the injured man. Bondarenko – still clinging to life – was immediately taken to the Botkin Hospital near the Cosmonaut Training Centre accompanied, it is said, by Gagarin. Having been burned over 90% of his body, he died eight hours later. The very next day, Gagarin and the other members of the team left for their first visit to the Baikonur Cosmodrome, where, just 18 days later, Gagarin rode Vostok 1 into history.

Cosmonaut Valentin Bondarenko fell victim to a tragic accident during an isolation chamber test just three weeks prior to Vostok 1.

For the next 25 years, the events of 12 April 1961 stood as a landmark in the history of exploration. But only a very few knew of the accident of 23 March that claimed Bondarenko's life and haunted the cosmonaut team for many years.

In January 1967, three American astronauts also perished in the flash fire of the Apollo 1 spacecraft. Whether their lives could have been saved by the knowledge of the Bondarenko incident, it is hard to say. But if the details had been known, then perhaps NASA would have considered the consequences of the atmospheric mix in the Apollo spacecraft more carefully, as well as more efficient escape techniques and faster-opening hatches.

## FLYING ACCIDENTS

The one element of the space explorers' training programme that has claimed more lives than any other is flying accidents in high-performance jets. As the majority of the people who have been selected to fly in space to date (especially in the early days) have come from a flying or military background, high-performance aircraft, along with their own unique dangers, have become part of their daily lives.

Most of the early astronauts were former test pilots or combat veterans, and the risks of flight testing were part of their job, stretching new aircraft beyond design parameters to test their performance every day. Such feats were both personally rewarding and inherently dangerous. Those who lived to tell the tale gave birth to the

'Right Stuff' legend, and along with the exploits of Chuck Yeager, Scott Crossfield, Bill Bridgeman and others, at what became known as Edwards Air Force Base, the mould for the astronaut-hero was cast.

Because space support facilities and contractor factories are deliberately distributed across the USA and the rest of the world, the need for crews to travel from place to place quickly led to the acquisition of a fleet of high-performance jets. American astronauts used T-38s, while the Russians flew MiGs. In their T-38s, the astronauts could cover 1,000 miles across continental United States between NASA field centres or contractors spread across the country. Some of the fabled 'tales from the astronaut office' include stories of astronauts trying to make a trip between two locations on one tank of fuel and glide in to the runway with only fumes left in the tank!

Another astronaut flying 'competition' in the early years was the 'Tire-Biter's Award.' Competition to score points off their peers was rife in the Astronaut Office. This even included getting the T-38 to its parking place first. An astronaut would fly in and (since the aircraft could be stopped in a very short distance) apply the brakes and swing into the allotted parking slot – a dangerous manoeuvre which could easily result in the plane rolling over. A blown tyre was usually the price to pay, and the astronauts created a special award for those who achieved this lesser-known flying skill. The origin of this award was credited to astronaut Dick Truly who, on landing at LA International Airport one day, tried to beat his colleagues, in a second aircraft, to the parking area. He applied all brakes almost on landing, and burned them off to the rim. He then quietly asked the control tower for a tow vehicle, as he had blown a tyre.

Most of the astronauts also enjoy the thrill of racing and aerobatics. Sometimes this has cost them the chance of a space flight, and sometimes their lives. In 1989, Shuttle astronaut Dave Griggs – an experienced NASA research pilot before he became an astronaut – died in the crash of his vintage North American T-6 aircraft during off-duty aerobatics, while rehearsing for an air show, when it slammed into a field near Earle, Arkansas. In 1990, three-time Shuttle pilot Robert 'Hoot' Gibson clipped the wing of his small Cassutt racer on a second aircraft piloted by Rocky Jones during a Formula One air race in New Braunfels, Texas. Jones went straight into the ground and was killed, but Gibson managed to land safely, due to his considerable flying skills. Coming one year after a new NASA rule, brought in after Griggs' death, forbidding astronauts currently assigned to a crew (Gibson was in training for a 1991 mission) to attempt 'recreational' flying, Gibson was taken off the flight, although his future career did not suffer too much. He was soon back in training, flew a fourth mission in 1992, was appointed Chief of the Astronaut Office, and then commanded the first docking of the Shuttle to Mir in 1995!

Over the years, astronauts' or cosmonauts' flying skills have been put to the test – and not every pilot has survived to tell the tale.

### 1963: Scott and Adams
In 1963, USAF pilots Dave Scott and Mike Adams were students at the USAF Aerospace Research Pilots School (ARPS) at Edwards Air Force Base, California.

In October 1964 Ted Freeman was the first NASA astronaut to be killed during training.

ARPS was an extension of the USAF Experimental Test Pilot School, and included space studies covering the fundamentals of rocket planes and spaceflight by winged vehicles.

In August of that year, both were completing a simulated X-15 approach in an F-104 aircraft when a propulsion system failure caused the aircraft to slam tail-first into the runway. Adams, in the rear seat, made an instant decision to eject. Scott, in the front seat, elected to stay with the aircraft and rode out the crash landing. Luckily the aircraft did not explode, and both men survived the incident.

In the post-flight investigation it was discovered that both men had made the correct choice, each relying on his instincts. Had Scott ejected he would probably have been killed, as his seat ejection system was found to be faulty and only in partial working order. If Adams had decided to stay in the aircraft he would certainly have been killed, as the engine of the aircraft rammed into the rear cockpit on striking the ground a split second after he had vacated it!

Two months later Scott was one of fourteen Group 3 astronauts selected by NASA, and he subsequently completed three spaceflights, one of them to the Moon. Adams remained with the Air Force, was selected for the USAF's MOL programme in November 1965, and was then transferred to the X-15 programme in July 1966. Tragically, he was not so lucky the next year when he was killed in a flight of the X-15 aircraft after attaining the 50-mile altitude that earned him the coveted USAF Astronaut Wings award, posthumously.

**1964: Freeman**

On 31 October 1964, 34 year-old astronaut Ted Freeman had just completed his first year as a NASA astronaut since joining NASA as part of the third intake in October 1963. He was just completing his landing approach at Ellington Air Force Base, near the Manned Spacecraft Center outside Houston, Texas, after a flight in his T-38. He had taken off at 10 am for a routine flight over the Gulf of Mexico and back to Ellington. He was a proficient T-38 pilot – more proficient than most of the astronauts, since the T-38s were new and recently acquired by NASA for the astronaut corps – and had completed some of the final flight-test work on the aircraft at Edwards AFB before coming to NASA.

As he lined up to Runway 4, the tower told Freeman that there was other traffic in the area, and waved him off for a second approach. At 10.50 am, as he climbed and turned to the east, rolling to the right, an 8-lb snow goose with a 4-foot wingspan struck the left side of the canopy and shattered it. Fragments of Plexiglas entered both the engine intakes, but the engines operated for a few seconds, allowing Freeman to level out at 400 feet and then attempt to climb while he continued his turn to the north-west. The engines flamed out at 1,500 feet.

Freeman, sensing the impending events, tried to glide the aircraft on to the runway. Unfortunately he did not have enough lift, and the aircraft dropped towards the ground. Nevertheless, he banked sharply to avoid hitting a barracks on the base, and punched the ejector button. Witnesses on the ground saw the T-38 descending lower and lower, and then heard an explosion. That 'explosion' was Freeman ejecting, but by this time he was less than 100 feet above the ground. Worse still, as a result of his last-minute manoeuvre, the nose of the T-38 was pointing down, so he ejected forward instead of upward and his parachute had time to open only partially.

Freeman's body was found still in his seat, near the smoking wreckage of his jet, only a mile from the air base. He was pronounced dead at the scene by one of the NASA doctors from the nearby Manned Spacecraft Center, and the subsequent post mortem revealed that he had died from a fractured skull and extensive internal injuries. He was buried with full military honours at Arlington National Cemetery. Freeman was the first American astronaut to lose his life during training. Ironically, one of his hobbies had been ornithology – the study of birds.

**1966: See and Bassett**

Astronauts Elliot See and Charles Bassett were named as prime crew of Gemini 9 in October 1965, and their back-ups were Tom Stafford and Gene Cernan. As their training progressed toward the May 1966 launch date, they followed the construction and testing of their spacecraft, and worked on the Gemini mission simulator at the McDonnell plant in St Louis, Missouri.

Early on Monday, 28 February 1966, the four astronauts arrived at Ellington Air Force Base, Texas, for a flight in their T-38s to St Louis, for rendezvous training in the simulator. Upon arriving at Ellington they received the latest weather situation at St Louis, but it was not very good. Cloud covered the area at 600 feet, with visibility down to 1.9 miles. It was raining and foggy, and the weather was not expected to improve before they arrived. The McDonnell plant was located right next to the

The original prime crew for Gemini 9, Elliot See and Charles Bassett, who were killed in an aircraft crash in February 1966.

large Lambert Field airport, and while See contacted Lambert Field tower for the latest forecast, Bassett and Cernan reviewed the runway options.

At 07.35 am Houston time, the four astronauts took off and headed for St Louis – a flight that would take them about 90 minutes. See was piloting the lead aircraft, with Bassett in the rear cockpit, while Stafford flew the second aircraft in wing position, with Cernan in the rear.

They arrived at St Louis just before 09.00 am, and See radioed Lambert Field tower to learn that, although the overcast cloud had lifted to 800 feet since the original call from Houston, the visibility had deteriorated to 1.5 miles. In addition, there were now light snow flurries mixed with the rain and fog.

Both pilots descended through the fog towards the runway, approaching on instruments. See and Bassett made a visual circling approach, while Stafford manoeuvred the second jet in behind them to follow in single file. As they descended below the cloud cover they found themselves too low, too slow and too far down the runway to make a safe landing. Stafford followed a missed approach procedure and climbed back into the clouds to circle for another instrument approach. See, however, elected to keep the landing field in sight, as he circled to the left at low level, to come around for a second landing attempt. With the south-west runway in sight, and approaching at 500 feet, he could not see McDonnell Building 101 shrouded in fog. He continued on full after-burner and headed straight for the building, in which technicians were working on the Gemini spacecraft he would be taking into space. See attempted to fly a nose-up attitude to try to miss the building, but misjudged it. As he banked he must have realised his sink rate was too high, and he cut his after-burner to attempt a sharp right turn. But it was too late, and the aircraft's belly struck the roof of Building 101, dislodging internal support frames that crashed down on the workers below. The T-38 slid across the roof, then fell and bounced in the car park beyond, exploding on impact and killing See and Bassett instantly. Fourteen people on the ground were injured, though none seriously.

Stafford and Cernan, in the second aircraft, could not understand why See had apparently continued his approach. They lost sight of See's plane as the fog bank and snow flurries closed around them. They were told to hold for landing, unaware of the tragedy below them, and thought that See and Bassett might have been diverted to another airfield. Running low on fuel, Stafford was authorised to land, still unaware of the fate of his colleagues. They landed safely at Lambert some minutes later, and were asked to confirm their names. This in turn named the two lost pilots in the still burning wreckage. They were told of the loss of their fellow astronauts as they climbed out of their aircraft. Hours later, Stafford and Cernan were told by Deke Slayton that they would take the place of See and Bassett and fly Gemini 9 into space.

A scant six hours after the accident, the first meeting of the Accident Board, chaired by Chief Astronaut Alan Shepard, began its investigation. The seven-man board reviewed the accident, the maintenance history of the aircraft, the weather conditions on the day, and the pilot experience and medical histories of the astronauts. They heard testimonies from several witnesses as well as examining the wreckage. The report found that there were no faults with the T-38. Indeed, it had functioned properly up to the moment of impact with the roof. Both astronauts were in good physical and mental health prior to and during the flight, as attested by recent routine medical examinations and by reported comments made during the flight, as well as by witnesses to the activities of the day. In addition, each had renewed his instrument flying certificate during the six months prior to the flight.

See was reported to be one of the better pilots NASA had in its astronaut corps at that time. However, Slayton, in contrast, indicated in his autobiography that this was certainly not the case. After flying with him a couple of times, he concluded that See was not aggressive enough and flew too slowly, which in the T-38 would easily result in a stall at a speed of less than 270 knots. The investigation board found that although the weather was a factor contributing to the accident, pilot error – prompted by See's desire to keep the field in sight – had brought the aircraft too low to effect a safe path around the building that it hit.

Despite damage to the roof of Building 101, Gemini 9 was unscathed. Stafford and Cernan replaced See and Bassett as the crew of Gemini 9, with Jim Lovell and Buzz Aldrin as their new back-up crew. Gemini 9 flew in June 1966.

On Wednesday, 2 March 1966, Gemini spacecraft No. 9 was moved to its aircraft transporter for the flight to Cape Kennedy for final preparations for its launch. It passed an American flag at half-mast and the courtyard that had been the final resting place of the T-38 wreckage. The next day, See (38) and Bassett (34) were buried at Arlington National Cemetery. They were laid to rest just 150 yards from the grave of Ted Freeman. Apart from the obvious loss of both men to their families and colleagues, losing both astronauts from the Gemini and early Apollo flight rota had a dramatic effect concerning crew assignments on the early Apollo lunar missions later that decade.

### 1967: Williams
4 October 1967 marked the tenth anniversary of the dawn of the Space Age. From the launch of Sputnik 1 to the imminent launch of the first unmanned Saturn V rocket, the first decade had seen remarkable strides made in the exploration of space. But it had also witnessed its first tragedies. Both America and Russia were recovering from the loss of crew-members in the Apollo 1 pad fire and Soyuz 1 landing accident.

In America, NASA was gearing up to resume flights in the Apollo programme that would take them to the Moon. At this time the crews for the first three missions were in training, and their back-ups would fly the next three missions in turn if all went according to schedule. It was from these 18 men that the crews for the first lunar landings would be selected.

In September 1966, astronauts Pete Conrad and Dick Gordon completed the Gemini 11 mission, and in December they were named Commander and Senior Pilot of the back-up crew of the third manned Apollo, with rookie Clifton C. (CC) Williams joining them as Pilot of the Lunar Module. For the next year the trio trained for the mission, in the hope of moving on to the sixth manned Apollo – a lunar landing flight – after completion of their back-up assignments.

On 5 October 1967, Williams' participation in the Conrad crew came to a sudden and tragic end. He was flying home to Ellington, from Patrick AFB in Florida, in a brand new T-38 (NASA 922) that had logged less than 50 hours. He had apparently planned to stop over at Brookley AFB near Mobile, Alabama (his home town), to visit his parents. His father was dying of cancer at the time, and Williams visited them whenever his training schedule allowed. The aircraft he was flying had a broken transponder, so he was limited to a ceiling of 24,000 feet during his flight home. As Williams approached Tallahassee he was to make a turn west toward a refuelling stop at Brookley AFB, before heading out to Houston.

The following sequence of events was never clearly determined, but as Williams banked – probably to visually locate Tallahassee – the controls locked and the T-38 began a rolling dive. Despite the low altitude he was still able to switch to an emergency radio frequency (with a radio set that took a second to warm up after switching between frequencies) to give a Mayday call, and was clearly heard by the tower at Tallahassee. At 12.34 pm EDT, some 33 minutes into the flight from Patrick AFB, controllers at Panama City, Florida, and Valdosta, Georgia, both reported a weaker, garbled cry of, 'Mayday, Mayday, NASA 922, am ejecting', followed by silence. Given this time, he could have ejected safely, but did not.

The T-38 was in a near-vertical dive when it slammed into a dirt road in a semi-wooded area of a private game preserve, two miles from the town of Miccosukee, at around 13.00. The subsequent investigation board, headed by Alan Shepard, reported that it was travelling at about Mach 0.95 when it hit the ground, and it was totally destroyed, leaving a large crater measuring 15–20 feet wide and 20 feet deep. Williams had ejected, but at only 1,650 feet, which was too low and too late for the parachute to open to save him. His body was found near the wreckage. Weather was not a factor, in that the clouds were at a 4,000-foot ceiling, with 12 miles visibility and light and variable winds. Witnesses reported hearing the plane 'sounding like breaking the sound barrier', and then rushing to the wreckage to find that it had not even singed or broken a tree. It must have come straight down.

Press reports indicated a suspected fault in the oxygen supply that deprived Williams of oxygen for a few seconds, causing loss of consciousness. When he recovered he was too low to implement a survivable ejection. The board was unable to verify the source of the accident clearly, although a kit bag left on the rear seat of Williams' cockpit had probably slipped and jammed the rear-seat controls. Deke Slayton always thought it was just bad luck and that Williams was flying a 'bum airplane.' Williams was a good pilot. A lesser one would probably have bailed out sooner, but he probably thought, in a true test pilot response to the situation, that no matter how bad the situation he would be able to get out. He was wrong.

Had he survived, Williams would probably have continued on Conrad's Apollo crew, backing up Apollo 9 and flying to the Moon on Apollo 12, to become the fourth man on the Moon. Al Bean informed Williams' widow of her husband's death. 35-year-old Williams was buried with full honours at Arlington National Cemetery on 9 October 1967. Bean later replaced him on Conrad's Apollo crew. He had backed up Gemini 10 with Williams, and it was his idea to put an extra star on the Apollo 12 emblem for the colleague he replaced on the flight to the Moon.

### 1967: Lawrence

Between 1963 and 1969 the USAF ran the Manned Orbiting Laboratory (MOL) programme. It was planned as a series of two-man, 30-day Earth orbital missions with an all-military crew, launched from an adapted Gemini spacecraft with a cylindrical laboratory. They would be launched into polar orbit by Titan III rocket from Vandenberg AFB, California. Three groups of military officers were selected to train for the programme as astronaut designees in 1965, 1966 and 1967. Major Robert Lawrence Jr was selected in the third group. He was the first African-American selected for astronaut training.

On Friday, 8 December 1967, during the final two weeks of the MOL pilot training course, Lawrence was completing a proficiency test flight in an F-104 Starfighter at Edwards AFB, California. Also aboard the F-104 was Major Harvey Royer, USAF, Chief of Operations at the USAF ARP School. Lawrence was simulating the very high speed and quick descent profile similar to the lifting body and the X-15 programmes then underway at Edwards.

Early in the X-15 programme it was noted that in a 'dirty configuration' (landing gear extended, speed brakes in the 'down' position and drag parachute out to increase aerodynamic drag) the F-104 closely resembled an X-15 in gliding flight. This would prove a valuable tool in training future X-15 and space pilots. But the F-104 was an unforgiving brute in these conditions. Having only small wings, the pilot had to divert engine bypass air over the wings to 'fool' the structure into thinking it that was going fast enough to support it, or the aircraft would oscillate so much that flight control would be lost and be unable to be regained. Flying too low and adding power gave too much lift to the wing, and sent it up like a rocket. When power was cut it lost all boundary air, and the wing acted as a high-speed elevator – down. The only way to fly this type of approach was to be perfect first time, or to abort early and try again.

Royer took the front seat of the aircraft, as he had to complete a proficiency flight as part of his Phase II curriculum programme, and Lawrence took the rear seat as co-pilot to gain more time in the aircraft. After setting up on Runway 4, the sleek Starfighter (labelled 'a missile with a man in it') roared down the runway and headed straight up, like a rocket. It was 14.58 PST.

The pilots were to fly standard ARPS lift/drag profiles, including two simulated X-15 approaches, until the fuel load recorded 3,500 lbs or less. It was then to fly two 'clean' wheels-up lift/drag approaches to a fuel burn of 2,500 lbs, and finally, two 'dirty' lift/drag approaches. The tests were to evaluate student performance under exacting conditions.

As the pilots completed their profiles that afternoon, something went wrong. During one of their approaches the aircraft 'appeared to contact the runway without appreciably breaking rate of descent', according to the post-flight inquiry. The F-104 hit the runway left of centreline, 2,200 feet from the approach end, its underbelly on fire. The landing gear collapsed on contact and the canopy shattered. For more than 200 feet the fuselage dragged on the runway before taking to the air again and continuing for another 1,800 feet. Royer ejected first, followed by Lawrence. Their aircraft veered left, and just after the 4,000-foot marker the wreckage left the runway and skidded to a stop on the sand.

Royer was seriously hurt, but he survived. Lawrence, however, was not so lucky. He died instantly and his body was found 75 yards from the wreck, still strapped into his seat, his parachute unopened. His fatal injuries included a crushed chest and lacerated heart. A few days later Lawrence was buried at his home town of Chicago. He was 32. Had he lived, it is probable that he would have transferred to NASA in 1969, when MOL was cancelled by the Air Force, and he could well have piloted one of the early Shuttle missions. Some 30 years after his death, and after a long battle to have him recognised as America's first black astronaut trainee, his name was added to the Astronaut Memorial near the Astronaut Hall of Fame, Titusville (next to the Kennedy Space Center), Florida.

### 1968: Gagarin

In 1961 cosmonaut Yuri Gagarin became the first person to explore space. His one orbit in Vostok 1 pioneered every subsequent step into space and, as a result, he became not only an international celebrity and national hero, but also too valuable to let him fly again (a decision also made by President Kennedy about American astronaut John Glenn). This annoyed Gagarin, who was never happy flying a desk. By 1967 he was wearing his space boots again as back-up to Vladimir Komarov on Soyuz 1, and after the loss of his friend he was assigned to fly the next mission in order to restore confidence in the Soviet programme.

By the early spring of 1968, Gagarin had not piloted a jet for more than five months. But as part of his preparations for his return to spaceflight, he had conducted a series of familiarisation training flights with top Soviet engineering test pilot and instructor Vladimir Seryogin. On 27 March 1968 he was making his last flight with Seryogin before going on to a series of solo flights.

The two men began their pre-flight preparations at 09.15 local time, and a little over an hour later, at 10.19, the two-seat UTI MiG 15 (call-sign 625) took off. By 10.30 it was reported that Gagarin had completed his assigned tasks and had turned back for landing. A minute later, the plane crashed, killing both pilots. The final flight of Gagarin had lasted just 12 minutes.

Rumours about the crash circulated in the West for many years. Why had they crashed on such a routine flight? – if it *was* routine! Seryogin was one of the most experienced Soviet pilots at that time, and Gagarin was a very competent pilot. Could they have been testing a new type of vehicle, and not flying the MiG? Were they so badly burned and injured that they were indeed alive, but only just? And so it went on. Whatever the reason, coming after the loss of Komarov on Soyuz 1 the year before, the loss of Gagarin came as a bitter blow to an already grieving Soviet nation. It was not until 1988, 20 years after Gagarin's death, that Professor Belotserkovski (whose detailed findings were later published in *Pravda*) instigated a detailed investigation. Belotserkovski was a professor at the Zhukovsky Air Force Engineering Academy, from which Gagarin had recently graduated.

The evidence presented revealed that just one minute after Gagarin and Seryogin had taken off, they had been overtaken by two faster MiG-21s as they broke through the cloud layer. The incident almost resulted in a mid-air collision, but this was avoided and the cosmonaut was able to continue with his training flight.

A further MiG-15 (call-sign 614) appeared on the scene and flew past Gagarin's plane at a distance of only 500 m. A review of the transcripts of conversations with flight control from the 614 pilot revealed that he had not observed Gagarin's plane. Gagarin had completed a programmed spin manoeuvre and was executing a turn when 614 shot past them as the aircraft passed through the upper border of a double cloud layer at 1,500 m.

Initial reports indicated that Gagarin had either struck a weather balloon, been caught in the turbulence of another aircraft, or encountered a powerful air current that caused the pilots to lose control and send the MiG into a dive from which they were unable to recover.

A search helicopter, dispatched to determine what had happened to Gagarin's aircraft, flew over a blackened patch of earth located 96 km north-east of Moscow. Originally, the pilot thought that the steam venting from the ground was a natural phenomenon. He found a large crater where the fallout from some sort of impact had distributed charred debris over a wide area. Broken trees were littered with twisted metal. He soon evaluated this to be the site of a recent high-speed plane crash, but there were no obvious signs of any part of the aircraft. Throughout the night and the next day, recovery teams sifted the crash site and found personal belongings of both Seryogin and Gagarin. As they dug into the crater the fate of the world's first space explorer became apparent.

The nose of the MiG had rammed hard into the ground to a depth of several metres and, combined with the weight and momentum of the engine, had severely mangled the cockpit area. The recovery teams faced a period of great distress as they recovered what they could find of both men's bodies from the mangled wreckage and surrounding area. It soon became obvious that the aircraft had hit the trees, severely damaging the cockpit area and the two pilots inside. The recovery teams were only able to identify pieces of the engine, one wing and a little of the landing gear. Everything else was unrecognisable. The grim task of identifying Gagarin fell upon his friend and colleague Alexei Leonov. He recognised a birthmark on a fragment of neck remains and knew they could stop searching for his friend.

Belotserkovski reported that Gagarin and Seryogin had in fact found themselves in the trailing vortex of MiG 614 and were sent into a flat spin. They did manage to recover after five full revolutions of the aircraft, but found themselves in thick cloud, which disorientated them. Opting not to eject but to regain control and try to restart the stalled engines, they fought for several seconds to pull out of the dive, enduring forces of up to 10–11 g. It was evaluated that if they had had another 250 m of airspace or 2 seconds of flight time, they could have pulled themselves out of the dive, to at least allow a safe ejection. But this was not afforded them, and the plane smashed into the ground in a forest near the airfield, close to the village of Novoselove, in the Kirzhach Oblast (district), Vladimir Raion (region) near Moscow.

Gagarin's wife, Valentina, was in hospital at the time, undergoing an operation for a stomach ulcer. On the evening of the 27 March she tried for 90 minutes to telephone her husband to check on his flight, but was told the telephone lines were down. She sensed something was wrong, and the next morning, quite unexpectedly, Valentina Tereshkova, Pavel Popovich and Andrian Nikolayev appeared at her hospital bed. Fear gripped her as she asked if anything had happened to Yuri. 'Yes … yesterday,' they told her. Initially she thought his body had been retrieved intact, but when the truth was learnt she became understandably hysterical.

For the next 30 years mystery surrounded the loss of the first cosmonaut. Was it pilot error? – and if so, which pilot? – or was it the fault of the ground controllers? In the late 1990s the files were reopened and a new investigation board was established, and there began one of the most in-depth air-accident investigations in all of Soviet/ Russian history. It included a complete recovery of the MiG wreckage and an analysis and study of the aircraft's past history. As the investigation board sifted the remains and completed their report, rumours grew that the two men had been drinking the night before the crash. They had been to a colleague's party a couple of days earlier, but nothing in the post mortem laboratory examination revealed anything other than insignificant levels of alcohol. Lactic acid levels revealed that both men were fully conscious and alert at the time of the impact.

Examination of the wreckage of the aircraft revealed that the flight controls were in the positions in which they would be in the event of the pilot's trying to control a faulty aircraft, and it was evident that they had tried to recover from the fatal spin. The aircraft hit the ground in belly-down pancake fashion, and not nose first. Even so, the momentum was enough to drive it several metres into the hard icy ground.

The Commission Report noted that the aircraft was 12 years old, and since its production in 1956 had undergone two major overhauls. The engine, also produced in 1956, had had four major overhauls. There were two 26-litre external drop-away fuel tanks under the wings, which were 'aerodynamically poor' and, according to cosmonaut Alexei Leonov, were a significant drawback with this particular configuration of the Mig, reducing the safety parameters of a flight. More importantly, the Commission also determined that the 'height-to-ground' indicator on Gagarin's plane was faulty, and could have shown an incorrect altitude reading.

The 30-volume investigation into the tragedy revealed that the aircraft, power unit and avionics (apart from the altimeter) were in working order, and that all onboard systems functioned normally up to the moment of impact. There was no information to support Western rumours that they had been sabotaged, shot down, suffered an onboard fire, or that either pilot lost consciousness. Nor had they suffered any breakdown, or been intoxicated at any time during the short flight. Both pilots apparently performed the required steps in attempting to save themselves, in the 'highest possible degree', and did everything possible to escape the situation.

Blame pointed to the flight controllers, who were seemingly unaware of the aircraft's presence in the area that they were controlling. The flight control on the airbase was unaware that Gagarin had taken off, and therefore did not prevent the subsequent take-off of the two MiG-21s and the second MiG-15, moments after Gagarin had left the runway.

Two days after the accident, on 29 March, nine fellow cosmonauts provided an honour guard as their friend and colleague was cremated and, along with Seryogin, was interred in the Kremlin Wall alongside Komarov. Almost 200,000 mourners filed past the biers of the two national heroes in the Soviet Army House in Moscow prior to the funeral. Communist Party leaders Kosygin, Podgorny and Brezhnev briefly joined the honour guard of soldiers near the biers during the State Funeral. To the cosmonauts and the Russian population, Gagarin had become an almost saint-like figure in Soviet space history. His office at Star City was preserved as it was the last time that he walked out of it, with the hands of the clock stopped at 10.31, the reported time of his tragic accident. It is tradition that new crews visit Gagarin's office prior to their spaceflight and lay fresh flowers on his statue upon their return. His portrait is often seen inside the crew compartments of spacecraft and space stations.

There is a monument on the site of the crash, where a major event is held by the cosmonauts of the Gagarin Cosmonaut Training Centre (TsUP) to mark Gagarin's death.

### 1968: Armstrong

Neil Armstrong's flying career was punctuated by several incidents before he flew to the Moon. In September 1951, during a combat-flying sortie in Korea, a wire stretched across a valley tore off the wing of his F9F-2 Bearcat, forcing an immediate ejection. He then went on to fly the X-1 and X-15 before selection by NASA. Following his Gemini 8 experience (see page 261) he trained for Apollo, and in 1968 ejected from a Lunar Landing Training Vehicle a split second before it crashed and

The Lunar Landing Training Vehicle used to train LM pilots for landing approaches on the Moon.

exploded. It has been said that his flying skills and coolness in dealing with the incidents helped him gain an early Apollo assignment. However, in the mysteries of crew selection and flight rotation, his assignment to Apollo 11 carried a strong element of luck.

In order to train potential Lunar Module Pilots in the skills of landing the vehicle on the Moon, NASA's Flight Research Center in California developed the Lunar Landing Research Vehicle (LLRV) in which a jet engine aimed downwards to take ⅚ of the vehicle's weight, while rockets lifted the rest, allowing a system of reaction control jets to control the attitude of the vehicle as it descended to the ground. Aerodynamics played no part in the design, which resembled an open framework supporting the pilot, engines and fuel system. This ungainly appearance gave rise to the term 'flying bedstead', derived from the early British flying simulators used to develop vertical take-off and landing (VTOL) technology in the 1950s.

Two LLRVs were built and flown at the FRC before shipment to Houston, to be used for astronaut training at nearly Ellington Air Force Base. These were joined by three more vehicles incorporating the modifications resulting from the experience in flying the two original LLRVs. These new vehicles were called Lunar Landing Training Vehicles (LLTV), and together the five vehicles allowed astronauts who were to fly the LM, the mission commander and his back-up, to practise landing techniques.

The two original vehicles were designed as LLTV A1 and A2, while the three new vehicles were LLTV B1, B2, and B3. A potential LM crewman would first go to helicopter school for a three-week course, then to the Langley Lunar Landing

Facility where a large gantry 393 feet long × 246 feet high supported a tethered test rig's $\frac{5}{6}$ weight, while rockets supported the remaining $\frac{1}{6}$. This was followed by 15 hours in ground simulators before finally going on to fly the LLTVs.

On 6 May 1968 Armstrong was piloting LLTV-A1 (the former LLRV-1) in simulated lunar landing approaches at Ellington AFB, near the Manned Spacecraft Center, Houston. At a height of only 196 feet, the vehicle suddenly lost helium pressure in the propellent tanks, leading to a shut-down of its attitude control system, and causing the vehicle to begin nosing up and rolling over before plummeting to the ground. Armstrong instantly ejected and parachuted to safety. The vehicle was totally destroyed. A five-person Board of Investigation was set up the same day to review the accident at MSC. In addition, on 16 May NASA Headquarters formed a Review Board to consider the implications of the accident for future training vehicles and LM design.

On 17 October, both Boards reported that a loss of attitude control (due to the early emptying of helium in the propellant tanks) allowed a fall in pressure-feeds to the thrusters and attitude-lift rockets. The subsequent warning displayed to the pilot was too late for him to take any corrective action. The board concluded that improvements in the design of the system, adequate caution and warning display controls and operating procedures should be implemented in the lunar landing training programmes. It was also determined that there was no need to adjust the LM spacecraft or the planned landing trajectories.

6 May 1968: astronaut Neil Armstrong descends safely after ejecting from a Lunar Landing Training Vehicle which went out of control and crashed to the ground during a training flight at Ellington Air Force Base, near the Manned Spacecraft Center, Houston. The burning wreckage can be seen in the background.

There were two further accidents on LLTVs. On 8 December 1968, MSC Chief Test Pilot Joseph S. Algranti was forced to eject from LLTV B1 when it became unstable and lost control. This was on its eleventh flight, prior to transfer to the astronauts for LM training. As the review board studied this accident, the previous incident involving Armstrong was re-evaluated. The problems were subsequently overcome, and the system provided an adequate training tool for all future LM Pilots. Finally, on 29 January 1971, MSC pilot Stu Present ejected safely following an electrical system failure on LLTV B2.

### 1971: Cernan

In 1971 Gene Cernan was in training as back-up commander for Apollo 14. He was in line for the Command of Apollo 17, the last scheduled Moon-flight, but that had not been publicly announced and was still open to change. With the cancellation of three Apollo missions and the second Skylab space station, and with the so-called Space Shuttle probably a decade away from orbital flight operations, there were too many astronauts chasing too few seats for space. Astronauts had to be careful in everything they did to ensure that they did not lose their seat due to an error of judgement or indiscretion.

On 23 January Cernan was flying one of the small H-13 Bell helicopters used for practising lunar landing approaches. At this time the astronauts flew helicopters because they handled a little like a Lunar Module. Very few of the former fighter pilots had previously flown rotary-wing aircraft. Both the prime crew and the back-up crew for Apollo 14 were at the Cape preparing for the final stages for the launch on 31 January. Cernan had decided to put in extra hours and took one of the little helicopters to practise a few miles down the Indian River. As he flew down the Atlantic Cocoa Beach, over the town of Melbourne and back up towards the Cape, he found the controls a little sluggish. This was due to the amount of fuel he was carrying, so he decided to 'loaf around' a little to burn off some of the fuel before starting the approach pattern. He was flying over one of the playground areas of the Florida waterways, and could not resist a chance to try a little mischief. He would fly a few feet above the surface between the boats and islands, to let off steam after all the intense training that had run, virtually without a break, almost since 1965, when he had been assigned to his Gemini mission. Demonstrating his piloting skills to the waving onlookers, he lost sight of the surface of the water for a split second. Momentarily distracted (some fellow astronauts like to joke that it was probably a bikini clad distraction), the tip of his left skid dipped into the water. His instant reaction was to try to twist the machine back out of the water by applying more power, but it was too late. The whole skid and then the cockpit struck the water as the helicopter, in an explosion of water, flame and debris, dropped from a speed of 100 knots to zero with one of the most severe lurches he had ever experienced. As the helicopter broke up, jagged fragments of still spinning rotor blades cartwheeled away from the wreckage.

The small helicopter disintegrated around him as he began to sink, still strapped to his seat. Cernan has never been able to remember if he lost consciousness, but he has a clear memory of still trying to fly the helicopter as it settled onto the bottom of

Apollo 14 back-up Commander Gene Cernan trains in a Bell helicopter in 1970. It was in a helicopter similar to this that he crashed during training in Florida in January 1971.

the river! Luckily, as the weight of the transmission had been separated in the crash the wreckage did not flip over. Unstrapping himself from the seat, he struggled out of the wreckage and headed for the surface, utilising his years of Navy and NASA submerged aircraft training.

As Cernan broke surface he gulped for air and found himself inside an inferno of heat as the wreckage burned around him. Already with singed hair and eyebrows, he gulped and dived, resurfaced, and dived again and again under the flames, heading for the shore. His flight suit and boots filled with water, weighing him down. He regretted his decision not to wear a 'Mae West' life preserver that day, and he was now tiring rapidly. Despite standing 6 feet tall, in water only 10 feet deep, he could not tread water.

Suddenly, a one-person fishing boat – one of a score of witnesses to the fireworks display he had inadvertently created – came to his rescue. He was taken to the shoreline, and transferred to the local Patrick Air Force Base for an initial examination and treatment of his superficial wounds. He was then taken by car to the crew quarters, where he churned the consequences of his actions over in his mind. Apollo 14 Commander Al Shepard – the so-called steely-eyed 'Icy Commander', and one of Cernan's bosses – just looked straight at him as he ate his breakfast. Cernan told Shepard and Deke Slayton that he 'just screwed up'. He thought he had lost Apollo 17, but it was probably his honesty in owning up to pilot error that allowed him to keep that seat.

On 18 October 1971 the report into the accident stated that the cause was a misjudgement in estimating altitude. The five-man board investigating the accident was established by Manned Spacecraft Center (MSC) Director Robert Gilruth on 25

January, and was chaired by astronaut Jim Lovell. Board members were astronaut Alan Bean, Harold Ream of MSC Aircraft Operations Office, Conway Roberts of the MSC Aviation Safety Office, and Dick Lucus of the Aircraft Quality Assurance Office.

The board listed several mitigating factors which could have contributed to the accident. These included the lack of familiar objects on the water surface to help him judge altitude, possibly focusing on a false water surface due to the water's millpond smoothness, or a change in the reflection of the Sun on the water caused by changing course just prior to the accident. The board also suggested that Cernan's extensive experience in flying fast high-performance jets could have been a contributing factor (the lower the flight path, the faster the ground appears to pass by). However, this effect is not so pronounced when flying a helicopter, and this could have led Cernan to think that he was flying higher than he actually was. The board concluded that Cernan's extensive water survival training, both as a Navy pilot and an astronaut, had helped his ability to escape major injury. Examination of the salvaged helicopter wreckage, combined with Cernan's statement, indicated no evidence of any mechanical malfunction.

The following year Cernan became the eleventh man to walk on the Moon – and he was the last person in the twentieth century to leave it. Tom Stafford – his Apollo 10 Commander – gave him a tough time for almost messing up his chances of a place on Apollo 17. Dick Gordon – who was back-up on Apollo 15 and had lost the Apollo 18 Command in the budget cuts – later joked with Cernan that he thought he had won the Command of Apollo 17 when Cernan crashed the helicopter. Cernan kept his battered and scorched flight helmet in his office to remind him that he had been able to escape from the crash alive when so many of his fellow astronauts and friends had not been so lucky. He later wrote that he could fly to the Moon without a scratch, but actually crashed a helicopter when flying only a few feet above the ground.

## COSMONAUT ACCIDENTS

On 23 August 1976, cosmonaut Leonid Ivanov was one of nine Air Force officers selected for cosmonaut training as the sixth Air Force cosmonaut enrolment since 1960. After a year of test-pilot training at the Chkalov Test Pilot School, Akhtubinsk, the group embarked on a two-year cosmonaut candidate training programme at the Gagarin Cosmonaut Training Centre (TsPK). Following completion of the cosmonaut training course, the candidates returned to operational test-flying until recalled to TsPK for assignment to a Soyuz–Salyut group to train for an assignment on a prime, back-up or support role on a space station mission. On 23 October 1980 Ivanov was killed in the crash of his MiG-23 after the aircraft went into a spin during a test flight from which he was unable to recover.

In 1977 the Soviet Ministry of Aviation Production (MAP) selected a group of five civilian test pilots for what was officially known as Vehicle 11F35 – more commonly known as Buran, the Soviet equivalent of the American Space Shuttle. At that time the Soviet military test flights were initially conducted by civilian pilots from MAP, and then assigned to military pilots to conduct acceptance flights later in the programme prior to operational deployment to air units. This resembled the US system in which pilots from the aircraft production companies performed the initial flight tests before handing the flying over to military test pilots. The group – based at the Flight Research Institute at Zhukovsky, near Moscow – consisted of Igor Volk, the commander of the group, and Oleg Kononenko, Anatoly Levchenko, Rimantas Stankyavichus and Alexandr Shchukin.

The five were not formerly enrolled into the Cosmonaut Detachment but did go on to complete basic cosmonaut training at TsPK during 1979–1980. After completion they returned to operational test flying as well as their Buran assignments. They became known as the 'Wolf Pack' ('wolf' in Russian is 'volk'), and along with other military pilots selected to train for Buran missions they performed a long programme of atmosphere approach and landing tests using a variety of aircraft and training versions of the Buran shuttle, including a jet-powered version allowing a runway take-off and an unpowered landing. The Buran programme was plagued with delays, and resulted in only one unmanned orbital flight test in November 1988 before being cancelled two years later as a result of the break-up of the Soviet Union and restrictions in the funding for human spaceflight programmes.

As part of the programme to provide potential Shuttle pilots spaceflight experience, the Wolf Pack was assigned to space station training groups to fly short week-long Soyuz missions to Salyut 7 or Mir. Shortly after landing they were to be taken to a nearby airport to fly, under medical supervision, a Tu-154 aircraft back to Moscow in a demonstration of piloting skills shortly after a week in orbit, to provide data on the difficulty of such tasks in returning Buran from orbit. Volk and Levchenko eventually flew in space, but unfortunately the three other members died before getting their chance to fly into space.

Oleg Kononenko was killed on 8 September 1980, just a few weeks prior to completing the Cosmonaut Basic Training Programme in November. At the time of his death he was on a test-pilot assignment in the South China Sea aboard the carrier *Minsk*. He was killed while testing a Yak-38A vertical take-off and landing (VTOL) from the carrier.

Between June 1986 and April 1988, Aleksandr Shchukin completed nine Approach and Landing Test (ALT) flights in the jet-powered version of Buran. In December 1987 he had served as back-up to Levchenko on the Soyuz TM-4 flight to Mir, and he and Levchenko had been assigned as the back-up crew to the first Buran manned spaceflight which was to have been flown by Volk and Stankyavichus. Tragically, he was killed on 18 August 1988, just twelve days after Levechenko had died of a brain tumour. Shchukin was killed in the crash of an Su-26M sports plane at the Zhukovsky Flight Research Centre while rehearsing for an appearance in an air show

Stankyavichus had flown thirteen ALT missions, including the first ALT flight between November 1985 and April 1988. He was assigned as Volk's back-up on the Soyuz T-12 mission to Salyut 7 in July 1984, and as second back-up to Levchenko on

Soyuz TM-4 to Mir in December 1988. He was also assigned to fly with Alexandr Viktorenko and Alexandr Balandin on the Soyuz TM-9 mission to Mir in 1989, but a suspension of flights due to delays in hardware forced a delay and reassignments of crews assigned to TM-9, and with them he lost his chance to fly in space. Stankyavichus became the fourth member of the Wolf Pack to die when on 9 September 1990 his Su-27 jet crashed during an air show in Treviso, Italy.

As with all pilots, the personal flying skills of cosmonauts were something of which they were very proud, and like their American counterparts they often tried to 'push the envelope' of the aircraft. Sometimes, however, they pushed a little too far.

# The Apollo 1 fire, January 1967

The exploration of space was considered dangerous long before man actually ventured out into the cosmos in the early 1960s. In those pioneering days, risk and danger seemed to be a major element in planning every mission. As the decade progressed, the relative ease with which space was being conquered by the pioneering Vostok, Mercury and highly successful Gemini programmes (as well as many unmanned successes) led many to believe, in error, that travel in space was becoming safer. It was thought that, following the achievement of the first landing, the American Apollo astronauts would eventually develop a routine programme of lunar exploration. To many, from both inside and outside the programme, Apollo was seen as one more great adventure into the cosmos.

As the Apollo programme gathered pace and the Soviet programme appeared to slow down, the stage was set for the first manned Apollo flight. This was to be a 14-day test flight of the Apollo mother craft (the Command and Service Modules) in February 1967. It would begin a series of 'shakedown cruises' designed to man-rate the hardware and confirm the procedures and methods devised for reaching the Moon by Kennedy's deadline of 1970.

The crew for this pioneering mission consisted of two experienced astronauts and one rookie. Commander Gus Grissom had previously flown on Mercury Redstone 4 in July 1961 and Gemini 3 in March 1965. Senior Pilot was Ed White, the first American to walk in space during Gemini 4 in June 1965. The rookie was Pilot Roger Chaffee.

The crew was announced on 21 March 1966. They were assigned to Apollo Mission 204 (AS-204), launched by the fourth Saturn 1B rocket – one of the most powerful rockets ever designed to carry a human cargo – and not the mighty Saturn V, which was still under construction for its first unmanned test flight. As this was also to be the first Apollo mission, it was more commonly known as Apollo 1. Originally, the crew was to have been Grissom and rookies Donn Eisele and Roger Chaffee. However, shortly before the announcement of the crew to the press, Eisele suffered a shoulder injury that forced him to be replaced. Deke Slayton, who chose the crews, brought Ed White forward from the Apollo 2 flight, to replace Eisele, who was reassigned to the second crew, led by Walter Schirra. The back-up crew was Jim

Apollo 1 prime crew: Ed White, Gus Grissom and Roger Chaffee.

The Apollo 1 mission emblem, showing the planned maiden test flight of the Apollo Command and Service module in Earth orbit.

McDivitt, Dave Scott and rookie Russell Schweickart. For the rest of 1966 the crew prepared themselves for their mission, but they found almost immediately, that both hardware and astronaut training was suffering technical and procedural problems as the Gemini programme wound down and the Apollo programme took precedence.

When the CSM arrived at the Cape in August 1966 it was found that a considerable amount of engineering work was still necessary – work that should have been completed at prime contractor North American Aviation prior to shipping to the Cape. The need to complete this work, plus several other problems and delaying factors, pushed the mission from a planned launch in the autumn of 1966 (and for a short time a possible joint flight with Gemini 12) into the first quarter of 1967. On the training front, Grissom often expressed his frustration at the fact that the constant changes made to his spacecraft were not applied as quickly to the simulator.

With new Apollo crews also coming into the training flow for later missions, a bottleneck was developing. Grissom was so frustrated with the situation that one day he hung a lemon above the simulator.

Despite these problems, progression towards the launch date continued. In November 1966, at Schirra's recommendation the second manned Apollo mission (Apollo 2) was cancelled as an unnecessary repeat of the first flight. As a result, in December 1966 some of the Apollo crews were reassigned.

The Apollo 2 prime crew of Walter Schirra, Walter Cunningham and Donn Eisele became Grissom's new back-up crew on Apollo 1. The McDivitt trio moved to the prime slot for the new second manned Apollo mission, featuring the first flight of a manned Lunar Module in Earth orbit. Their back-ups were to be Tom Stafford, John Young and Gene Cernan. Both of these missions were to be launched by Saturn 1B launch vehicles, while Apollo 3 would be the first manned launch of the Saturn V Moon rocket and feature a high-Earth orbit into deep space test of the CSM and LM combination. The crew for this flight would be commanded by Frank Borman, with Mike Collins and Bill Anders, and their back-ups would be Pete Conrad, Dick Gordon and 'CC' Williams. It was planned that the three back-up crews would fly Apollo 4, 5 and 6. The first lunar landing would probably be made by Apollo 5 or Apollo 6, at the earliest.

As 1967 dawned, so did expectations for the beginning of manned Apollo

A pristine Apollo Spacecraft 012 (Apollo 1) during pre-flight preparations.

operations with the two-week Apollo 1 mission in Earth orbit. Late in January, the crew were putting the final touches to their training and progressing through a series of manned systems tests inside the Command Module, sitting on top of their Saturn 1B booster on Pad 34. The training schedule for Friday, 27 January 1967 consisted of yet more routine tests which would last most of the day. As the three astronauts moved around the crew quarters early in the morning, completing final assignments prior to suiting, they appeared to several observers to be preoccupied and somewhat quieter than normal. Perhaps the strain of the final weeks of training for their all-important maiden flight on Apollo was beginning to tell on them. They suited for the test by 10.00 am, and departed for the day's work inside their capsule, which they entered a short time later.

The scheduled test for the crew was part of the normal pre-launch test procedure. Known as a 'plugs out' test, it was designed to verify each electrical system of the Apollo spacecraft during the planned launch phase. Since the spacecraft would be configured as for the actual launch, the capsule would be pressurised with 100% oxygen, and all hatches would be sealed and locked. In order to minimise the risk of leaks, the inside of the CM was pressurised to 10% above atmospheric pressure, which would ensure any leaks would be from the inside out and not from the outside into the crew compartment.

As the test was considered a 'non-hazardous' exercise with a non-fuelled launch vehicle, medical attendants and fire crews were at ready stations only, rather than full alert. Safety, however, was important to the crew, and Grissom had asked for the inclusion of a simulated crew escape at the end of the test. The procedure would have seen the crew open the hatch, run suited along the gangway through the White Room to waiting lifts, and make a hasty exit to comparatively safe areas. As with many Apollo procedures and techniques, training was an important element of each mission, and all the astronauts were aware of the potential dangers that awaited them on the launch pad.

The planned test was an important stage in the road to the launch. All personnel were fully aware that the success of the test would point the way to a high probability of a launch the following month. The three astronauts entered the spacecraft and settled in their respective seats in the CM, as appropriate to their roles during launch. First to enter was Grissom, who slid over to the left seat with the flight control instruments he was to monitor during the test. The CM had been on internal power for about five minutes when he entered the vehicle, and he noticed an odour apparently originating from the suit oxygen loop in the environmental control system. He reported it, and the countdown simulation was immediately halted at 13.20. Samples of the gas were taken and examined and, when nothing serious was found, the test continued, although it was to be more than an hour later.

Next into the CM was Chaffee, who entered the right-hand seat where the communications equipment was located. During the actual mission it would be his job to handle communications during ascent. For this test, all actions would be simulated. Last to enter was White, who he took the centre couch, where he was responsible for navigation during the mission. On that day's test he was to perform

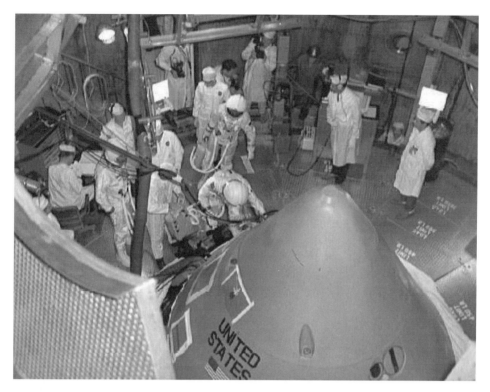

The Apollo 1 prime crew enter the Command Module for a pre-launch simulation.

monitoring of important procedures and actions and assist Grissom in operating flight controls in the spacecraft as well hatch opening and closing duties

The day before, Grissom and Slayton had discussed having Slayton, in shirtsleeves, sit in the capsule during the test to enable him to listen in to the communications problems that had plagued previous simulations. Over breakfast the next morning, Grissom suggested that Apollo Spacecraft Manager Joe Shea should also join them in the test, sitting in the lower equipment bay to monitor the operation. Actually achieving this, however, would be difficult, as there would be room for only one other person, who would be forced to stand up for hours or crawl under the three crew couches. The idea was abandoned because technicians would have to jury-rig extra communication devices, which would not have been the same as the crew were using. There was also no easy way for the visitor to join the communication loop once the hatches were closed. Without communication there was no real value in being inside the spacecraft, so Slayton monitored communications between the spacecraft and the Control Centre from the Blockhouse near the pad, along with CapCom astronaut Stuart Roosa.

With the three men inside the spacecraft, and the odour problem solved, the hatches were sealed, and at 2.45 pm the countdown resumed. All three astronauts were now on their suits' ECLSS, with their faceplates closed.

The design of the Apollo hatch was complicated, but it could be opened in about 60 seconds, with all three astronauts safely out 30 seconds later. The original concept for the Apollo CSM hatch was a simpler design affording both pressure integrity and thermal protection. It would be a one-piece design that would be explosively released after landing. After Grissom himself nearly drowned after his 1961 Mercury 4 flight, when his explosive Mercury hatch blew prematurely and the capsule shipped water and began sinking (see page 346), the design for the Apollo hatch was changed. The new hatch was a three-piece design, which none of the astronauts particularly liked, partly because it precluded any EVA from the CM. Realising this and looking towards possible EVA from the CM, NASA had already begun development of a simpler hinged hatch. However, this would certainly not be ready for the early Apollo 'Block I' manned orbital test flights. It would be incorporated in the 'Block II' missions, which would fly the lunar distance missions. For Apollo 1, the crew had to train with the older hatch design.

The inner hatch was the first to be installed. It was designed to retain the internal atmosphere and could be opened only from the inside by the astronaut in the central couch (White), by using a special ratchet to individually release each of six sockets. Once free, the hatch had to be stowed on the floor, under his seat. The next hatch was the thermal shield, designed to integrate into the ablative coating on the surface of the CM, forming a smooth surface on the outside of the vehicle. It could be opened simply by one handle on the inside. The final hatch covering was not actually attached to the CM. It formed part of the Launch Escape System's Boost Protective Cover (BPC), which sat over the CM during the early phase of the ascent through the atmosphere. Made from fibreglass and cork (as was the rest of the BPC), it was designed to open along with the outer hatch and yet also be jettisoned with the LES during the launch. With both the inner and outer CM hatches installed and locked, the BPC hatch was installed, but bundles of wire around the area meant that this hatch could not be locked without distorting its shape, so it was left unlocked.

For almost three hours, the crew and ground controllers ran the test. One problem which plagued the test was consistently poor communication links between the CM and ground control crews. Further problems were associated with communications between facilities. At 17.40 Chaffee took the opportunity of a further hold in the proceedings to try various switch positions in an attempt to locate a fault the crew were experiencing with an 'open' microphone that could not be turned off. After eight minutes of troubleshooting, an extremely busy period of communications between the astronauts and the ground would see a series of system checks and verification of switch positions. However, at times the astronauts' comments could not be heard or understood on the ground. Grissom, typically, barked his disapproval of the state of the communications, asking how they were expected to fly to the Moon if they could not even communicate properly on the ground.

The primary test of the day was the evaluation of the operation of several electrical circuits on internal power, isolated from ground connections. However, fears were being voiced about the growing list of problems and setbacks that had plagued the test all through the afternoon. Darkness was approaching over the Cape, and Grissom continued to express concerns for the integrity of these obviously faulty

The back-up crew for Apollo 1 – Schirra, Cunningham and Eisele – prepare for an altitude chamber test at the Kennedy Space Center on spacecraft 012 (Apollo 1) during training in December 1966. This view shows the positions of the astronauts, similar to those of the prime (Grissom) crew during the fatal test in January 1967.

systems and connections. He had been openly critical of such faults for some time prior to the launch test. In addition, it was suspected that the Environmental Control and Life Support System (ECLSS) had sprung a leak, and that a water–glycol mixture soaking the wires had probably caused the communication problems. A further hold was ordered, to provide a chance to clear the communications problems. Meanwhile, a review of the countdown checklist continued to a point planned at 18.20 (T–10 minutes), where the spacecraft would go to simulated fuel cell internal power. Once again, awaiting clearance for this event, a hold was called. The White Room technicians (who would 'pull' the plugs) took a break.

Arc lights illuminated the Pad area, as 27 personnel watched and waited for the 'GO' to continue the countdown simulation. The simulation had already lasted 5½ hours. TV cameras and telemetry devices continued to record the activities of the spacecraft, crew and ground teams. Biomedical recorders logged every move of the astronauts, and gimbals in the spacecraft sensed their movement. Ground technicians were casually looking at pictures and data as tiredness began creeping in after a long day and a difficult simulation. It was now 18.30, and thoughts that the test would soon either be completed or abandoned were on the minds of most at the Cape as they looked forward to heading home for the weekend. Events then began to move rapidly towards a tragic conclusion.

18:30:21.0. Biomedical telemetry indicated a small rise in the pulse rate of White, followed seconds later by an increase in his heartbeat. Data also recorded Grissom moving about in his couch. This was hardly surprising, after nearly six hours lying on his back. For several seconds, data showed significant movements in the vehicle. They were nothing violent, but the crew was definitely active. An increase in the flow rate of suit oxygen demand was recorded.

18:30:54.25. A momentary power surge was recorded originating in electrical AC Bus 2, which was still connected to ground sources through inverter number 1. The trace on the recorder graph barely moved, but it was enough of an uncharacteristic shift to identify a major short-circuit, somewhere in the 20 miles of electrical wiring within the CM.

18:31:04.7. '*Fire! We've got a fire in the cockpit!*' Grissom screamed over the intercom, as the effects of the fire in the pure oxygen environment rapidly took hold. The fire apparently started below Grissom's couch, expanded past his foot well, along his left side and into the instrument panel in front of all three astronauts. As the fire spread it ignited netting restraints, and these dripped flames down into lower storage areas, in which there were flammable lockers, bags, wiring and logbooks. As Grissom alerted the startled ground controllers, White was already frantically trying to release the six bolts retaining the inner hatch. In front of them the instrument caution and warning display lit up, and an alarm sounded, warning the astronauts of a serious situation. They hardly needed such a warning, however!

18:31:12. A significant rise in pressure and temperature was recorded in the

Apollo 1 commander Gus Grissom (*left*) inspects spacecraft equipment inside an Apollo Command Module during a visit to North American Aviation Inc., Downey, California, in June 1966.

spacecraft as the fire took hold. Sheets of flame rapidly engulfed anything that could burn. In the Apollo Block I design there was a considerable amount of flammable material. The flames licked around the oxygen relief valve to Grissom's left. He was apparently unable to activate the vent to relieve the rapidly building pressure in the cabin, but even if he had done so it would still not have extinguished such an oxygen fire quickly enough to save the crew.

18:31:16.8. 'We've got a bad fire – Let's get out. We're burning up!', cried Chaffee. But before he could complete the sentence the pressure build-up inside the spacecraft reached its limit, ruptured the cabin and sealed the fate of the three astronauts as the fire spread uncontrollably across the cabin.

18:31:19.5. The exact determination of the pressure at the moment of cabin rupture is not certain, but it ranged between 2 and $2\frac{3}{4}$ that of sea level. As the load built up to around twice that of normal sea level, which exceeded the design specifications, the pressure surge tore the hull of CM 012 around the lower right section. This resulted in a diagonal sweep of gas, smoke and fire across the three astronauts and the instrument panels. With a final roaring sound, the flames and smoke billowed around the outside of the spacecraft. After one last cry of pain filled the intercom, there was silence on the communication circuits.

18:31:25. The most intense phase of the fire had occurred as the pressure split the hull. Then, with the reduction in the supply of oxygen, the inferno turned into a cloud of billowing black smoke and soot, which ruptured the oxygen feed lines. Deadly poisonous fumes entered the environment system. Just ten seconds after the spacecraft tore apart, the atmosphere inside was lethal. It had been just 25 seconds since Grissom's first call of a fire in the spacecraft.

18:31:30. Thirty seconds after the probable start of the fire, the three astronauts were beyond help from the outside. All three had suffered extensive burns, but remained alive, falling into unconsciousness within three or four seconds. After a further three or four minutes, their unconsciousness, brought on by lack of oxygen to the brain, led inevitably to heart attack and death.

Outside the spacecraft, onlookers were helpless to do anything. The White Room TV, monitoring the CM's hatch, suddenly filled with a blinding light, and at the sound of splitting metal the technicians in the White Room ran from the area, thinking the spacecraft was about to explode. When they realised this was not the case they rushed back with fire extinguishers, but they were initially forced back by the flames and choked by the choking dense black smoke that filled the White Room.

Over the intercom, Roosa constantly called the spacecraft in a vain attempt at receiving some sign that the crew were alive, while Slayton bellowed for medics. There were two in the blockhouse, and they set off for the Pad. Back in the White Room, five members of the ground crew finally reached the spacecraft and reported that their attempts to remove the spacecraft hatch had begun. Some five minutes after the tragic sequence of events began, and working in a dense cloud of smoke and soot, they finally removed the hatches.

As the last hatch came off, a wave of intense heat greeted the rescuers. One of

The Apollo 1 capsule, with part of the booster protective cover removed, showing the effects of the fire.

them likened it to opening the door of an oven that had been on for hours. Soot and smoke billowed outwards, adding to the discomfort, but no one retreated. All that was on their minds was to get the crew out.

To the amazement of the hopeful rescuers, their initial inspections seemed to show that the crew was not in the cabin. It was a charnel house, and each of the astronauts' bodies was located by sight and feel. What they found, in the first seconds, was too much to absorb. In the original reports from the gantry, less than 5½ minutes after the incident, it was reported that all hatches were open, but the scene within was totally indescribable.

Fire crews arrived eight minutes after the start of the sequence, three minutes after the hatches were removed. Attempts were made to extract the astronauts, but this proved to be impossible because they seemed to be stuck to the structure. Three minutes or so after the firemen arrived, the medics entered the White Room, examined the bodies and confirmed what was already obvious: all three astronauts were dead. The positions of the bodies were recorded, as were their injuries, and in an extensive effort taking more than 1½ hours they were removed.

Initial examination of the positions of the crew provided evidence of the actions of the three astronauts during the final seconds of their lives. Grissom's couch was in the 170° position, basically flat out. Grissom himself was found with his feet on his own couch and with his body sprawled underneath White's centre couch. His legs were under the raised leg support of White's couch, and his faceplate was closed. Of the three, Grissom had received the most injuries from the heat and flames and his suit was the most extensively burnt. It was apparent that Grissom had disconnected his harness and oxygen hoses in a vain attempt to seek some protection from the fire as it swept up and over the couches.

The interior of CM 012 (Apollo 1), showing the effects of the intense heat of the flash fire which claimed the lives of the three Apollo 1 astronauts, Grissom, White and Chaffee.

White was found lying across the aft bulkhead of the capsule, perpendicular to the couches, under the headrests and just beneath the side hatch. His couch was found in the 96° position with the leg support section bent upwards. His faceplate was closed, and he had apparently decided to give up the effort to open the hatch and had tried to crawl under the couches with Grissom to escape the flames. His efforts must have caused extreme pain as the harness restraining him to the seat was found still closed and locked, but with the webbing burnt through. His decision to give up trying to open the hatch was the right action, as with increased cabin pressure it would have been impossible to open the hatch inwards.

Chaffee, on the right of the cabin, was the furthest from the intense heat of the fire. Lying on his couch in the 264° position, his seat harness was undone, but his suit hoses were still connected and his faceplate was also firmly closed. It was Chaffee, still on station in his couch, who managed to make the last communication from the spacecraft. His was the least burnt suit. Both Grissom and White had cut communications with the outside world in their attempt to escape the fire.

All three spacesuits were extensively burnt through. Very little of Grissom's suit remained unaffected by the flames. The most severe damage was the melted material, which had welded the astronauts to the cabin by combining with the liquid nylon at the very height of the fire. As the rescuers tried to remove the bodies, the now solidified portions of suit debris made this task all the more difficult.

The news was soon flashed to the media and thence around the world. All America mourned the passing of its astronauts. What could not be understood was that the accident had happened on the ground and not in space.

Fellow astronauts and NASA employees immediately went to comfort the families of Grissom, White and Chaffee. Within hours it was announced that a board of enquiry would be set up to investigate the tragedy, that all further flights were cancelled, and that all crews would be stood down.

The following day, 28 January, the Apollo 204 Review Board was established by NASA Deputy Administrator, Robert Seamans, to investigate the accident that had claimed the lives of the three astronauts. The nine-member team included astronaut Frank Borman and Max Faget, Director of Engineering and Development at MSC, a leading spacecraft designer. The board submitted its final report on 5 April 1967, and basically concluded that the CM and the entire Apollo system had (inadvertently, perhaps) been geared towards mission success rather than safety. This had to change. Attention focused on the complicated hatch, as it was also determined that protection from fire, especially once the spacecraft left Earth orbit, was even more important than quick egress during a ground emergency. Primarily concerned with the CM and the fire itself, the findings and recommendations also covered other areas of the programme, especially the Service Module, Lunar Module, Saturn V and the astronauts' spacesuits. Hardware came under the scope of the review board, and subsequent personnel changes reflected the findings of the investigation. Key personnel in NASA and its prime contractor, North American, were changed during 1967, as America tried to recover from the loss of the three astronauts and resume the race for the Moon in order to achieve Kennedy's goal by the end of the decade.

Fellow Mercury astronauts and John Young walk beside the coffin of Grissom during the funeral ceremony in February 1967.

On 31 January, funeral services for the three astronauts were held with full military honours. Grissom and Chaffee were buried in Arlington National Cemetery, Virginia, while White was buried in West Point, New York. The ceremonies followed memorial services at Houston on 29 and 30 January. As the board was formulating its actions, news came in of a similar accident on 31 January, the day of the funerals, in which two USAF airmen were killed in a flash fire in an oxygen-filled altitude chamber at Brooks AFB, Texas.

For the next two months the review board examined the spacecraft, visited NAA and the Cape, interviewed key personnel, and initiated tests and experiments to find the cause of the accident. They proposed a series of recommendations into future spacecraft and Apollo programme activities.

During the investigation, several accounts of events were recorded from the launch pad technicians, who provided eyewitness statements, having been closest to the vehicle at different levels on the service structure. This allowed a subsequent reconstruction of events from different perspectives.

Between 18:31:00 and 18:31:15 EST, 27 January 1967. Witnesses on launch work-platforms felt two definite rocking or shaking movements of the vehicle aft inter-stage Level A-2 before the 'Fire' report. They were unlike vibrations experienced in the past from wind, engine gimballing or equipment input. Witnesses on Levels A-7 and A-8 heard the 'Fire' or 'Fire in cockpit' transmissions, and heard a muffled explosion and then two loud whooshes of escaping gas (or explosive venting). They observed jets of flame from around the edge of the CM and beneath the White Room. TV monitors heard the 'Fire' or 'Fire in cockpit' transmissions and observed astronaut helmet, back and arm movements through the hatch window. There was an increase of light in the spacecraft window and tongue-like flame patterns within the spacecraft. They observed flame progressing from the lower left corner of the window to the upper right. Then the flames filled the window, burning around the hatch openings and around the lower portion of the CM and cables.

Between 18:31:15 and 18:33 EST. Witnesses on Levels A-7 and A-8 reported attempts to penetrate the White Room for egress action. They fought fires on the CM, SM and in the White Room. TV Monitors observed smoke and fire on Level A-8. There was progressive reduction of visibility of the spacecraft hatch on the TV monitors because of increasing smoke.

Between 18:33 and 18:37 EST. There were repeated attempts to remove the hatch and to reach the crew. North American Rockwell personnel J.D. Gleaves and D.O. Babbitt removed the boost protective cover hatch, and J.W. Hawkins, L.D. Reece and S.B. Clemmons removed the Command Module outer hatch. They then opened and pushed down the inner hatch at approximately 18:36:30. No visual inspection of the spacecraft interior was possible, because of the heat and smoke. There were no signs of life within the cabin.

Between 18:37 and 18:45 EST. The remains of the fires were extinguished. Fire and medical support crews arrived. Fireman J.A. Burch Jr and North American technician W.M. Medcalf removed the inner hatch from the spacecraft. The crew was examined and their status confirmed.

Between 18:45 27 January and 02.00 28 January. The service structure was cleared and photographs taken to thoroughly document the scene. The bodies of the crew were removed, and the launch complex and surrounding area were placed under secure conditions. Personnel from Washington and Houston arrived and assumed control.

Eyewitnesses and TV/audio recordings from 18 agencies and contractors were examined, and 572 written and 40 recorded statements were filed from 590 persons (some submitting two or more statements).

On Sunday, 29 January, three representatives of the press were allowed into the White Room to record their observations and to relay to the outside world the state of the room and the now-empty Apollo 1 capsule. Forbidden to touch anything in the CM, they had to rely on only visual sightings of the area. For security reasons they were escorted from the road block 1½ miles from Pad 34, through every gate and level. A film cameraman took TV and film coverage, and a still photographer took pictures for newspapers and magazines. George Alexander provided the reporter's point of view, and what he described in a post-examination press conference is summarised below. It is perhaps the most graphic account of the events, written only hours after the accident.

Describing the inside of the capsule, Alexander noted that bare metal was visible and that the wiring cables were 'jagged' where the fire had eaten away the insulation covering. The floor of the CM was littered with debris, including pieces of the green restraint harness used to strap the astronauts to the couches. The simulation flight plan rested between Grissom's and White's seats, and it was severely burnt and brittle. There was one readable page, near White's headrest, with only the edges of the page scorched. The interior was a uniform slate grey, with an off-white colour on one side, to the right. The head supports for both Grissom and Chaffee were in the 'up' position, but with the support sides folded down. White's was folded down and below the frame in order to provide access into the spacecraft after the fire. Also noted were handprints in the soot of the overhead rail. Instead of being shiny white, the hatch was also covered with a film of soot. The boost protective cover was outside the White Room, in the service room, and it was white with a soot border, with several finger marks in the soot. The window was totally dark from the effects of the fire. The internal hatch, originally brownish with an aluminium border, was blackened with soot along its border.

Alexander also described the very evident presence of an odour as he walked into the service room as the door was opened. 'A bitter smell of smoke', he recalled, which increased as they went into the White Room and reached its repugnant strongest as they reached the hatch area. It was similar to the smell of an electrical fire. The guide identified it as the smell of burnt insulation wires.

On the outside, as they walked into the White Room, charred items of polyurethane were found in the entrance, but Alexander was unable to identify where these came from. He also pointed out that they were not allowed to ask questions of anybody in the White Room: 'This was taboo.'

Astronaut Frank Borman (rear) and spacecraft designer Max Faget (foreground) inside a mock-up of the Apollo 1 Command Module during the investigation into the fire.

A crumpled and discarded expansion hose, used for gaseous transportation, was also on the White Room floor. Apart from the hatch, the main conical boost protective cover was still on the CM. It was white in colour, and about 270° of panels were removed, revealing CM 012 beneath. In addition, a sheet of black plastic, originally placed over the CM boost protective cover to protect it during the test, had either been torn away or had burst, as there remained only a short strip interwoven into the escape tower. It was like a 'black mourner's ribbon'.

Three control panels were visible in the White Room, in the base of the CM, with ground support instruments hooked up to them. The first was the fuel control panel. All the wire bundles in this test equipment appeared moderately burnt and a triangular smoke trace extended up from the panel along the side of the CM. The second panel examined was about four feet away, and was identified as the oxidiser panel. There was a great deal of charring and burning on this panel, and it was extremely difficult to identify individual components or even wire bundles. The third panel – the most damaged of the three, with nothing left apart from mounting brackets for electronic components – was not identified. This third unit was immediately adjacent to the spacecraft power umbilical, which at this time was still plugged into the vehicle. It was here that the greatest amount of damage was visible on the exterior of the spacecraft. In relation to the crew access hatch, this panel was to the immediate right and over Chaffee's shoulder. Small blisters were found on the heat shield and paint had flaked off in several areas to reveal the next layer of the heat ablative coating. The American flag on the CM was partially obscured by grey soot, and 'looked very dingy'. The letters 'UNITED STATES' were also covered in soot. This whole burn area was wedge-shaped, extending upwards to the boost protective cover.

All around the spacecraft was carbonised material which resembled paper. It could have been the effects of the fire, or items of debris from the boost protective cover's black sheeting. About four feet away there was a table littered with further carbonised material. At least twelve fire extinguishers were around the spacecraft, some of which, discarded in one corner, had obviously been used.

As the time approaches for launch, a fairing is placed around the base of the CM and is attached to the SM. On Apollo 204 this had yet to be attached. It exposed the top of the SM, and mylar covered some of the visible areas. Where the fire had not touched it, it remained very bright and shiny, but by the umbilical section it had burnt through and was very discoloured. The top surface resembled a crumpled cigarette paper. It was also noted that several used gas-masks, thin rubber gloves and an asbestos jacket were discarded around the White Room.

The group then descended to the seventh level (the floor immediately beneath the CM) and examined the SM. An access hatch, marked 'Cryo Storage Access Hatch' was removed from the SM. This area displayed cabling and obviously burnt insulation hanging in shreds, with blackened and jagged edges.

Looking upwards to where the umbilical joined the CM, smudge marks were very evident on the white surface of the vehicle. Several items of debris littered the floor area, but fire extinguishers on this level appeared not to have been used. Oil drips were noticed, with 12–16 strings of oil running the length of the CM–SM, and by now dry. This was the only visible evidence on the outside of the SM.

In general, cables were in good condition, but burnt wires and fire damage was very evident behind the plates. The greatest amount of fire damage was located in the area of the umbilical connection to the CM. There was no visible evidence of structural failure or collapse. Controls in the CM were almost universally covered in soot, and instrument readings were indistinguishable. Two technicians were working on the CM during this macabre tour.

After Alexander's tour, fellow pressmen asked for his impressions. 'The only thing I have seen like it [the cockpit] is the cockpit of a jet fighter that burnt. No signs of over-pressurisation as you would get in an explosion. It's just a gutted shell. It looked like a bunker from World War Two or the Korean War that had taken a direct hit.' He stated that almost everything was burnt apart from a few odd bits and pieces. The couches had been removed to provide easy access to the capsule. The open panels on the side of the CM were very evidently burnt, and their covers were removed as part of the normal test procedure when the CM is linked to ground test equipment. Normally, after the test, these panels would be replaced for flight after the cables were unhooked. Alexander emphasised that the fire did not blow off these panels, as they had already been detached prior to the

At the witness table before the Senate Committee on the Apollo 1 accident are (*left–right*) Dr Robert Seamans, NASA Deputy Administrator, James E. Webb, NASA Administrator, Dr George E. Mueller, NASA Associate Administrator for Manned Spaceflight, and Major General Samuel C. Philips, Apollo Programme Director.

accident. From what he could see in the poor lighting conditions, switches did not appear to be melted.

According to the information Alexander was given, nothing had been removed or changed after the crew and couches had been removed early on Saturday morning, 28 January. The tour was the only press access to the spacecraft. It provided a harrowing insight into the state of the vehicle, in stark contrast to the crisp, clean machine proudly displayed by NASA in the months leading to the accident.

The Subcommittee on NASA Oversight of the House Committee on Science and Astronautics held a series of hearings on the Review Board report during 10–12, 17 and 21 April and 10 May. The Senate Committee on Aeronautical and Space Sciences hearings were held on 11, 13 and 17 April and 4–9 May.

The final report of the Review Board had been published on 5 April 1967:

- A momentary power failure had occurred at 18:30:55 EST, as determined by evidence of electrical arcs found in the investigation. No single source of ignition was conclusively identified. The most probable source was thought to have been where DC power cables crossed over aluminium tubing under a lithium hydroxide access door, near the floor in the lower forward section of the lower equipment bay.
- The CM was packed with several types of combustible material in areas close to possible ignition sources. At 100% oxygen atmosphere, the test was,

belatedly, considered extremely hazardous, and the Board recommended a severe reduction of combustible material in the CM and stricter controls in the future.

- The rapid spread of the fire led to a dramatic increase in both the temperature and the pressure, rupturing the CM and creating a toxic atmosphere. The autopsy determined that death from asphyxia was caused by inhalation of the toxic gases from the fire and that thermal burns were a contributing factor. It was determined that the crew rapidly slipped into unconsciousness.

- Internal pressure prevented the CM hatch from being opened, and as a result the crew had no time to react to their situation and effect any emergency procedure before they fell into unconsciousness. The board recommended that a complete review of egress procedure times be initiated, and that they be simplified and speeded up.

- There were no procedures for this kind of an emergency, and rescue equipment was inadequate to deal with the results of the fire, the smoke and the heat. Rescue attempts were hindered by lack of provision for emergency crew access.

- Frequent interruptions and communications failures had hindered the whole test, and these should be improved to ensure reliable lines of communications in future tests.

- Late revisions to the checkout procedures, while not directly contributing to the accident, did hinder personnel from becoming familiar with the latest test procedures before use.

- Full-scale CM mock-up fire tests were conducted as part of the investigation. It was determined that such tests could be used to provide realistic appraisal of in-flight fire risks. It was recommended that further tests of flight configuration spacecraft be conducted.

- The CM ECLSS provided a severe fire risk and would continue to do so if the amount of hazardous material in the CM were not reduced. Studies and tests of a diluted gas mixture should be pursued.

- Deficiencies in the design, workmanship and quality control of the CM were cited, together with a long troubled history of CM 012 hardware removals and technical difficulties. All elements should be reviewed, and the design of joints, wiring, ducting and other issues (such as coolant loop leakage and spillage) addressed. A full vibration test should be conducted and the most effective methods and supplies of equipment for fighting cabin fires investigated.

- Too many open items were still an issue when the spacecraft was delivered from the prime contractor. Problems with programme management and relationships between NASA field centres resulted in confusion and a lack of communication or understanding of responsibility and quality control issues.

With the results of the inquiry available, but with no clear indication of the exact cause, NASA, North American Rockwell, and leading contractors, sought to impose procedures and safeguards that would prevent such a tragic accident from happening again. Many thought the Moon landing would be delayed as a result of the hundreds of changes the fire had imposed on Apollo. Not all of the recommendations were implemented, as it would have taken many more months to incorporate them in the Apollo design. By the summer of 1967 new crews had been formed, and in November the first unmanned flight of the huge Saturn V rocket was accomplished, beginning its eventual 100% success rate.

Manned Apollo flights resumed with the Schirra crew in October 1968, and by December, Apollo 8 was in lunar orbit for Christmas. The outstanding success of Apollo 11 in July 1969 achieved President Kennedy's goal of landing on the Moon. Despite the failure of Apollo 13 with an inflight explosion on the way to the Moon (see p. 277), the Apollo programme achieved six manned landings out of a planned ten (including the last three which were cancelled due to budget restrictions). Apollo hardware flew successfully on the Skylab and Apollo–Soyuz mission through to 1975.

It has been said that the loss of the Apollo 1 crew cost NASA and the Apollo programme several months in delays. Others reported that the extra safeguards built into the system and those already developed for Block II spacecraft, such as the new hatch design, contributed to the outstanding success of the Apollo programme and helped put NASA into world history and headlines. Although there was a gap of 21 months between Apollo 1 and Apollo 7, the achievement of landing on the Moon with Apollo 11 would probably not have come much earlier, due to the difficulties that were becoming evident with the Saturn V and the Lunar Module.

Alongside the successes of the Apollo programme, the loss of Grissom, White and Chaffee can never be forgotten. A duplicate of the Apollo 1 emblem was left on the Moon in the dust of Tranquillity Base, and perhaps one day the names of the three astronauts will be given to manned lunar outposts.

Nineteen years and one day after the loss of the Apollo 1 astronauts, seven more astronauts were lost in the *Challenger* accident (see page 169), forcing NASA to soul-search for a second time. In the aftermath, NASA realised that its plans for a space station serviced by the Shuttle risked having astronauts stranded in space in the event of another Shuttle accident; so, with the development of the International Space Station there was increased emphasis on crew rescue vehicles. Early design configurations reviewed the Apollo CM design, and some CM conical features were incorporated into the shortlist of design studies. Ironically, a modern configuration of the spacecraft that claimed the lives of three astronauts, and a redesign that allowed man to reach the Moon and return alive, was being considered as a possible crew rescue vehicle for the space station.

# Summary

In reviewing the first 40 years of spaceflight crew training it becomes clear that training can be as challenging and risky as completing the mission itself. Working in a field that is dependent on technology working at incredible speeds and in extreme environments is difficult on its own, as many a failed automated space vehicle has demonstrated. Adding a human to the equation increases the level of risk and complexity.

There is currently no opportunity to train for spaceflight in space itself, and all training has to be carried out on Earth, with very little opportunity to simulate the lack of gravity other than brief periods in aircraft or submersion in water. The use of life-like simulators and complex training cycles helps prepare the crew for what they may encounter on the actual mission.

The skills of simulation engineers and training staff have put countless flight crews through more life-threatening situations than the astronauts or cosmonauts could expect to encounter in a career in space, let alone during one mission. It is the mundane, repetitive training that provides the often forgotten added edge that can be the difference between life and death. Extensive training in simulators certainly helped the crew on Gemini 8 in 1966. In addition repeated simulations of normal and problematic launch profiles resulted in the almost second-nature actions of the crews of Apollo 12 and 13, Shuttle 41-D and 51-F, and others. However, some of the problems that these crews encountered were not always recognised at the time, and the experience of Shuttle 51-L *Challenger* clearly demonstrated that a sequence of events can sometimes happen much more quickly than the crew's response.

The 1970 Apollo 13 situation and the 1997 Mir incidents also highlighted that there are always going to be incidents, and that no matter how good the training programme can foresee possible risks, there are some things that will happen unexpectedly. The benefit of crew training is in being able to react to a situation instantly and effectively without worsening the situation, provided that the crew reacts in time and that there is provision to survive the effects of the accident. On Gemini 8, Apollo 13, and Mir these provisions were in place, but on Soyuz 11 and *Challenger* they were not.

At the beginning of the crew-member selection programmes one of the primary skills required to become an astronaut or a cosmonaut was flying experience. Though this has changed over the years, the requirement for flight hours remains essential. The early exploits of astronauts and cosmonauts in demonstrating their flying prowess varied between demonstrations of natural flying ability to the very edge of risk-taking, and sometimes beyond. The legends of the Right Stuff test pilots gave the impression that it was possible to escape from any situation. In fact, taking a risk was not part of a true test pilot's agenda, and those that *did* take risks did not live long.

A review of flying incidents from NACA/NASA's High Speed Flight Station/ Flight Research Center/Dryden Flight Research Center between 1954 and 1975 reveals that in the 21-year period almost 28,000 hours were logged in more than 23,000 research and test flights, during which ten accidents were recorded by the civilian agency. There were others from the military programmes, but the figures show that the skills of a test pilot include awareness of safety and risk while at the same time pushing the vehicle to its design limits and sometimes beyond. As many of the nation's top pilots moved into spaceflight training, the love of flying continued. The T-38s became the astronauts' runabouts, and added to these were fast cars and sports aircraft. In 1991 – after several instances of astronauts being involved in air crashes or near misses involving sports aircraft – NASA issued a ban on 'recreational flying' when an astronaut was assigned to a flight crew. To ignore NASA flying and safety rules risked losing the seat on the mission, as happened to veteran astronauts Robert Gibson and Dave Walker. Though grounded, their space careers were not affected for long, and both later again flew in space.

Training for space is not an individual role, as even single-seat missions had replacement or back-up crew-members available to take the place of a sick or injured prime crew-member. Selection for a crew had always been a mystery, not only to spaceflight observers but also to many of the cosmonauts and astronauts directly involved. One of the criteria for selection was medical fitness to train for and complete the mission safely and successfully. Compatibility with other crew-members is also very important, especially during long-duration missions, in which being able to distance oneself from a confrontation is almost impossible.

The problem of training for stressful situations on Earth is that no matter how real the simulation and how instinctive they might feel or react, the crew knows that in the final analysis it is just a practise and is not life-threatening. But all this changes in space. Being told that there is risk and danger in spaceflight is not like experiencing it first-hand.

By only the fifth flight of the Shuttle it was declared 'operational', and by the 25th mission, plans for flying private citizens into space had been achieved. One of the first pioneers was teacher–payload specialist Christa McAuliffe – one of seven who paid the ultimate price. The Shuttle was never, and will never become, an 'operational' vehicle, as by the nature of its design and flight profile every mission carries equivalent risk. After 199 free flights, the X-15 was still considered to be a research aircraft for later 'operational' high-performance hypersonic aircraft that have yet to be designed. The Shuttle is only a step towards a totally reusable craft with safer access to space, which is reflected in the training for the flights. All crews train extensively to deal with various malfunctions, aborts, emergencies and rescue scenarios, but a crew-member's reaction to a real event can be known only when it happens.

Early in the programme the Americans developed a system in which one crew was backed up by a second and supported by a third. The back-up crew offered redundancy in the event of a sick or injured crew-member being unable to make a flight. The reassignments of the Gemini 9 and Apollo 13 crews demonstrated this system. These two alternate 'crews' then rotated, sometimes as a unit, and sometimes being individually reassigned to later missions. The system changed early in the Shuttle programme, in which an adequate pool of trained flight crew-members were available to replace sick or injured crew-members on a flight crew. Changes in crew assignments during the Shuttle programme have proven much more complex, and do not follow a pattern as in the case of the Gemini and Apollo era, reflecting the pool of adequately trained personnel.

The Russian system has a different approach in that from the cadre of trained cosmonauts a small training group is established, made up from the ranks of the Air Force cosmonaut team, or from cosmonautically-trained engineers and specialists seconded from the institutes or bureaus. Only a short time before the flights are the prime and back-up crews selected, based on their training course results and medical examinations. It has been seen that a cosmonaut could still fall at the last hurdle by not achieving adequate training examination results or by failing a medical examination.

The nature of the two different training systems of the American and the Russian space programmes has been highlighted in the difficulties encountered by members of both teams during training in the other country's methods and procedures. This is an important hurdle to overcome during the early International Space Station missions, and a combination of both training scenarios will be required for total crew safety and compatibility.

Early examples of restricted communications were the fire-related training accidents of cosmonaut Valentin Bondarenko in 1961, the Apollo 1 fire in America in 1967, and the later incident of Alexandr Viktorenko in 1979. Due to the nature of coverage of American spaceflight programmes, the Apollo fire was highlighted in the world's media shortly after it happened, whereas the reporting of the Bondarenko incident was hidden by the Soviets for 25 years. While the Americans learned from the tragic loss of three astronauts, the incident of Viktorenko revealed that the Soviets had not learned from the Bondarenko accident.

# Launch to space

# Overview

Once a flight crew completes mission training, the next step is the launch from Earth to space. Since the only way to reach space is to ride a rocket system filled with potentially dangerous gases, liquids and other explosive fuels, this is the most nerve-wracking and hazardous part of any spaceflight. Most crews breathe a sigh of relief once they have entered orbit, and usually consider the most dangerous part of the mission to be over.

The vehicles providing human access to space evolved initially from adapted military missiles, through huge purpose-built rockets such as the Apollo Saturn V, to the current Shuttle system. The pioneering rockets did have a tendency to explode as they lifted off the pad, but as technology improved and experience was gained, so the systems matured. Since men would be sitting on top of these rockets, technicians put a little more dedication and effort into their contribution to the space programme, to ensure as far as possible that their part of the rocket would not fail. Quality and reliability became driving factors – standards which have gradually filtered through to the rest of us in the form of hundreds of space spin-offs we all take for granted.

Even with this faith in the systems and their designers, there was still a need for contingency planning. Escape methods were incorporated into the launch systems as a back-up to provide the crew with an escape option. For many years, ejector seats had been fitted on high-performance jet aircraft and rocket research planes, and had saved the lives of many pilots. These were adopted for use in the Vostok and Gemini spacecraft for crew ejection in the event of a malfunction during ascent, or in the latter stages of descent. On the first four Shuttle flights, such seats were also available to the two-man orbital flight test crews, and a similar system was planned for the Soviet Buran shuttle flight tests.

In 1964 and 1965, Soviet Voskhod missions were launched with no escape system. There was no room for one, as the additional crew-members or equipment took up the space for the ejector seat in the modified Vostok capsule. They were lucky to have no mishaps – a fact of which the Soviets became well aware following the two Soyuz launch aborts. The events of 28 January 1986 clearly demonstrated, to the entire world, the ever-present dangers of a launch to space.

For Mercury, Soyuz and Apollo, an escape tower was attached, and was fitted

Voskhod 1, October 1964: the first crew to launch into space without the safety of an escape system (*left–right*), K. Feoktistov, V. Komarov and B. Yegorov. (Courtesy Novosti Press Agency via Astro Info Service Collection.)

with high-impulse motors to pull the crew capsule away from an exploding rocket to parachute to recovery in the normal way. In more than 100 manned spaceflights, these systems were needed only twice – on the Soyuz missions in 1975 and 1983.

With the Shuttle's capability of carrying up to eight crew-members, an escape tower, ejection pod or multi-ejection seats were impractical, and prohibitive in both development costs and payload weight. Safety equipment on commercial airliners does not include individual parachutes, so why have a personal escape system on the 'operational' Shuttle missions? The twin ejector seats on the flight deck of Columbia for STS-5 – the first Shuttle mission to carry a crew of four – were disabled. This was a relief to the two Mission Specialists, who did not like the idea of being left behind if the Commander and Pilot suddenly ejected!

With no effective escape system for the entire crew – such as the ejection cockpit fitted to the B-1 bomber or the F-111 – other contingency modes were developed in the event of launch mishap. The Shuttle countdown and launch sequence featured milestones where a launch can now be halted for a few minutes or hours, or the ascent can be modified if one or more of the main engines loses power or shuts down completely. Even after main engine ignition, aborted countdowns on the Pad are possible right up to the moment of sending the command to fire the twin solid rocket boosters. When these are started, there is no stopping them until burn-out, and when the crew feel the SRB ignition they know they are going somewhere.

There are four basic abort modes available to an ascending Shuttle, and are options only after separation of the SRB, some two minutes into flight. The first is Return to Launch Site (RTLS). This is available with the loss of one engine, from separation of the solid rocket boosters (SRB) until 4 min 20 sec into the ascent. At this point there would be insufficient fuel left to accomplish the return trajectory. If an engine is lost while the SRB are active, the crew selects RTLS ABORT on the ABORT MODES switch at the earliest possible time and then presses the ABORT button. A three-engine RTLS Abort (following a major systems failure such as cabin depressurisation) would occur as late as possible (GET 3 min 34 sec) in order to burn as much fuel as possible before attempting the return trajectory. This requires the orbiter to fly further down-range to burn propellant, then turn around under power from the main engines, before cutting the propellant flow and discarding the external tank. A controlled descent to the landing strip at the launch site would follow, some 25 minutes after leaving the pad.

The Trans-Atlantic Launch Abort (TAL) is available to a vehicle that loses an engine or has a major systems failure after passing the last RTLS opportunity and before reaching the Abort To Orbit threshold. This allows a controlled gliding descent with the orbiter providing lift but with a limited cross-range capability, and without the need for additional firing of the orbital manoeuvring system, to head for a runway landing on designated sites in Spain or coastal Africa, some 45 minutes after launch.

Abort Once Around (AOA) is the option to continue towards orbit, complete one circuit of the Earth and then make a nominal re-entry and runway landing at the earliest opportunity, either in California or Florida, some 90 minutes after lift-off. This is used where power is sufficient to allow a vehicle to achieve a trajectory that can take it around the Earth once, but is insufficient to attain a long-term orbital trajectory.

The final option is Abort To Orbit (ATO), where power is sufficient to achieve a temporary low Earth orbit, to allow the Orbital Manoeuvring System OMS engines to increase orbital height over the course of one or more burns. This allows the crew and ground control to plan for an early return at one of many certified landing sites around the world, following the ground track of the orbiter, or plan for using the OMS to raise the orbit.

Following the loss of *Challenger* and her crew of seven, a personal 'slide-pole' escape method was designed for use in an emergency. This depends on the orbiter being in a reasonably controlled descent (which it probably would not be), giving time to get everybody out and safely past the wings and probable debris trailing the vehicle. No one wants to test-fly the Shuttle in-flight abort modes, but they would call upon them if required, as in the case of Shuttle 51-F in 1985. The same view is held of the slide-pole. It is good to have it, but would a crew have time to use it – and would auto-pilot control be sufficient once the last person had left the flight deck? The *Challenger* crew had no time to react to their sequence of events (as discussed later). The likelihood of being able to use a slide-pole escape system in a real emergency is, in reality, quite low.

The actual moment of launch requires millions of working parts to reach a specific point at precisely the same moment in time, to move the vehicle off the pad. The countdown to launch time is the process to prepare hardware and the crew for the predetermined moment of launch (lift-off). This is determined by the requirements of the mission, correct positioning of the target (either another spacecraft or a planetary body), recovery, landing and emergency situations.

Launch must occur within a set period of time and still meet these constraints. This period, (the launch window) can vary from minutes to hours. From the scheduled time of lift-off in this window, the programme of preparations is scheduled over a period of days, weeks and months prior to launch, and addresses the milestones as the preparations progress in the countdown to zero and the moment of launch (T, time). From then, the mission progresses and becomes Mission Elapsed Time until return to Earth at the end of the mission.

After lift-off this synchronisation needs to continue to work for the next ten minutes or so until orbit is achieved. America's first woman in space, Sally Ride, called the Shuttle launch the 'best theme-park ride ever', and onboard mission films have recorded that launch as a 'shake, rattle and ride' event. The separation of rocket stages during the launch is a further milestone to overcome. The crew, pinned to their seats by the acceleration during powered flight, are suddenly thrown forward at engine cut-off, and are then slammed back into their seats at the ignition of the new stage. Six-time space veteran John Young has called this the 'great train wreck', and it is an apt description; g forces and staging aboard the Shuttle is gentler than the pioneering flights strapped to the top of a missile, but is nevertheless dramatic. From the outside, a launch looks smooth enough, but on the inside it is far from being so.

**1965: GEMINI 6**

Despite the pioneering nature of the early years of manned space flight, launches in the Vostok, Voskhod and Mercury programmes were relatively free of trouble. By the summer of 1965, three manned Gemini spacecraft had also been launched without serious problems being encountered. However, getting off the pad was not always straightforward, and crews sometimes spent hours flat on their backs in cramped capsules, awaiting the 'GO' for launch. This caused its own problems – such as the need for Al Shepard to relieve himself in his suit during a lengthy launch delay on 5 May 1961. He had been lying in the capsule, waiting for technicians to clear several small problems before he became the first American in space. When told he could empty his bladder, the liquid pooled at his lower back and set off the suit's environmental control warning system as it tried to cope with a sudden flood of body liquids!

The Gemini programme was designed to provide crews with experience in rendezvous, docking, long-duration flights of up to two weeks and EVA (space-walking) techniques, which would be used in the forthcoming Apollo lunar missions and beyond. Gemini 6 was planned as the first rendezvous and docking mission, with a previously launched unmanned Agena docking target. The morning of 25 October 1965 saw the crew of Walter Schirra and Tom Stafford in their spacecraft, waiting for the Agena launch from a nearby pad. When news came that the Agena had been lost in an explosion during launch, their mission was scrubbed. Gemini 6 was using the last of the battery-powered spacecraft and was limited to a two-day mission. Rendezvous and docking were the only objectives, and with no target in space the mission looked to be delayed several months.

However, almost immediately, plans were devised to allow the next mission, Gemini 7, to be launched first on a two-week space marathon in December. Towards the end of that flight, Gemini 6 would be launched to perform the first space rendezvous between two manned spacecraft. They would close to within a few feet, but there would be no docking capability. Both craft would fly in formation and then perform separate re-entry and recovery profiles. They would also need to avoid inadvertent contact, which might create electrical charges between the spacecraft. Small dissipation rods were installed to prevent charge build-up. The Soviets had claimed the first dual flights with Vostok 3 and 4 in 1962, and again in 1963 with Vostok 5 and 6. But they flew different orbits, and their closest approach had been to within only a few miles.

On Sunday, 12 December 1965, Schirra and Stafford slid into their Gemini 6 capsule for the second launch attempt. Above them, in orbit since 4 December, were Frank Borman and Jim Lovell aboard Gemini 7. All four were looking forward to the historic meeting in space. Gemini 7 passed over the Cape 26 seconds before the planned launch of Gemini 6 at 09:54:06 local time. Schirra and Stafford prepared for a spectacular space chase. As the clock ticked towards zero the crew received a 'GO' for launch. The crew felt the whining roar of the ignition of the first stage booster rocket, signalling the start of their mission. The onboard mission event clock had been activated, indicating launch, and the blockhouse began the famous final countdown as the engines built up to full thrust for liftoff: '. . . three, two, one, LIFT-OFF.' Then abruptly, after only 1.2 seconds, both engines automatically shut down to complete silence.

'We have a shut down on Gemini 6', explained the launch control commentator. According to mission rules, Schirra should then have pulled the D-shaped handle between his legs and triggered the ejection seat system that would have catapulted the astronauts out and away from an expected exploding launch vehicle. But he did not. Schirra later wrote that he knew there was no lift-off, as he had not felt the same sensation 'in the seat of my pants' as on his previous Mercury 8 mission in 1962. He elected to sit tight, knowing the vehicle was not going to topple as it was still firmly held in the support mountings. Stafford had his own D-ring, but he looked at Schirra and also sat tight. According to Schirra, Stafford *did* announce his reaction – but the commentary was being heard live over TV and radio, and a slight time delay allowed NASA to 'bleep' over Stafford's comment and replace it with 'Oh shucks!'

None of the astronauts had much faith in using the ejection system. It was designed to pull the astronaut clear of the vehicle on the pad or during early stages of launch, for descent by their own parachutes. In any event, it would have been a rough ride due to expected high g forces, although how rough was unknown, as it had not actually been flight-tested, and no pioneering test-pilot astronaut wished to try it just for the experience! During unmanned tests of the Gemini ejector seat, astronaut John Young witnessed an unfortunate event. The hatch was supposed to be jettisoned, and the seat containing the dummy astronaut mannequin fly clear of the vehicle. However, on this occasion the hatch did not open and the seat and dummy ploughed right through it. This prompted Young to comment that if this happened for real it would result in 'one hell of a headache, but a short one!'

There was surprise from some (particularly the training team) that Schirra had not ejected, as in a simulation and by flight rules there was no choice but to eject. However, this was no simulation. They were sitting on 150 tons of highly explosive propellant, which could have exploded, engulfing the astronauts, vehicle and pad in a catastrophic ball of flame. But his gamble on his experience and instinct paid off, and the vehicle sat safely on the Pad, with Schirra commenting, 'Fuel pressure is lowering.' In space, the Gemini 7 crew, looking down at the Cape area, commented, 'We saw it ignite. We saw it shut down.'

At 11:33 – some 99 minutes after the event – both astronauts were pulled from their capsule to return to the crew quarters to evaluate the situation. With disappointment clearly etched on their faces, two tired and frustrated astronauts walked from the pad to begin debriefing and recycling for a new launch date.

It was soon established that Schirra's actions had saved the vehicle for another attempt. Had they ejected, then the seat rockets would certainly have destroyed the capsule, cancelling the mission permanently. Schirra later stated this was not a factor in the decision at the time, 'but it was satisfying to realise it later.'

With just six days remaining in the Gemini 7 mission, engineers planned to launch Gemini 6 in four days. Teams of technicians began investigating the cause of the engine shutdown as soon as it was safe to approach the booster. These investigations revealed that a malfunction detection system had sensed no upward movement, and closed valves to prevent fuel gushing into the engines. It was also discovered that an umbilical connector separated prematurely from the base of the booster, activating the mission programme that was to have started at lift-off.

Later that day, analysis of telemetry from the abort indicated that a decaying

Gemini 6 astronauts walk dejected from the launch pad after ignition and shut-down of the Titan II launch vehicle on 12 December 1965.

thrust level in one first-stage engine sub-assembly had been identified before shutdown. Engineers of Aerojet General worked throughout the night to discover that the fault was a protective cover dust cap inadvertently left in the engine's gas generator oxidiser injector inlet port, which would have shutdown the engine a second after the computer. If this had not been found, a second abort would have been initiated, and this time, Schirra wrote, 'I suspect I would have ejected.'

Months before, the gas generator had been removed for cleaning at the Martin Marietta plant at Baltimore. A plastic dust cover was put into the gas generator port when the check valve at the oxidiser inlet was removed. When the unit was reinstalled the dust cap was overlooked, and since it was in such an inaccessible place it was not seen in later work on the unit prior to the abort. Once removed, the generator was found to be undamaged, and it was cleaned and reinstated.

The umbilical tests revealed that the plug was fitted correctly, but it was also discovered that some plugs did not fit as well as others and could be dislodged fairly easily. Improved plugs – with the addition of a safety wire – became standard features for all future Gemini launch vehicles.

The crew displayed a remarkable level of calmness in what could have been a disastrous event. Schirra's heartbeat rate rose to only 98, and Stafford's to 120. Stafford later expressed his concern for the expected 20 g that he and Schirra could have experienced during their ejection. The sheer force needed to throw the astronaut clear of the exploding booster, and the speed at which this needed to happen, could have severely injured even the fittest astronaut. Had he experienced the ejection, he expected that at best he would have been walking around with a crick in his back for months, as had many pilots who had ejected from aircraft using similar high-impulse ejection seats. In fact, in one such incident in which a new ejector seat was being tested, the subject's eyes had to be surgically repaired when the force of the ejection ripped them from the muscles as they rotated inwards!

Schirra also stated that had he needed to use the system, then he would have done so. Had the vehicle left the launch pad and indicated that the booster was about to blow up by settling back down on the pad again, there was no option left to the astronauts. 'It's death or the ejection system', Schirra later wrote. For the event, Schirra was awarded the NASA Distinguished Service Medal, with the citation 'For his courage and judgement in the face of great personal danger; his calm, precise and immediate perception of the situation that confronted him and his accurate and critical decisions.' Schirra had relied on his test-pilot training to address the situation that he and Stafford faced in a split second, fully living up to the image of the Right Stuff.

Gemini 6 lifted off on time at the third attempt, and the subsequent rendezvous with Gemini 7 provided one of the milestones in manned spaceflight, together with photographs that over the years have been used in countless books.

## 1969: APOLLO 12

When any mission is assigned to a launch window, every effort is made to ensure that the launch time is met on time. Without affecting the safety of the crew or the success

The launch of the Saturn V marks the beginning of the Apollo 12 mission on 14 November 1969, 11.22 am EST. The mission was almost aborted when the rocket was struck by lightning within seconds of launch.

of the mission, technical problems usually are the main hold-up. Any delay in launch moves this window a few minutes, or shortens its for several days, weeks or months, depending on the mission objectives. For Apollo lunar missions this window was governed by correct lighting conditions at the lunar landing site three days after launch, and for recovery conditions on Earth at the end of the mission.

For Apollo 12, the second manned lunar landing mission, this happened to be a window of 3 hours 5 min on 14 November 1969. Delays after this time would push the launch back two days and reassign the landing site to a back-up, further west. This was not ideal, as the crew had trained hard for a pinpoint landing near the unmanned Surveyor 3 spacecraft in the Ocean of Storms.

In addition, President Nixon was to witness an American manned launch for the first time, with several other distinguished guests. At the time, NASA was fighting Congress for funds for the rest of the Apollo programme and promoting the lunar landings for all mankind, as a way to secure generous funding for new programmes. These were to include the follow-on Apollo programmes such as a fleet of Skylab space stations, a lunar base and a huge space station in Earth orbit supplied by a vehicle called Space Shuttle. It would not be politic to cancel a launch when the President was in attendance.

A rain belt had covered the Cape the day before the planned launch and, following a clearing of the clouds during that night, a launch on schedule was decided. As the crew of Pete Conrad, Al Bean and Dick Gordon breakfasted and then put on their suits, two weather patrol aircraft monitored the growing fronts heading for the Cape area at the very time the Saturn was due to launch. As the crew entered the spacecraft, rain lashed down all around the vehicle. Conrad noticed rain had even seeped in under the protective launch shroud around the Command Module (CM) and was running down his observation windows on the left of the spacecraft. Though this did not infringe launch rules he thought then that he and his all-Navy crew were not going to have a comfortable ride through the storm front. The fact that they were heading for the Ocean of Storms on the Moon also seemed appropriate. Three miles away, launch management confirmed a 'GO' for launch as the clock ticked towards zero.

Weather was marginal bordering on a 'hold' conditions, with winds light at all altitudes (16 mph at ground level), and there were no problems recorded by the lightning potential meters. Light rain showers were expected from broken cloud cover at 820 feet and above. With six highly successful Saturn V launches behind them over the past two years, the decision was given, at the Cape, to proceed with the lift-off. At that time Mission Control Center in Houston, Texas, had no live TV link with the Cape, only voice and telemetry from the vehicle, and they therefore assumed that launch conditions were acceptable. Launch control: '10, 9, 8, ignition sequence start, 6, 5, 4, 3, 2, 1, ZERO. All engines running. Commit. LIFT-OFF! We have a lift-off at 11:22 EST.'

Despite a declining interest in the Apollo mission due to more pressing needs at home and in south-east Asia, more than 3,000 invited guests and onlookers witnessed the five huge Saturn V F-1 engines ignite, build up thrust and lift the monstrous vehicle. President Nixon and his group – which included the then NASA Administrator Dr Tom Paine and former astronaut Frank Borman – stood impassively under a canopy of umbrellas as NASA launched its first mission in the rain. Twenty seconds later the vehicle disappeared from view into the cloud deck, some 800 feet above the ground.

Conrad: 'Roger, cleared the tower. We have a pitch and roll programme and this baby is really going. That's a lovely lift off. That's not bad at all.'

The Saturn was then travelling at 235 mph, just 36 seconds after leaving the pad. As the spacecraft completed a programmed roll manoeuvre to steer on the proper heading, Conrad monitored his instruments and the '8-ball' artificial horizon indicator slowly rolling with the vehicle. Suddenly he noticed a bright flash of light outside and a loud burst of static in his ears. He felt the whole spacecraft shudder and asked, 'What the hell was that?'

All three astronauts aboard the Command Module were drawn to the bank of caution and warning displays, which had 'lit up like a Christmas tree', with so many lights showing that the crew could not read them all. As the vehicle built up speed, lightning continued to discharge. A few observers on the ground saw the lightning streak down the launch tower steelwork along the Saturn's exhaust plume. A second strike, unseen by the crew, also wiped out the navigational platform. Back in the

CM, warning lights and buzzers indicated the loss of the onboard guidance system – one of the most critical navigational devices onboard – followed by another system, then another. It appeared that Conrad and his crew were riding a dying bird, the 8-ball now tumbling aimlessly as the guidance platform lost the inertial fix.

At Mission Control the control of the mission had begun when the Saturn had cleared the tower, and the problem facing the flight controllers was that they had no way in which to control the vehicle. For 26 long seconds all telemetry from the Command and Service Module to MCC Houston had dropped off line. The CSM was essentially dead, with all electrical power cut except the emergency batteries in the CM used for re-entry at the end of the mission.

Conrad then relayed, with only a minor strain in his voice, 'OK, we just lost the [guidance] platform gang. I don't know what happened here, we had everything in the world drop out. I've got three fuel cell lights, an AC bus light, fuel cell disconnect, AC bus overload 1 and 2 out. Main Bus A and B out.'.

For a few long seconds it appeared that the crew would have to abort the mission. The massive surge of electrical power tripped the circuit breakers and shut down the main electrical supply (Main Bus A and Bus B) to the vehicle and switched automatically to the reserve emergency batteries. At GET 52 seconds, as a fork of lightning hit the launch pad far below them, electrical strikes on the Saturn ceased, the vehicle resuming power from its back-up battery system on the instrument unit on top of the third stage. Had the launch vehicle guidance platform failed, located in the Instrument Unit, the Saturn V would have deviated out of control, and the launch escape system would have automatically triggered an abort mode, separating the CM from the pending destruction of the rogue Saturn to begin a parachute recovery sequence ending with the dumping of the crew in the Atlantic Ocean

At Mission Control, Houston, Flight Director Gerry Griffin, on his first assignment as Lead Flight Director, warned his flight controllers on the Flight Directors' communication loop to watch for deviations in launch profile as the crew worked to restore full power in the ascending vehicle. He also thought to himself that he would soon be ordering the abort of the flight before it had a chance to begin.

By now, Apollo 12 was breaking out into clear sky above the rain clouds. The crew – relieved not to be experiencing the tremendous jolt of the escape tower firing to pull them clear of a potentially exploding launch vehicle – focused their attention on corrective commands as they headed for staging. They still had the escape tower attached if they needed it, and as the vehicle approached 33 miles altitude and 45 miles down-range, the first stage shut down as programmed. The crew was slung forward, then slammed into their seats as the second stage ignited to propel them higher into the atmosphere.

Back in Mission Control, Griffin called upon his flight team for answers. John Aaron, the mission Environmental, Electrical and Communications Flight Controller (EECOM), had no usable data from his display screens, but remembered that he had actually been faced with a similar problem in a simulated launch more than a year before, and thought he knew how to get out of it. In order to cope with a low-voltage power supply, the crew could flip the signal condition equipment (SCE) switch in the CM panel to 'auxiliary' mode. The Saturn was now looking fine, but the spacecraft appeared to be dead, or at best dying, so the instruction to reset the SCE switch was passed from Aaron through to Flight Director Griffin, himself a former EECOM, who knew of the correct switch but watched for Aaron to make the call. When he did so, Griffin instructed CapCom astronaut Jerry Carr to instruct the flight crew.

Houston CapCom (Jerry Carr): 'Apollo 12, Houston. Try SCE to auxiliary [and] try to reset your fuel cells now.'

Conrad had not encountered this request before, and queried NCE to auxiliary' with CapCom. As Carr repeated the call, Al Bean recognised the command and flipped the required switch. As the second stage of the Saturn V powered them higher into the atmosphere, Conrad began cycling the fuel cells back on line. At the end of the mission, after the Service Module was separated prior to re-entry Conrad reported scorch marks from lightning strikes on the outside module.

Conrad: 'Got a good SII [second stage] gang. We are weeding out our problems here. I don't know what happened. I'm not sure we didn't get hit by lightning. OK, Al has got the fuel cell back on and we're working our AC buses. We've got cycling $CO_2$ pressure high, which doesn't bother me particularly. We have reset all the fuel cells, have all the buses back on line, and we'll just square up the platform when we get into orbit. I think we need to do a little more all-weather testing!'

Houston CapCom (Carr): 'Amen to that!'

Conrad: 'That's one of the better sims [simulations], believe me.'

CapCom (Carr): 'We've had a couple of cardiac arrests down here too, Pete.'

Conrad: 'There wasn't any time for that up here.'

Although the Command Module guidance system had lost its alignment, the independent Saturn V guidance system was working properly, so the immediate danger had passed. It could have resulted in the vehicle tumbling out of control and breaking up in the upper atmosphere, forcing the crew to use the escape tower. The Saturn continued to orbit, with both the second and third stages functioning normally. As they restored systems, the warning lights went out one by one. Conrad began to chuckle, and Gordon and Bean laughed with him all the way to orbit. It was a close call.

Safely in orbit, the crew had time to check the spacecraft and realign the guidance platform. The Saturn itself had never faltered. 'She was chugging along nicely, minding her own business', as the crew expressed it. As the crew settled down in orbit, Conrad reflected on the lightning striking the vehicle. He thought unstable air might have jolted the vehicle, as the Saturn was a large electricity conductor flying through the atmosphere and they could well have discharged themselves. He later added, 'I knew we were in the clouds, and although I was watching the gauges I was aware of a white light. The next thing I noticed was that I heard the master alarm ringing in my ears, and I glanced over to the caution and warning panel, and it was a sight to behold.'

In orbit, Gordon went to the lower equipment bay to realign the navigation platform with star sightings – never an easy operation – and once his eyes had become accustomed to the darkness he was able to reset the platform, allowing the CSM to once more take over the guidance role from the Saturn instrument unit (IU). The undervolt had robbed power from several vital systems, including the

CSM inertial guidance platform. This platform was preset prior to launch, and provided information on which way the CSM was heading during any point in the mission. Resetting it required a deliberate sequence of computer key inputs by the crew to realign the platform for a full check prior to the TLI burn. During the orbital pass they chatted about probably scaring their wives and giving the press something to write about, and for themselves, an all-Navy crew, a 'sea story' to retell for years. 2 hrs 28 min into the flight they received the 'GO' for the engine burn to take them to the Moon. 'Whoop-de-do. We're ready. We did not expect anything else', replied a happy Conrad.

During the first orbit, Director of Flight Crew Operations Chris Kraft and Apollo Program Manager Rocco Petrone reviewed the available data and approved the recommendation to proceed with the TLI burn, fully aware that the lightning strike could have seriously damaged vital systems for which they had no data. They were both fully confident in the design of the Saturn and the CSM, and in the skills of the astronauts and ground controllers in dealing with any eventually that might face them. They also knew that the LM, protected inside the spacecraft launch shroud during ascent, was probably unaffected by the event. What they could not be sure of was that the lightning may have damaged the pyrotechnics used to deploy the CM landing parachutes. Kraft and Petrone decided not to tell the crew, as they could do nothing about it. The astronauts would be as dead if they tried to use the parachutes now in an aborted mission, as they would be in ten days after landing on the Moon.

Launch Director Walter Kapryan later compared the power surge to a blown fuse in a house, and said it protected vital instrumentation. In post-flight examinations, NASA field centres in Houston, Alabama and Florida studied the telemetry, film and witness reports as well as crew debriefings. The theory they formed was that the whole 363-foot Saturn V, plus its ionised exhaust trail, acted as one huge lightning rod which triggered static electricity in the heavy cloud cover. Just seconds after leaving the pad, a bolt of electricity discharged right through the vehicle and down the exhaust column to the tower 6,000 feet below.

Post-flight evaluation of the recorded weather showed that the minimum requirements for launch were met, and were well within pre-launch criteria. Unlike the Shuttle, the Apollo system was designed to cope with launch in rain, but the possibility of lightning had concerned the Launch Director and a hold was considered. USAF weather stations reported only minor turbulence and no indications of lighting within 32 km of the Saturn on the pad. US Air Force One, bringing in the President and his party for the launch, had also experienced no turbulence while flying through the front. With this information and a 'launch now or scrub' situation, the 'GO' was given for lift-off.

The second strike, 16 seconds after the first, had been the one that had sent the inertial platform on the CSM tumbling, and both strikes recorded 60,000–100,000 amps passing through the metal exterior of the stack to the bottom of the exhaust plume. This damaged external instrumentation and reaction control system thruster reserves, neither of which were critical. Fortunately the direct charge did not penetrate inside the CM, but it did induce voltage and current charges which raced around the electrical circuits, powerful enough to trip systems off line but not severe enough to destroy the main circuits in the CSM.

Apollo 12 was saved by the early design decision to incorporate guidance for the Saturn launch vehicle independent of the Apollo spacecraft. Early designs of the circuitry had direct control of the Saturn V from inside the CSM, providing the crew with more hands-on control of their launch vehicle if required. The Saturn V designers at Marshall Space Flight Centre in Huntsville, Alabama, had always resisted this, and eventually won the argument against overburdening the crew with a role that the computers in the instrument unit could handle automatically. The instrument unit on the top of the Saturn IVB third stage would handle the Saturn guidance, and on Apollo 12 the software continued to work in the IU despite increased current, and prevented the Saturn gyro platform from tumbling. If an extended connection from the Saturn to the CM had been chosen, and probably routed along the outside of the vehicle for simplicity then this would have made it even more susceptible to a lightning strike.

In fact, the reports of a lightning strike were disputed for a while, due to speculation that lightning damage would have probably been more severe. Post-flight analysis of the film and photographs of the launch recorded two lightning hits. The whole event triggered a more scientific analysis of the threat of lightning hitting an ascending vehicle. In December 1969, after the crew had returned from their lunar landing mission, the American Geophysical Union discussed the NASA data and agreed that the Apollo 12 stack had itself triggered the recorded lightning discharges. For six hours before and following lift-off, not a single incident of a strike was recorded.

As a direct result of the Apollo 12 incident, additional instrumentation was incorporated into the ground equipment to monitor electrical charge around the pad area. A lightning conductor was also placed on top of the launch tower to divert any lightning from the Saturn V (or the Shuttle in later years). A NASA Lightning Investigation Team strongly opposed any major modifications to the Apollo vehicle itself, although minor changes were included on Apollo 13 onwards, to obviate the possibility of serious damage to components resulting from a future strike. Flight rules were amended so that no launch flight path would take a vehicle to within 8 km of a thunderstorm, cumulus clouds with a 3,050 m-plus ceiling, or 'through a cold front' where seasonal weather could also have a 'moderate' effect on launch plans. Temperature and wind guidelines were also checked. NASA's management decision was that never again would a manned vehicle be launched in such adverse weather conditions. That rule was enforced for 15 years, but temperature and weather conditions would again be a major feature in the Shuttle programme – dramatically so for *Challenger*'s tenth mission in 1986.

On 11 December, the flight crew visited KSC to thank the launch team for their efforts on Apollo 12. Conrad commented, 'We forgive the weather man for his job, but had we to do it again, I'd launch under exactly the same conditions. We had such fine equipment [despite] the little difficulties we came up with on the flight.'

## 1983: STS-8

By 1983 the Shuttle had flown seven highly successful missions. Not since Apollo 12 had a serious launch event threatened the safety of an American crew. The danger was always present, and several minor incidents had been recorded on a couple of launches (see the table on p. 149), but confidence was growing in the reliability of the Shuttle system.

On 31 August 1983, STS-8 became the first American night launch since Apollo 17 had set off for the Moon in December 1972. The twin solid rocket boosters (SRB) and three Space Shuttle main engines (SSME) turned night into day as the Shuttle *Challenger* roared to orbit. As it was a smaller and much lighter vehicle, the Shuttle climbed more rapidly than the old Apollo, and at 2 min 30 sec the SRB burned out, separated and headed for an ocean recovery, to be refurbished and used again. *Challenger*, meanwhile, continued upwards and flew a successful mission in Earth orbit.

It was not until the boosters had been recovered and examined that anyone realised how close STS-8 had come to an SRB burn-through of the type that later contributed to the loss of STS-51L (mission 25).

Following parachute descent into the Atlantic Ocean, the twin SRB were recovered and towed back to the Cape, where they were disassembled, dried, cleaned and prepared for reassembly and reuse on a later mission. On 27 September – a full month after the launch of STS-8, and three weeks after the return of the crew – Morton Thiokol, prime contractor for the units, discovered excess corrosion during routine inspections of one of the carbon–phenolic resin linings. These linings were used to protect the nozzle's aluminium and steel structures from the 5,800° F exhaust gases.

Pre-ignition thickness was 3 inches, designed to be ablative and burn down to 1–1.5 inches at SRB cut-off. Post-flight examination of the left-hand set from STS-8 recorded a burn down to 0.2 inches in various parts of the nozzle area. No details were released on exactly how long the area was away from complete burn-through, but some reports indicated just 14 seconds.

Examination and tests of nozzles for STS-9 (and later missions in processing at that time) revealed mixed results. Some showed no problem at all, others a near duplication of the STS-8 boosters. The anomaly was eventually pinpointed to the curing of the batch of resins used on this set of boosters.

Had the burn-through occurred, the nozzle itself would have been breached a short time later. Hot gases would have spewed out from the side of the nozzle, possibly sending *Challenger* into an uncontrollable cartwheel that would have caused aerodynamic loads beyond the vehicle's tolerance and probably resulted in the loss of the crew.

At the time, the media's reporting of the incident and the deterioration of the lining was minimal. They again attributed it to the long history of setbacks and niggling problems that had plagued the Shuttle programme for years. Despite other launch mishaps and anomalies, this spectre would come back to haunt the programme in 1986.

**1984: STS 41-D**

On 26 June 1984, six Americans were once again seated aboard a new orbiter, *Discovery*, awaiting launch. The day before, their maiden launch had been halted at T–32 sec, due to errors on the back-up computer system. Ground controllers had been unable to solve the problem, and so the launch attempt was postponed for 24 hours.

This time, all progressed smoothly towards lift-off. At T–6.6 sec, the normal staggered main engine sequence started with the ignition of No. 3 engine. 120 milliseconds later No. 2 engine was ignited, but before No. 1 engine was started computers located a failed signal in No. 3 engine. All three engines automatically shut down, and the launch was aborted just 2.6 seconds after the initiation command and 4 seconds before the SRB command was due, from which there would have been no turning back from lift-off.

Launch Control: 'We have a 'GO' for main engine start. 7, 6, we have main engine start. We have a cut-off. We have an abort by the onboard computers of *Discovery*.'

At 2 seconds the confirmation of a Redundant Set Launch Sequencer (RSLS) abort was recorded, followed by commands to ensure the safing of the ground launch sequencer and three engines to prevent the SRB ignition. Readings indicated that engine No. 1 was still alive and running, and quick communications between controllers in the 30 seconds after shutdown revealed that it was not actually ignited. Commander Hank Hartsfield radioed, 'We have a red light on engine two and three in the cockpit, not on one.' After verifying that engine No. 1 was indeed safe, some 5 seconds after the abort, the command to make *Discovery* safe was given to the crew.

At 3 min 30 sec, a report of a fire around the engine compartment of the orbiter was relayed, and there began a series of rapid discussions about the possibility of a fire on the vehicle. Suspected to originate around engine No. 3, a 10-foot high flame was seen on TV coverage of the orbiter body flap, away from the engine in question. Despite negative indications of a fire by detectors, several fire sensors in the SSME area were activated, prompting the controllers in Launch Control to initiate water spray nozzles to extinguish any potential fire hazard.

For several minutes the fire sprinklers deluged the pad with thousands of gallons of water, and flames were seen on TV pictures used to investigate progress. Some 20 minutes after the abort, cameras revealed heat waves and a possible fire around the engine area. The fire then gradually petered out, allowing removal of the crew 40 minutes after the abort.

Support astronaut Mike Smith (who would later be killed onboard STS 51-L) closed out *Discovery* as the flight crew made it down to pad level. They were not in any danger, and it was deemed unnecessary to use the slide wire systems available to evacuate a crew from the pad quickly in such a scenario. Almost immediately, NASA initiated a board of inquiry to investigate the abort and reschedule the mission a few months later.

All three engines had been used before, including test firing. Engine no. 1 had flown twice, logging 1,752 seconds of burn time; engine No. 2 had flown once and logged 1,032 seconds of total firing time; and engine No. 3 had completed three flights and logged 2,658 seconds of burn time.

A NASA/Lockheed inspection team works beneath *Discovery*'s main engine following the aborted launch attempt of 26 June 1984.

A review of the sequence of events revealed that engine No. 3 started on time at T–6.6 sec. During the start-up sequence, two checks, 20 milliseconds apart, by the Main Engine Controller (MEC) of Engine No. 3 on channel A, which were to verify the position of the main fuel valve (MFV), found that it did not open. Upon 'seeing' this, the MEC automatically switched to channel B, which began to drive the MFV open. However, due to the lack of redundancy (no back-up system) for this valve, this violated standard launch commit criteria. The MEC signalled a 'failed' flag to the orbiter general purpose computers (GPC), which in turn issued an immediate shutdown command to all three engines. Engine No. 3 never achieved internal ignition, nor rose above an approximate pressure of 18–23 psi. Engine No. 1 never received the start command, and engine No. 2, which received its start command, was at about 20% thrust when the shutdown command was issued by the GPC.

In the post-abort press conference, Commander Hartsfield explained the situation from the crew's perspective inside *Discovery*:

'The major impression from the crew was that of disappointment. We'd had one of the smoothest counts I'd seen. When we got under 31 seconds I told the crew, 'Hang on gang, here we go.' We felt a large bump as engine ignition started and just a fraction of a second later we got the master alarm and the red light on two of the engines indicating they had shut down, or had shutdown commands. At this point I knew we weren't going anywhere. We had seen this in our simulations. We were interested to see the count stopped and made sure we did not go into flight mode, which would indicate that maybe the solids were going off, and we didn't want that to happen. The ground recognised the problem and began working on it. We kept very quiet in the cockpit, as we wanted to hear what the ground had to say. There was no concern on our part because the [safety] system had done exactly what it was supposed to: detect a problem and shut the engines down. It picked us to test itself and it worked fine. For that we were certainly grateful. We heard the launch team say they saw what appeared to be a small fire under the bird [which was found to be burning residual hydrogen] and we listened very attentively to that conversation. When things calmed down, one of the crew commented, 'Gee, I kind of expected to be a little higher than this at MECO [Main Engine Cut Off]!' The mid-deck people exited first. There was quite of bit of water around from the spray-down, so I knew we were going to get plenty wet. We got very wet making the dash to the crew van, [which] was our biggest discomfort.'

Hartsfield added that they briefly thought of using the slide wires, but they were not anxious to try it out. They could not see or smell anything out of the ordinary during their exit. For the whole crew, the feeling was one of extreme disappointment at having to abort twice in two days. Hartsfield's first reaction was 'rats', as he knew that, having had an engine start, they would not launch for some time after this second abort.

When a repeat of the failure was accomplished in subsequent tests, it was determined that microscopic contamination had clogged the valve in the engine. The whole engine was replaced with a new unit in the VAB, and *Discovery* was back on the pad by 5 July. It was not until 31 August that *Discovery* finally left the ground, with the same crew, but with a revised payload.

## 1985: STS-19 (5I-F)

As described earlier, NASA opted for several in-flight abort modes. These were determined to be the best options for the survival of both the orbiter and the crew in the event of certain types of emergency during lift-off and ascent.

By late July 1985, eighteen Shuttle launches had been successfully completed without the need to rely on these contingency plans. Shuttle 19 (designated 51-F), carrying the Spacelab 2 astrophysical payload, had already experienced a pad abort situation similar to, but not as dramatic as, STS 41-D. On 29 July 1985, *Challenger* finally left the pad for its mission in Earth orbit. The first minutes of the mission through the separation of the SRB progressed smoothly enough. As the boosters fell away the crew received the call that their first stage performance was 'low', which necessitated throttling of the SSME by Commander C. Gordon Fullerton and Pilot Roy Bridges to compensate for the lack of thrust.

At GET 3 min 30 sec, the SSME Flight Controller observed a warning light on her display console that indicated a failure in one of two temperature sensors in the No. 1 engine, the centre engine's high-pressure pump. She immediately advised the Flight Director on the controller's communications loop. Just ten seconds later, as *Challenger* continued to ascend through the upper atmosphere, the vehicle passed the point of no return, 110 miles down-range, travelling at 6,600 fps.

At GET 5 min 5 sec the crew received the call 'Press to ATO', meaning that the crew should be ready to dump fuel should an engine fail, and then proceed with the Abort-To-Orbit procedures if necessary. *Challenger* was by then travelling at 11,000 fps, 58 nautical miles altitude and 275 nautical miles down-range. At GET 5 min 45 sec, a loud warning alarm sounded in the headsets of the flight crew. They immediately noticed the SSME status light on engine No. 1 shining bright red, and correlated this with the rapid fall to zero of the engine chamber pressure reading.

Fullerton: 'We have a Centre Engine Failure!'

CapCom: 'Roger. We copy, standby. ABORT ATO. ABORT ATO.'

Public Affairs Officer: 'Mission Control Houston, we have a centre engine down on the *Challenger*. The crew have been instructed to Abort To Orbit.'

Onboard *Challenger*, Fullerton selected the Abort To Orbit (ATO) switch on the instrument panel, and pressed the switch 'ABORT'. This immediately instructed the onboard computers to configure their software to the selected abort status and commanded the OMS to use 4,400 lbs (1,995 kg) of fuel. The OMS were gimballed to compensate, to some extent, for the lost centre engine, thereby relaxing the gimbal angle on the two SSMEs. This allowed the two still-functioning SSMEs to operate a more efficient thrust, pushing *Challenger* onwards and upwards towards orbit.

With the two OMS firing in conjunction with the SSMEs, the additional 12,000 lbs of thrust, combined with the rapidly diminishing mass of the vehicle, would allow a safe entry of the spent ET over unpopulated areas.

In Mission Control, the flight control officer responsible for monitoring the performance of the SSMEs saw one of the two temperature sensors on engine No. 3 record a high temperature at GET 8 min 13 sec. Further indications suggested that a second sensor was about to fail in a repeat of the sequence of events experienced on the No. 1 engine a few moments before. At her recommendation, the Flight Director passed up instructions to the crew to inhibit the SSME controls. This meant that automatic shutdown of the two SSMEs was cancelled. The flight controllers were convinced that the signal was in error. Inhibiting shutdown allowed orbital insertion on the remaining engines. The crew would now shut off the engines at the appropriate time, either upon achievement of orbit, if all went well, or if a fault really did develop and the flight controller deemed it necessary to abort.

After burning for some 60 seconds longer than a normal mission, the shutdown of the remaining SSME occurred at GET 9 min 42 sec. Even so, *Challenger* was travelling 114 fps slower than the planned 25,760 fps, and with cut-off achieved at 70.25 miles altitude. The ET was separated normally, but re-entered without incident over Saudi Arabia instead of over the Pacific Ocean.

The OMS-1 burn was cancelled, and the designated OMS-2 burn at apogee of 194 fps increased the velocity of *Challenger* to raise the perigee into an orbit of 164.7 × 124.4 miles. With just a little extra effort and an amount of good luck, *Challenger* had reached orbit for the eighth time. Almost immediately, ground controllers reviewed the situation and mission planning teams soon determined that it would be possible to carry out the full mission duration.

Over the next 24 hours, the orbit of *Challenger* was gradually raised to a height that would enable the flight crew to accomplish most of the preplanned mission objectives. Three additional burns of the OMS system ensured that they were in a safe enough orbit to sustain a week-long mission. By conserving onboard supplies, almost completely rescheduling of orbital activities and rewriting the flight plan as the mission was flown, the mission concluded with a safe landing at Edwards AFB, California, eight days later. In fact, this even included a one-day orbital extension to allow collection of additional scientific data. With *Challenger* back on the ground, engineers were able to take apart the faulty engine and evaluate what exactly had gone wrong, using the recorded data obtained during the launch from both onboard the orbiter and on the ground.

On 6 August 1985 – the day after the landing – technicians removed the faulty sensors from the engines and returned them to their manufacturer for what NASA termed 'expedited analysis.' Fullerton stated that the crew recommended that, *Challenger* 'be sent back to the shop for an engine tune-up.' He also added that he was thankful that the failure had not occurred seconds earlier, as that would have resulted in a difficult Trans-Atlantic Abort landing at Zaragoza, Spain, 'undoubtedly setting a new Trans-Atlantic crossing record!'

NASA indicated that the cause of the malfunction would be traced before the

The inspection team views the area of tiles on the orbiter body flap (at left) where the tile bonding caught fire, discolouring the thermal protection system tiles, during the abort of STS 41-D on 26 June 1984.

launch of the next mission, planned for later that month. On examination of the sensors it was soon determined that all the data were correct and that it was the sensors, not the readings that were faulty. The engines had been fine.

In September 1985, NASA issued the post-flight Mission Report for 51F, which reported that all SSME parameters appeared normal during the pre-launch and countdown activities, and compared well with data recorded on the previous eighteen flights. At T + 120 sec, data from SSME No. 1 channel A high-pressure fuel turbo pump (HPFTP), measuring discharge temperature, displayed characteristics that indicated the beginning of a failure of that sensor, and drifting measurements. In channel B the sensor failed at T + 221 sec. The channel A measurement continued drifting, and exceeded redline limits at T + 343 sec, resulting in the premature shutdown of the SSME. All other parameters recorded on this engine appeared satisfactory.

During main stage operation of SSME 3, the channel B sensor data measurement of the HPFTP discharge temperature also began to drift upwards, and exceeded the red line value at approximately T + 493 sec. However, the measurement did not at this time exceed the 2,900° F limit. After a drift downward of a few seconds, and then back up, it exceeded the limit at T + 496 sec. Measurement of channel A on this engine remained within prescribed limits. With this information and other data recorded, operation of the SSME was satisfactory. There were no anomalies recorded with SSME 2. Further ascent performance before and after the premature shutdown of SSME 1 was satisfactory.

When the warning levels of the failing sensors were received at MCC-H, they were ignored, because the back-up sensors recorded normal levels. However, two minutes later the back-up alarm triggered and the computer signalled a shutdown of SSME 1. This was only the second time in an American manned spaceflight that a mission had lost an engine in the launch phase. (The previous incident of this type was recorded on the first stage of the Saturn V of the ill-fated Apollo 13 launch in April 1970.) Using well-designed procedures, and drawing upon the hundreds of hours of simulations and contingency training, it was soon possible to assess the situation and initiate adequate abort procedures to not only overcome the initial problem but adapt the flight plan and complete *Challenger*'s mission successfully.

It was also subsequently reported that the SSME sensors had developed faults in numerous ground tests for some years, and on receiving the data from *Challenger* the action was to switch off the faulty sensor until the vehicle aborted to Spain or Earth orbit. When the No. 3 sensor also indicated an impending failure, with a reading so high that the Flight Director could not believe it, he ordered the crew to isolate the circuits to prevent automatic cut-off, before computers shut off another engine. His actions proved correct, and the mission was saved.

Once in orbit, the problem posed no threat to the crew, as the SSME are used only during the ascent. But troubleshooting of the problem would have to wait until *Challenger* was back on the ground again at the end of the mission. NASA had always tried to provide a sophisticated computer monitoring system with redundant back-up circuits and a series of warning signals to prevent a build-up of multiple failures, but this does not ensure complete foolproof monitoring. *Challenger*'s computers actually mistook the information they were receiving. Inside each sensor was a wire of tungsten and rhodium that altered its ability to conduct electricity as the temperature changed. Computer analysis determined that when the wire broke, becoming unable to conduct electricity, it sent an erroneous signal of infinite temperature. Computers onboard *Challenger* then turned to the back-up sensor at a time when, apparently, insulation compound around the wires leading to the sensor had deteriorated. Instead of a sudden leap to an infinite temperature, the computer recorded a gradual climb towards a dangerously high temperature, and thus quite correctly shut down engine No. 1.

The multiple failure of an already suspect sensor design prompted NASA to supply the already designed replacement sensor type on *Discovery* for STS-20 (51-I) and for all subsequent missions, preventing a repeat of the STS-19 (51-F) scenario.

The abort mode controls on the flight deck of *Challenger*, set to the Abort To Orbit (ATO) configuration required during the ascent of 29 July 1985.

In a 1988 interview, STS-19 (51-F) Flight Engineer Story Musgrave related his personal recollections of the activities on the flight deck during the ATO profile. He stated that with this, his second launch into space, he was much more 'frightened' than on the first flight, where had had faith that everything would work correctly. However, he added that since the start of the Shuttle programme in its current format (with twin SRBs), he was 'scared to death of the entire process'.

The pad abort of 12 July had only supported his fears of this phase of the mission, and he went into his second launch with more apprehension than on his first flight. Despite selection as a scientist astronaut and qualification as a Shuttle mission specialist, Musgrave was a qualified jet pilot (as were the entire 1965 and 1967 scientist astronaut candidates). He flew as a member of the 'flight deck crew' on 51-F, with more responsibility for launch, landing and orbital operations of the Shuttle than for the actual payload, which was attended to by the 'science crew' members. That is, on this ascent, he was the third seat Flight Engineer, Mission Specialist Two (MS2) who assisted the Commander and Pilot, sitting between and behind them during flight deck activities for ascent and landing, rather than 'just going along for the ride,' as on his first flight.

Recalling the ATO situation, Musgrave stated that the flight crew (Fullerton, Bridges and himself) knew the cause even before the abort call came from the ground. Training simulations gave a very good indication of what they were actually experiencing, with a decrease of thrust as they lost an engine. Having gone through many hypothetical simulations of similar situations, the actions of the crew were almost instinctive on seeing the red light on the console.

The reaction onboard was inevitably linked to this training, and Musgrave's first remarks were 'What the hell are they [the controllers] doing that [inserting an abort situation] for today?' He soon realised that it was the real thing, however, and commenced procedures to configure the vehicle for the ATO profile. With that completed, Fullerton and Bridges handled the ascent of the Shuttle to orbit. Musgrave said he was busy studying the flight procedures malfunction books to review his actions if a second engine went down, so that he would be ready with the all important information that his fellow astronauts would need to complete a safe landing in Spain. He continued to say that he was mentally in that book of procedures, reviewing them one after another, when Karl Henize – the Mission Specialist, sitting in the fourth flight deck seat, who had no launch or entry duties – looked over and saw the malfunction procedure book. Seeing the page open at Zaragoza, he casually asked Musgrave, 'Story, where are we going?' Musgrave, without looking up from the page said, without thinking, 'Spain!', at which point Fullerton and Bridges spun their heads around to glare at him in amazement. Musgrave looked up immediately, realising what he said and added, 'Forget it ... I didn't say that', waving the looks away. Being so engrossed in what he had to do if called upon, he never heard Henize's comments properly. Musgrave concluded that at least it gained everyone's attention for a couple of seconds.

In a later interview, Henize added, 'I was sitting on the flight deck immediately behind Bridges when the computer gave an alarm that there was a problem and it was going to shut down an engine. I didn't quite believe it, but when it finally sank into my head that we really did have an abort, my first feeling was 'Oh my God, we're going to land in Spain. I'm not going to make it into space yet!' (He had waited eighteen years for this, his first and only chance.) When his initial feeling of disappointment subsided after seeing that no one else was unduly concerned, he asked what ATO meant (as a science crew-member he had no direct training in ascent procedures or abort simulations). He was told by Musgrave, 'Relax, Karl, we're going to orbit.'

Mission 51-F provided the unwanted, but useful, experience of an actual Shuttle launch abort mode. Despite the loss of one engine, it proved that a Shuttle could reach orbit and, with close co-operation between the crew and ground control, successfully complete its mission. Once again, NASA had proved that well-tried procedures and training, with back-up contingency, were effective if the astronauts had the luck and opportunity to use them.

Pilot Bridges later added, 'Although an engine failed during ascent, the crew – both air and ground – responded appropriately, and the mission continued to a highly successful conclusion.'

**1999: STS-93**

Following the loss of a main engine on the 1985 *Challenger* launch, the next major launch incident was the 'major malfunction' of mission 51-L, with the loss of seven crew-members, in the 1986 launch explosion (see p. 169). In 1988 the Shuttle returned to orbital operations and, despite numerous launch pad aborts and setbacks with ground and vehicle systems and hardware, no real emergency occurred for more than a decade.

The minor incidents went largely unnoticed by the media and the public outside the space community. This indifferent attitude to the programme was interrupted only by 'event' missions. One such event was the return to orbit, in October 1998, of Senator John Glenn who, as a Mercury astronaut in 1962, became the first American to orbit the Earth. His second, long awaited flight aboard STS-95 was always going to be a high-profile mission, and of course the launch marked an historical milestone for both the 77-year-old astronaut and the space agency itself, which was eager to gain public and political support for the troubled International Space Station programme, which was then preparing for the launch of its first elements.

Apart from the loss of the compartment cover panel for the deceleration parachute – introduced in 1992 on STS-49, and used at the end of the mission – which fell off in the first few seconds of the mission, the ascent of STS-95 was free of trouble. Vibrations from the SSME had shaken the panel loose and, for the first time in more than forty launches, it fell off, hitting the vehicle, bouncing off and falling on the pad area. Luckily there was no serious damage, but the decision was made not to deploy the landing parachute to slow the vehicle after touchdown in case some damage had been inflicted on the area or system.

Less than twelve months later the launch of STS-93 was to carry the third of NASA's Great Observatories into orbit. The Hubble Space Telescope had been deployed in 1990, followed by the Compton Gamma-Ray Observatory in 1991 and then, in 1999, the Chandra X-Ray Observatory was deployed. The mission also featured the first female crew commander, Eileen Collins, who had earned this command on merit from two previous missions as Pilot. Inevitably the media focused more on the role of Collins as the first woman in charge of a US space mission, as she made spaceflight history on 23 July 1999, taking *Columbia* off the pad and into orbit for its 26th mission. The world's media had already hyped the event out of all proportion. They had covered all the expected headline angles on the mission and the pioneering role that Collins was undertaking, and had exhausted the stock of links to the skills of female drivers and comments about who to blame if something went wrong. There was almost no reference to the importance of the payload or that she was just one of a group of suitability qualified women astronauts in a pool of astronauts available to NASA. The role of an astronaut was certainly changing, but launch was as risky as always.

There had already been two aborted launch attempts. The first, on 20 July (the thirtieth anniversary of the Apollo 11 lunar landing), was cancelled at the T–7 sec mark when the orbiter's hazardous gas detection system recorded unacceptable hydrogen concentration levels in the aft engine compartment of *Columbia*. There was a risk of an explosion in the bay. SSME ignition was programmed for T–6.6 sec, and the cut-off was initiated at T–8 sec after the sensor reading recorded an upward spike at T–16 sec, peaking at T–10. At T–8 sec, other hydrogen readings indicated that the earlier data were in error – but by then it was too late, and the command came to shut down the engines.

NASA technicians later reported that an ion pump located in the gas detection system had caused the fault. This had become saturated and had actually counted some ions more than once. This fault had been noted in past launches, but never so late in the launch cycle. This was an impressive safety demonstration, but NASA altered its strategy to ensure that it was not repeated. It was decided that any decision to halt the countdown at such a late stage would be made after the last reading had been taken. This would follow ignition of the SSME, and would require a pad abort if readings did not return to normal levels just prior to SRB ignition.

Two days later, on 22 July, after the area had been cleaned of hydrogen contamination, the second launch attempt was cancelled due to unacceptable weather conditions (lightning storms) at the Cape. NASA was not prepared to risk a repeat of the Apollo 12 lightning strikes.

It was a case of 'third time lucky' for STS-93. Mission Specialist Steve Hawley – on his fifth flight into space aboard the Shuttle – was taking no chances, and decided to cover up his smiling face as he prepared to board *Columbia* by wearing a brown paper bag over his head. He had already suffered launch pad aborts on his first three missions, and he was fast becoming an expert on them! Finally, after a short delay due to communication problems, *Columbia* left the launch pad and headed for orbit.

Just five seconds after lift-off a voltage drop was recorded in one of the vehicle's electrical circuits. This resulted in the shut down of one of two redundant main engine controllers – one that served two of the three main engines (1 and 3). Engine No. 2 was unaffected, and the other controller compensated for the loss with no impact on the ascent.

Then, nearing the peak of the ascent, the MECO command occurred seconds before it was programmed, due to lack of fuel being fed to the engines. It was reported to be about 4,000 lbs short, and this left *Columbia* seven miles lower than its planned orbit. Using the Orbiter manoeuvring engines, *Columbia* eventually made up the difference and completed a highly successful five-day mission, which included deploying the Chandra X-Ray Observatory.

When any Shuttle crew is formed, the delegation of responsibilities for activities, systems and hardware is divided amongst all crew-members. These responsibilities are the primary concern of one or more astronauts, with others serving as a back-up. The Commander is responsible for all aspects of crew safety and mission success, and has ultimate responsibility for everything that happens on a flight. However, he or she cannot have primary responsibility for everything, and even the Pilots takes the prime role in some aspects of ascent and landing, with the Command 'backing up' this role. For the STS-93 ascent, Pilot Jeff Ashby was primarily responsible for malfunctions, and was monitoring flight deck controls for these. Commander Collins had trained in a back-up role to Ashby in this area. Mission Specialist Steve Hawley, sitting between and behind them and acting as Flight Engineer, also participated in the monitoring and assessment of the situation in communications with Flight Controllers at Mission Control, Houston.

When the event was reported in the post-launch press conference, NASA's Shuttle Programme Manager Donald McMonagle, himself a former astronaut, compared the shortage of 4,000 out of 1.2 million lbs in the external tank to a 20-gallon car fuel tank running short by a half pint. It was also reported that a review board would investigate the shortage. However, this did not stop the tabloid media from deriding the programme with frivolous headlines such as 'First woman commander blasts into history. NASA asks: "Who forgot to fill the tank?"'

It was not until the results of the investigation were released that the seriousness of the situation became evident and it was appreciated how close Collins had come to being forced to attempt the first, very risky, Return To Launch Site (RTLS) abort.

The loss of the primary controllers seconds after launch forced the vehicle to fly on the back-up systems. Due to the early requirement to use the these systems, almost the entire ascent was flown with a much-reduced level of redundancy available. Had a two-engine failure occurred early in the ascent, the crew would have had only one option – an attempt to bail out over the ocean.

NASA engineers initially thought that the most probable cause of the 13.3-amp short-circuit was the primary controller of engine No. 1, which was connected electrically to engine No. 3. These faulty units were taken back to the prime contractor, Honeywell, for evaluation. Upon further investigation it was discovered that the short originated inside the orbiter's body for the top centre SSME (No 1). This tripped the circuit breaker, which in turn cut power to the back-up controller on the lower right engine (No. 3). It was a cascading fault.

Finding a new source for the electrical shortage moved the focus of the problem from the Engine Programme Branch to the Vehicle Engineering Office, and had further implications for the rest of the orbiter fleet. However, the controllers used on the launch of STS-93 were of an older design that was unlikely to be used again after that flight because they were scheduled for replacement by newer, more robust units.

Further inspection of the 217 feet of wiring inside the orbiter actually revealed a far more significant problem than was at first thought. It was found that a single electric wire had lost its insulation, apparently rubbed away by repeated ground processing over two decades, including contact from numerous workers and equipment. A full wiring checkout was planned for *Columbia* during its 12-month maintenance and refurbishment programme at the Boeing facilities in Palmdale, California, following the STS-93 mission. Wiring and connector inspections on *Endeavour* revealed 38 damage locations, and inspection of *Discovery* revealed 26. This was a major problem. The whole fleet would have to be inspected to ensure that a complete electrical failure would not occur during launch.

The results of the examinations established that the problem was not related to the age or operational use of the wiring. The small 'nicks' were apparently related to processing between missions. This gave a new priority to wiring safety, including flexible plastic conduit, smoothing and coating of rough edges and added protection to areas where ground turnaround activities are more frequent.

The shortage of fuel during the STS-93 ascent was finally traced to a hydrogen leak in engine No. 3. This also occurred early in the ascent and, according to NASA, never posed a risk of a catastrophic failure. However the leak affected the $H_2$/LOX mix ratio in the engine, and resulted in higher temperatures in the pre-burners. Had this situation grown worse and overheated the engine, it could have initiated automatic shut-down and forced an abort.

The other problem of the 2,500 lbs of hydrogen lost during the ascent was traced to engine No. 3. The loss of the hydrogen was sensed by the engine controller, which recorded lower pressure in the engine combustion chamber. The controller commanded an opening of liquid oxygen (LOX) valves to consume more of the oxidiser and compensate for this. This action saw *Columbia* run out of oxidiser, which automatically triggered an early engine shut-down.

This leak was not actually detected during the ascent. However, when the engines shut down 1 second early and data revealed that an additional 4,000 lbs of LOX had been consumed, ground controllers knew something was wrong. It was not until *Columbia* was back on the ground that access to the engine compartment and the unit itself was possible. Recorded data and photographic evidence from long range cameras monitoring the ascent performance were also evaluated to determine the cause of the leak.

Had a single engine failed, a flight profile involving an 'outside loop' attached to the external tank (ET) would have been attempted. This would have meant flying on the fringes of the atmosphere at 400,000 feet and 6,000 mph to bring the vehicle to 180°, and then head back for a runway landing at the Cape. This would have been a high-risk and untried (except in simulation) profile, and would put great demands on the flight crew (Collins, Ashby – on his first flight – and Hawley) and on the vehicle's structure.

In the event of an engine failure between 3 min 14 sec and 5 min 17 sec, the recovery option would have been a much easier TAL abort on a runway at Ben Guerir in Morocco. After that, an ATO profile would have been their option. This would have been similar to STS-19's abort in 1985.

When the Chandra satellite was manifested to fly on *Columbia*, the combined weight and bulk of the payload and inertial upper stage (IUS) required the use of engines which were lighter and offered better range of performance. When the engines were stripped down after the flight it was revealed that a liquid oxygen post pin had probably become detached from the main ejector during ignition and struck the hot wall of the engine nozzle, rupturing three of the 1,000 ⅜-inch diameter nozzle loops. NASA considered it unlikely that the tubes had ruptured by themselves, and had probably been hit by debris originating from the engine power head.

The post pin was a gold-plated inconel pin the size of a nail that had been used to plug suspect oxygen injector posts for most of the Shuttle flight history. They are inserted during pre-launch testing when it is determined that LH might leak from a tube from a place other than intended. Chilled in liquid nitrogen to shrink it, the nail is tapped into the tube. It expands as it warms and seals and strengthens the suspect weak area.

The SSME used on position 3 launched with two pins, but returned with only one. Electron microscopic examination of the three ruptured tubes actually revealed traces of gold, indicating that the pin had indeed struck the damaged area.

Analysis of engines and wiring circuits on the other vehicles preparing for launch grounded the fleet for most of the remaining months of 1999 and seriously delayed launch operations for the third servicing of the Hubble Space Telescope and further construction of the International Space Station. What was missed in the media was the more important factor that, during ascent, Collins and her crew had used up all their redundancy and had only just reached orbit. Had they not done so, then perhaps the next day's headline may not have been so frivolous.

## Space Shuttle launch abort chronology

Aborts within 60 seconds of main engines ignition or following SRB ignition

| | |
|---|---|
| 1981 Nov 11 | STS-2 aborted at T–31 seconds after clogged fuel filters in APUs cause over-high temperatures. |
| 1984 Jun 26 | STS 41-D launch aborted T–4 seconds due to loss of redundant control over a main fuel valve in SSME No. 3. SSME No. 2 has just ignited and SSME No. 1 is about to ignite when abort when shut-down occurs. *The first On-pad After-ignition Abort.* |
| 1985 Jul 12 | STS 51-F aborted at T–3 seconds when a hydrogen coolant valve in SSME No. 2 fails to close. |
| 1985 Jul 29 | STS 51-F suffers a one-engine shutdown 350 seconds into flight. *Abort To Orbit* profile flown for the first time. |
| 1985 Dec 19 | STS 61-C launch aborted at T–14 seconds due to out-of-tolerance readings on the right hand SRB hydraulic system. |
| 1986 Jan 6 | STS 61-C launch aborted at T–31 sec due to an accidental drain of 14,000 lbs of oxidiser through a faulty fill and drain valve in the liquid oxygen system. |
| 1986 Jan 28 | STS 51-L explodes 73 seconds after launch, *killing all seven crew members.* |
| 1989 Apr 28 | STS-30 launch aborted at T–31 seconds due to a fault with SSME No. 1 liquid hydrogen recirculation pump. |
| 1993 Mar 22 | STS-55 launch aborted at T–3 seconds due to incomplete ignition of SSME No. 3. |
| 1993 Apr 6 | STS-56 count aborted at T–11 seconds when an $LH^2$ high-point bleed valve in the Main Propulsion System (MPS) indicates 'off' instead of 'on'. |
| 1993 Jul 24 | STS-51 launch aborted at T–19 seconds due to problems in the right-hand SRB APU turbine assemblies on one of two hydraulic power units. |
| 1993 Aug 12 | STS-51 launch aborted at T–3 seconds due to faulty sensor monitoring fuel flow in SSME No.2. |
| 1993 Oct 14 | STS-58 launch aborted at T–31 seconds due to failed range safety computer; part of the Eastern Test Range network. |
| 1994 Aug 18 | STS-68 launch aborted at T–1.9 seconds due to detection of a high discharge temperature in SSME No.3 engine high-pressure oxidiser turbopump turbine. All three engines shut down by GPC. |
| 1995 Oct 7 | STS-73 fourth launch attempt scrubbed at T-20 seconds when master events controller 1 fails to operate properly and needs replacing. |
| 1998 Dec 3 | STS-88 countdown held at T–31 seconds to access situation of a master alarm in crew cabin; at T–4 minutes a fault in the hydraulic system number 1 is indicated; system engineers quickly attempt to assess the problem, but are unable to restart the countdown within the allotted time remaining in the launch window. |

1999 Jul 20     STS-93 launch aborted at T–7 seconds when the orbiter hazardous gas
                detection system records a 640 parts per million concentration in the aft engine
                compartment – more than double allowed level; manual cut-off by systems
                engineers in KSC Firing Room 1 less than $\frac{1}{2}$ second before SSME ignition.

# Soyuz launch aborts, 1975 and 1983

For more than forty years the Soviet/Russian space programme has used the launch vehicle that first flew in 1957, and which was developed initially as an ICBM. Development of the R (Rakyeta, or Rocket) series began in the late 1940s with the R1 based on German V2 technology. Variants of the original launch vehicle, designated the R7, have launched hundreds of spacecraft into orbit, to the Moon and to Mars and Venus, and since 1961 the R7 has also boosted more than 100 individual cosmonauts into space.

For many years an inconsistent and confusing designation system was used for identifying Soviet launch vehicles. In the media the names of the primary payloads were used to identify the launch vehicles. Therefore the launch vehicle used to launch the first satellite was called Sputnik, and the launcher used to launch the first cosmonauts into space was identified as Vostok, when in fact it was the same type of launch vehicle but with added upper stages.

In the early 1960s the US Department of Defense designated each new type of Soviet launch vehicle as SL (Soviet Launcher) followed by the next numerical sequence, and so the R7 was also known as the SL-4. In 1967 the US Government adopted a different type of designation system which identified the main core stage as well as add-on or upper stages. This system used an alphanumeric in which a family of launch vehicles was identified by an upper case letter followed by numbers and lower case letters for the number and type of upper stages added to the core vehicle. In this system – termed the Sheldon system, after William Sheldon, who devised it – the R7 was designated an A-2 class vehicle.

With the R7 variant, the Soviets had a production line-type launch vehicle that was capable of supporting a wide variety of automated as well as human cargoes. Over the past four decades, several hundred vehicles have been launched with it.

In 1966 the R7 lifted the first, automated, Soyuz (Union) spacecraft, which was originally designed as part of the Soviet human lunar programme in the 1960s, but which was subsequently cancelled. The Soyuz spacecraft evolved over the years into perhaps the most durable spacecraft in history, fulfilling both human and automated missions, with scientific and military objectives. During more than thirty years of manned Soyuz operations, more than fifty cosmonaut crews have been launched into

The Soyuz launch vehicle, the R7, leaves the pad at Baikonur Cosmodrome, Kazakhstan, in 1975. During the latter stages of the Soyuz 18-A launch in April of that year a rocket stage failed to separate correctly, forcing an abort and emergency return after only 20 minutes of flight.

space onboard a spacecraft that has been compared in part to the technology of the American Gemini spacecraft flown in 1965/66. The compatibility of the R7 and Soyuz spacecraft has stood the test of time.

Despite the secret nature of the early Soviet programme and the lack (unlike their American counterparts) of live coverage of launches (except the 1975 Soyuz 19 Apollo–Soyuz Test Project), they did acknowledge suffering at least two identified

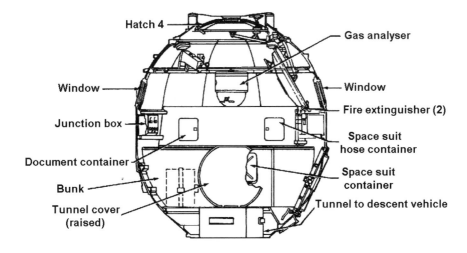

Cutaway of one side of the Soyuz Orbital Module (OM) (1975 era).

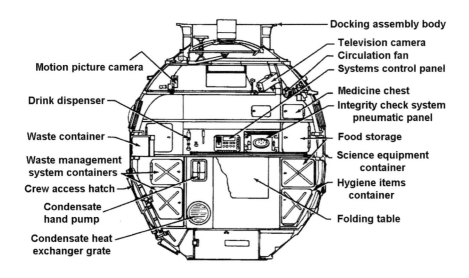

Cutaway of the other side of the Soyuz Orbital Module (OM) (1975 era).

launch aborts, in 1975 and 1983. Both were due to faults in the launch vehicle and not in the Soyuz spacecraft itself. That the cosmonauts survived is due to the robust construction of the spacecraft.

The R7 launch vehicle configured for Soyuz launches consists of a central core and four strap-on boosters which, on top of a lattice structure, was the upper stage. The Soyuz manned craft was mounted on top of this within a protective aerodynamic shroud, also supporting the launch escape tower system.

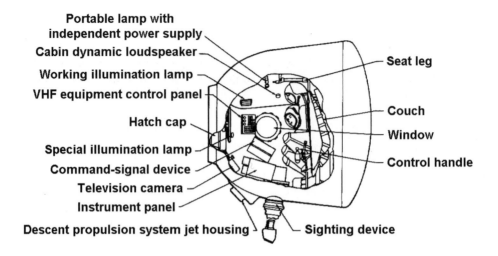

Portable lamp with independent power supply
Cabin dynamic loudspeaker
Working illumination lamp
VHF equipment control panel
Hatch cap
Special illumination lamp
Command-signal device
Television camera
Instrument panel
Descent propulsion system jet housing

Seat leg
Couch
Window
Control handle
Sighting device

Cutaway of the Soyuz Descent Module (DM) containing the crew (1975 era).

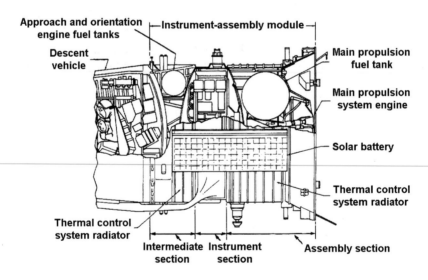

Approach and orientation engine fuel tanks
Instrument-assembly module
Descent vehicle
Main propulsion fuel tank
Main propulsion system engine
Solar battery
Thermal control system radiator
Thermal control system radiator
Intermediate section
Instrument section
Assembly section

Cutaway of the Soyuz Equipment Module (EM) (1975 era).

The design of the Soyuz spacecraft posed problems for the safe recovery of the cosmonauts in the event of a launch abort. The 2,800 kg Descent Module (DM), shaped like an elongated dome, carried a crew of up to three cosmonauts during the launch and landing phases of the mission, and served as a control module for the operation of the vehicle in space. It also carried the descent landing equipment, such as parachutes and the re-entry heat shield. The cylindrical Equipment Module (EM) – containing systems consumables for orbital operations, manoeuvring engines and solar wings or chemical storage batteries for power (depending on the mission requirements) – was attached to the rear of the DM. Located on the front of the DM was the egg-shaped Orbital Module (OM), which provided extra storage space for the crew, carried crew consumables and served as a valuable habitable area for research tasks in orbit. The OM also housed the hygiene facilities, an EVA hatch and in some versions the docking and transfer tunnel hardware.

This design, with the OM above the DM, hindered the use of a launch escape tower to pull away only the DM from impending disaster, as for the American Mercury and Apollo vehicles. Flying a three-man crew also prevented the use of ejection seats as on the Soviet Vostok and American Gemini vehicles. For Soyuz, a different method of crew escape was required; but at least there was an option, unlike Voskhod, which had no seats, no tower and no chance of escape.

The Soyuz escape system was based on solid-fuel rockets attached to an aerodynamic shroud encasing the Soyuz spacecraft. It was designed to pull the cosmonauts clear of a rogue launch vehicle either on the pad or in early stages of powered flight. In a normal mission the escape tower is jettisoned at 160 seconds into the mission where, a split second later the launch shroud splits into two along its longitudinal axis and falls away, exposing the Soyuz spacecraft.

In the event of an prior abort on the pad or before the 160-second mark the escape system is triggered and the shroud splits across the plane of the Soyuz DM and EM to take the cosmonauts to a safe distance from the rest of the launch vehicle. Four aerodynamic control flaps lower to stabilise the configuration prior to the DM dropping out of the rear of the shroud and achieving a controlled parachute recovery. Elements of this system were tested automatically in the early stages of the programme. In more than thirty years of Soyuz flight operations the escape tower has been used during only one manned mission, in 1983. An earlier event during the launch of a Soyuz in 1975 occurred after the escape tower had been jettisoned and the shroud separated, exposing the Soyuz vehicle and allowing spacecraft separation and controlled parachute recovery.

*Soyuz launch abort, 1975*
By the spring of 1975 there was feverish activity at the Baikonur Cosmodrome in the Soviet Central Asian Republic of Kazakhstan in support of preparations for two manned flights during the summer. Two separate teams of cosmonauts were in training – one for a two-month stay on the Salyut 4 space station and one for a joint mission with an American Apollo crew in July. It was the time of détente between the two superpowers.

The Apollo–Soyuz Test Project (ASTP) originated in 1970 as a joint Soviet–American manned flight to either the US Skylab or Soviet Salyut space station. By 1972 it was decided to pursue a more modest programme which was about to reach its climax in the summer of 1975 with the docking of a Soyuz and an Apollo spacecraft in Earth orbit. The planned few days of joint activities would, it was suggested, offer a global demonstration of how the two leading spacefaring nations could work effectively together, leading, hopefully, to joint space station operations and mutual rescue capability with common docking systems. These future projects were soon cancelled due to changes in administration and foreign policy in both countries. However, at the height of the work on the ASTP, the Soviet human space programme was for the first time displayed more openly than ever before.

It would, however, be another twenty years – with renewed international co-operation in space following the demise of the Soviet Union and development of the International Space Station – before the openness and pride of the Russian national space effort was once again seen in the West.

From the start of ASTP, NASA insisted upon open and frank discussion of Soviet manned spaceflight hardware and technology before America would consider allowing her astronauts to work inside Soviet spacecraft in space. It produced new and hitherto undisclosed information about the accidents of Soyuz 1 and 11 (see pp. 369 and 389) that resulted in the fatalities of the cosmonauts on board, and the more recent docking failure of Soyuz 15 in 1974. The Americans gained important insights into the Soyuz system and its hardware, and in April 1975 witnessed an unplanned demonstration of one of its safety features.

As teams of engineers, scientists, astronauts and cosmonauts worked towards ASTP, final preparations were also underway to begin the second phase of manned occupation of the Salyut 4 space station. This station had been in orbit since December 1974, and had already hosted one pair of cosmonauts for a record-breaking 30-day stay, earlier in the year. This would hopefully be the first time that a Salyut received a second crew (although this had been tried, but not achieved, on Salyut 1 and Salyut 3).

A two-man crew was assigned to fly Soyuz 18 to Salyut 4 for a 60-day mission returning in June, just weeks before Soyuz 19 was to fly the joint ASTP mission with the American Apollo 18. The Commander of Soyuz 18 was Lieutenant Colonel Vasili Lazarev, an Air Force test pilot and qualified physician. His Flight Engineer was civilian engineer Oleg Makarov. This team had already flown in space together on Soyuz 12 in 1973, when they man-rated the improvements made to the Soyuz ferry craft following the tragic loss of the Soyuz 11 cosmonauts in 1971.

Soyuz 12 was a two-day test flight of a new type of space station crew ferry craft, which relied on batteries instead of solar wings for power. Following Soyuz 11 – in which a sudden cabin decompression had claimed the lives of the three unsuited cosmonauts – the need to prevent a repeat of the disaster meant that the space formerly assigned to the third seat in the ferry was taken up by additional life support hardware. Soyuz would now become a two-person spacecraft.

Now, they were to fly Soyuz 18. At around 12 noon on 5 April 1975, Lazarev and Makarov arrived by bus at the launch pad from their suiting quarters. Following the traditional speech and wave of farewell to onlookers, they climbed the steps to the elevator that would take them to the access platform and the Soyuz craft. The two cosmonauts entered their Soyuz through the side hatch in the orbital module, inside the launch shroud, and descended into the cramped command module beneath. Makarov, who had to strap himself into the right-hand position, entered first, followed by Lazarev in the central position. With all checks completed and the crew safely in their seats, the hatches and shroud were secured and the countdown clock ticked on towards launch just two hours away.

Soyuz 18-A cosmonauts Vasily Lazarev (foreground) and Oleg Makarov (background) in a Soyuz simulator during training.

Shortly after 14:03 Moscow time, 5 April 1975, the main engines of the R7 burst into life. The vehicle rested on the pad until the engines built up enough thrust to overcome the vehicle's weight. Lazarev and Makarov then began what they thought would be a two-month mission in space. Typical flight speeds of a Soyuz were 500 m/sec at 70 seconds into the flight, 1.5 km/sec at 120 seconds, and 6 km/sec at 450 seconds.

As the vehicle rose into the sky, all seemed well as the onboard computer guided the vehicle towards space, pitching it over as it climbed and built up speed. When the four strap-on boosters had depleted their 'fuel' they were shut down and explosively separated from the main vehicle, which accelerated upwards along a north-east trajectory, using the central core stage. At 2 min 40 sec after launch, the launch escape tower and launch shroud were jettisoned, the crew no longer needing the system to ensure a safe re-entry and landing in the event of a failure. With the tower and shroud gone, the crew must have relaxed a little as another milestone and danger point had been met and passed safely; after all, they were flying a vehicle that had about twenty years service behind it. Reliability built on regular use of proven hardware was a major factor in the Soviet programme, and these vehicles were reliable.

Five minutes into the flight, the central core stage of the launch vehicle was shut down and the upper stage ignited to continue the push for orbit. It was at this stage that problems began to develop for Lazarev and Makarov. Normally, the final stage is ignited seconds after two sets of pyrotechnic couplings fire to separate the spent core stage. Each set of couplings contains six latches on the upper and lower attachment points of the interstage lattice structure, enabling exhaust gases from the firing upper stage to be vented. The lattice structure usually separates seconds after the central core, and splits into three elements that fall away from the climbing combination. The latches are activated by two sequencers – one located in the core stage controlling the lower set, and one in the upper stage of the vehicle controlling the other set. Both sequencers are electrically joined to ensure combined separation and a clean staging.

However, on this flight the correct operation of the staging resulted in the abort of the mission. In a post-flight mission failure briefing, issued by the Soviets as part of their agreement with NASA, the events that cut short the mission of Lazarev and Makarov were revealed. Apparently, a high level of vibration caused the relay in half of the upper sequencer to close down and to signal three out of six latches to fire prematurely, with both the upper and lower stages still firmly attached. The malfunction of the latch sequencer occurred seconds before planned separation was programmed, but when the latches were armed. The connections that fired were located in the same region of the structure, along with the electrical link connecting the upper relays to those on the lower part of the structure. This electrical connection was also severed, and all links with the lower latches were cut. The premature firing of the upper latches caused an uneven link between the upper and central core stages as the vehicle sped onward.

As the final stage was ignited, the core stage was still attached and the rest of the latch separation devices failed to operate. The vehicle was dragging an empty, but heavy, spent stage through the upper reaches of the atmosphere. After a 4-second burn, the extra weight of the spent stage and the increase in the aerodynamic drag from the unstable booster caused the onboard gyroscope to detect a deviation to the planned flight path in excess of the 10° safety limit. It activated the automatic abort programme.

The Soyuz spacecraft received the abort signal at about 90 miles altitude, and the spacecraft was automatically separated from the failing booster to complete a fully automatic re-entry sequence. Separation was triggered by the Soyuz propulsion system, and after a safe distance had been reached the orbital and service modules were separated. The computer programmed the DM, containing two surprised cosmonauts, to automatically orientate for entry and landing from what was no more than a sub-orbital space hop similar to the US Mercury flights of Shepard and Grissom in 1961. During re-entry the high-altitude low-speed configuration combined to impart 18 g on the two cosmonauts. The aerodynamic lifting capability of the Soyuz DM, and the posture of the form-fitting crew couches designed to absorb 3–4 g, possibly helped the crew remain conscious, though they still had to endure a painful 14–15 g force during re-entry. For comparison, a fighter pilot in a high-performance jet will encounter loads seldom in excess of 4 g for short periods. The forward acceleration of a Boeing 747 Jumbo jet at take-off is only $\frac{1}{4}$ g, and the launch of a Space Shuttle produces little more than 3 g for a few minutes.

As the normal parachute descent programme was initiated, all the cosmonauts could do after recovering from their high-speed entry was to ask for reassurance from ground control that they would not land in the unfriendly country of China. Their flight path crossed the north-west part of the town of Sinkiang, a region where recently two Soviet helicopter pilots had reportedly landed in error and had been captured by a Chinese patrol. With this on their minds and the hope that they would survive the unplanned emergency landing, all thoughts of disappointment at not reaching Salyut 4 were, for the moment, put aside. The crew played no part in the re-entry sequence, as instrumentation and controls required for doing so would have added an unacceptable weight and volume to the Soyuz. They remained passive, pressed deep into their couches, as they dropped like a stone, until the drogue and single main parachute were deployed, gently lowering them to what they thought was the end of their eventful mission.

The region in which they landed was the very rugged terrain of western Siberia, south west of Gorno-Altaisk, about 1,600 miles from Baikonur and 515 miles north of the Chinese border, inside Soviet territory. However, they hit a snow covered mountain, and almost immediately the capsule began to roll down the side. As it approached a sheer drop, the parachute line snagged on scrub trees, halted the spacecraft and held it safely. The cosmonauts were bruised, shaken and dizzy. Their flight time, planned at 60 days or around 1,500 hours, lasted just 21 min 27 sec from launch to landing.

Dazed but otherwise unhurt, the two cosmonauts scrambled out of their spacecraft to prepare for rescue. They had been tracked by radar and the capsule had a radio beacon, but they had to light a fire and prepare for a long wait, as the region they landed in was inaccessible except by helicopter. Conflicting reports indicated that they were either reached by local villagers within an hour or that they had to spend the night in the region awaiting search parties from the launch centre. In any event, it was reported that the cosmonauts were found to be in good health following their unplanned emergency landing.

The two men were airlifted back to Baikonur for medical check-ups and then on to Moscow. They made their first public appearance on 10 April at a meeting to mark the fifteenth anniversary of Star Town, and on 11 April they arrived in Moscow to celebrate Cosmonautics Day on 12 April. On 28 April, both men were presented with the Order of Lenin, for courage displayed during the Soyuz launch, but not with the order Hero of the Soviet Union.

The aborted flight was not mentioned by the Soviet authorities until 7 April, two days after the event and after the crew had been recovered. In the official announcement the Soviets stated that the planned mission to Salyut 4 failed when 'parameters of the carrier rocket's movement deviated from pre-set values. An automatic device produced the command to discontinue the flight programme and detached the spacecraft for return to Earth.'

Initial reports from the Soviets stated that both men were in good health and suffered no ill effects from their ordeal. Cosmonaut Vladimir Shatalov, the Director of Cosmonaut Training, reported that after meeting both cosmonauts shortly after the mission, both were 'ready to make another spaceflight.' Shatalov also stated that the cosmonauts had told him that their landing was very gentle because of the soft-landing rockets – as though they had landed on water. Their landing in the Altai mountains was, in their opinion, much gentler than their Soyuz 12 landing on low, flat land. This was in contrast to later reports indicating a rough landing.

In another report issued some years after the event, it was stated that Lazarev informed ground control of the failure of the clear separation of the spent stage. He was not believed, as the upper-stage engine readings on the ground were normal, and the 10° off the planned flight path of the vehicle was so high that it was off the scale, leading to the belief that the meter, not the vehicle, was at fault. Apparently, some choice comments by a rather concerned Lazarev shook ground control into action, and the automatic activation of the Soyuz abort system put paid to any disbelief on the part of the controllers.

Shatalov's comments on the failure indicated that even in a difficult situation the reliability of the Soyuz spacecraft, the operation of all its systems and the actions of the crew during the emergency proved to work well when called upon. This would induce confidence in the forthcoming ASTP. However, the Americans were not convinced.

On 8 April, during a routine telephone link between Glynn Lunney, a NASA Flight Director and the US ASTP Technical Director, and Konstantin Bushuyev, his Soviet counterpart, concerns were expressed over the possibility of another launch failure during the Apollo–Soyuz mission in July. The Soviets, although unsure of the exact cause of the abort, assured the Americans that when their investigations had isolated the problem, the information would be transmitted to US ASTP officials immediately. Bushuyev also informed Lunney that the booster used in the launch abort on 5 April was of an older design, and that the ASTP prime and back up launch vehicles were newer, and contained none of the suspect components. NASA was confident that the Soviets had enough time to conduct the failure analysis and make any necessary corrections to the allocated launch vehicles for ASTP well before 15 July.

When full details of the accident were released to the Americans, the ASTP progressed towards the target launch date with no delays. The ASTP boosters were indeed a newer version of the booster with heavier payload lifting capability and improved pyro-lock circuitry. However, due to the launch abort, both ASTP and all subsequent Soyuz boosters were to receive a series of staging circuitry changes. A redesigned relay system and wiring paths to activate the stage separation system were implemented. The Soviets attested that should any one of the latches fire prematurely, then the rest would be triggered to ensure a clear separation.

The Soviets also stated that the launch vehicle that was used in the April mission was 'less diligently checked' than that being prepared for the joint mission – indicating that it was perhaps not as sophisticated as the ASTP version. This aspect of pre-flight checks – or rather, the lack of them – had been of concern to the Americans since the beginning of ASTP. One account of a tour of American ASTP officials to the Soyuz vehicle processing facilities at Baikonur summed up this concern. A NASA launch processing engineer asked his Soviet counterpart why there were no super-clean 'white room' facilities available to protect the hardware from the visible layers of dust and sand he saw lying on the surface of the launch vehicle. The reply, in typical Soviet fashion, was that when the vehicle was taken to the pad and then launched, 'The dust just falls off!'

On 7 April NASA commented on the launch abort and expressed confidence in the Soviet system. They were informed that the aborted mission had nothing to do with the ASTP flight programme, as it was part of the Soviet domestic space station programme. The statement added that nobody was more concerned with the reliability and safety of the Soviet Soyuz system than were the American astronauts.

Soyuz was considered safe enough for the portion of the flight with which the astronauts would be dealing (in orbital rendezvous), and NASA remained satisfied with the design. American planners for ASTP had spent hours meticulously reviewing past missions with the Soviets, including the Soyuz 1 and 11 failures and the Soyuz 15 docking failure. NASA insisted on the right to inspect the design of the Soyuz in detail before the fine points of the mission had been determined. If this was not the case, NASA would have refused to fly with them. But not all Americans were convinced.

Several critics of the joint mission and of the space programme used the launch abort to once again blast the value and usefulness of such a vehicle. Senator William Proxmire, a longstanding opponent of the US space programme, revealed in a speech on the Senate Floor on 9 April that he had asked the CIA to produce a report assessing the safety of Soviet manned spaceflight programmes and current technology. 'The in-launch failure of another Soviet manned satellite reinforces my deep concern that the upcoming joint Apollo–Soyuz experiment may be dangerous to American astronauts. The history of Soviet manned programmes shows an appalling lack of consistency. As soon as one severe problem is solved, another occurs.'

True, the Soviet programme did have its share of setbacks with bad luck, unreliable and 'crude' (by US standards) technology and constant battles for funding and support, but so did the Americans in the early days. Unlike the Americans – who developed new equipment and spacecraft for a given programme, and then discarded it and started again for the next – the Soviets continued to use production line components and adapted programmes to fit the hardware. The Soyuz failure – officially called 'the 5 April anomaly' by the Soviets (but Soyuz 18-A in the West) – was, of course, a failure and a setback to the programme, and it could have cost the lives of another crew. However, all elements of the onboard safety system proved to work as designed under real flight emergency conditions. This reassured ASTP planners that the Soyuz launch abort system really had functioned as planned with men onboard – a contingency that had fortunately never been called upon by a manned American Apollo-type vehicle.

The standard Soyuz launch vehicle was used again just nine days after the aborted launch, when a Molniya 3 communications satellite was successfully launched into orbit using an added fourth stage. In May, Soyuz 18 (18-B in the West) was launched with the back-ups for Lazarev and Makarov onboard, and it successfully docked to Salyut 4 for the long awaited 60-day mission. In fact, the mission was flying during the ASTP flight itself, which prompted Senator Proxmire to voice concern over the ability of the Soviets to control two separate missions at one time. However, using two different control centres – one in Moscow, and the other the old centre at Yevpatoria in the Crimea – they demonstrated a growing maturity in the Soviet programme.

The crew of the Soyuz T-10A pad abort, Vladimir Titov and Gennedy Strekalov. (Novosti Press Agency via Astro Info Service Collection.)

### Soyuz launch abort, 1983

In April 1983 the Soviets launched a new three-man crew to Salyut 7 for what was planned as a flight of several months on the space station. Cosmonauts Vladimir Titov, Gennedy Strekalov and Alexandr Serebrov flew Soyuz T-8 to rendezvous with the Salyut station, only to find that their rendezvous radar was damaged when the Soyuz separated from the shroud. Their later attempts at manual docking proved unsuccessful. Although the rendezvous was completed, the final approach had to be called off at the last minute when the vehicles flew into the Earth's shadow.

The three cosmonauts aborted their mission and returned to Earth two days after launch. Despite this setback, the two-man back-up crew – cosmonauts Vladimir Lyakhov and Alexandrov – were launched to Salyut 7 on 27 June 1983, and docked with the station to continue the research programme originally assigned to the Titov crew. At the time, Salyut 7 was suffering a severe lack of power caused by a voltage drain on the supply due to an increase in experiments. The original Soyuz T-8 crew was to have conducted two periods of EVA to erect additional solar panels on one of the three 'huge' solar arrays outside the station, in order to increase electrical output for experiments and station operations. Other arrays would be added by later missions.

Lyakhov and Alexandrov had not received the same detailed training for the EVA construction work as had Titov and Strekalov, and, it was therefore decided to recycle the two former Soyuz T-8 cosmonauts to be launched to Salyut 7 again to conduct the EVA and to take over from Lyakhov and Alexandrov on the station. Lieutenant Colonel Titov was a former Air Force test pilot, and Strekalov was an experienced civilian engineer.

In September 1983, Soyuz T-10A became the first acknowledged pad abort when the carrier rocket exploded on the pad. The two-cosmonaut crew escaped by using the escape rocket system, mounted on the nose of rocket, which carried the crew module safely away from the burning pad area.

In early October a launch window opened for a mission of eight days, the nominal visiting mission duration. However, the cosmonauts needed more time to complete their planned EVA tasks due to the more complex work with the solar panels. This, plus the requirement for adequate lighting conditions at the landing site at the end of the planned mission, dictated a launch on 26 September.

Two hours before the scheduled launch, the two cosmonauts clambered aboard their Soyuz capsule (which would receive the T-10 designation upon entering orbit) and awaited the final stages of the countdown, a completely automatic sequence. Strapped into their couches, the crew faultlessly followed pre-launch procedure.

Soviet reports on the morning of 26 September stated that the scheduled launch time for the crew (call-sign 'Okeany', or 'Ocean') was 11.37 pm Moscow Time. The temperature was 27° C, but had fallen at launch time to 10°. Wind gusts were recorded up to 40 fps. Pre-launch preparation continued smoothly. The launch vehicle had been prepared horizontally, like all previous R7 vehicles, in workshops linked to the launch pad by rail tracks. When delivered to the pad area some two days before launch, it had been raised into a vertical position over a flame trench and surrounded by access towers. Several tests and inspections had not revealed any serious problem with the vehicle, so it was declared ready for flight and assigned to the Soyuz T-10 mission.

The pad was in darkness as launch time approached. Searchlights illuminated the pad area and picking out the ghostly shape of the launch vehicle with the Soyuz on top, inside the launch shroud. Some 90 seconds before the planned launch, a valve failed to close and a leak of the launch vehicle fuels (liquid oxygen (LOX) and kerosene) started a fire at the base of the R7 carrier rocket. The fire could have been started by an electrical spark or by faulty ground equipment, but once it spread it soon became apparent that the automatic firefighting equipment would not be sufficient to extinguish the flames. There remained only seconds before flames ruptured the main fuel tanks holding 270 tons of kerosene, and the whole launch pad, including the booster and two-man Soyuz, was turned into a raging fireball.

Watching these events unfold from the safety of a nearby bunker, Aleksey Shumilin – head of the ground testing group at Baikonur – saw, through a periscope, a fire break out at the base of the rocket while a fuel line was being purged with nitrogen. Observers also noticed that the umbilical cord did not detach as intended, and at 25 seconds before launch an unusual reddish-brown flame was observed. Black smoke began to rise around the vehicle as the seconds ticked away. Ground controllers knew that the only way to save the crew was to activate the Soyuz launch escape system (LES) – a system of small rockets located at the top of the launch shroud covering the spacecraft. In normal circumstances the firing of the solid rocket LES should have occurred automatically, but as the fire spread rapidly around the vehicle, wiring designed to command the LES to operate was severed and the automatic ignition failed to materialise. The controllers quickly noticed this and had to initiate the command from their own stations. This awkward procedure involved two commands to activate the LES from two different rooms by two different controllers at the same time. This took all of ten seconds, but must have felt like a lifetime before the signal to the LES was finally completed. It was the first time this system had been called upon in a real emergency.

As soon as the command was received by the escape system, events moved swiftly as the launch vehicle's explosion was imminent. Within one second, the pyrotechnic separation of the DM (carrying the crew) and of the EM was achieved. At the same time the 176,000-lb thrust of the LES fired and pulled the combined OM and DM away from the rocket, still inside the shroud. The flight path took the cosmonauts away from the pad at an angle, at high velocity and 10 g profile. The capsule was blasted away from the rocket, which suddenly exploded in a massive fireball, severely damaging the pad area. There was no sound of engines inside the capsule, and later, Titov recalled feeling a rocking motion (which he thought was a wind gust), two unusual waves of vibration and suddenly a strong jolt. Observers noticed that a cloud of yellow, red, black and orange smoke rose to the top of the vehicle, and that an object suddenly shot upwards with sparks shooting from it, announcing that the LES had been activated.

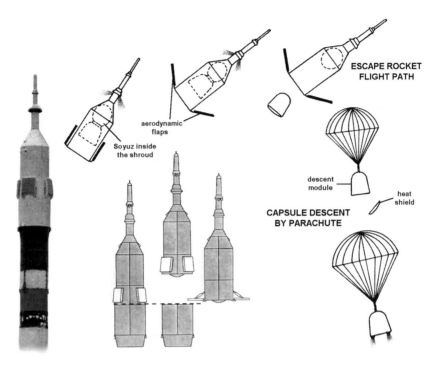

aerodynamic
flaps

Soyuz inside
the shroud

ESCAPE ROCKET
FLIGHT PATH

descent
module

heat
shield

CAPSULE DESCENT
BY PARACHUTE

The Soyuz Launch Abort profile first used operationally by Soyuz T-10A in September 1983.

Three seconds after separation, the cosmonauts had reached Mach 1 vertically. At 3,117 feet, five seconds after ignition, the solid rockets burnt out and the four aerodynamic, petal-like airbrakes were deployed on the shroud to slow the ascent. The smaller set of rockets on the top of the LES was fired, and the command to separate the OM from the DM was given. The DM then dropped out of the base of the shroud to begin the descent to Earth. As they were too low to activate the main parachutes, the reserve emergency parachute was deployed. The crew was shaken around inside the capsule, unable to do anything. As the parachute slowed the rapidly descending capsule, the heat shield automatically dropped to expose the soft-landing retrorockets at the base of the DM. At a height of 5 feet above the ground the retros fired to slow the landing of the capsule. Despite the rockets, the landing was still much rougher than normal, but the cosmonauts touched down safely 2.5 miles from the blazing pad area. Observers had beamed searchlights into the darkness, to see the parachute open and descend with the capsule beneath it, and the dust-clouded soft landing.

The rescue teams found both cosmonauts shaken but uninjured after their ordeal, within minutes of their explosive and sudden departure from the launch vehicle. The cosmonauts were each given a glass of vodka before a brief medical examination revealed that they did not need to be taken to hospital. From the vantage point of their unplanned landing site, the cosmonauts could clearly see the flames licking skyward around the launch pad they had just vacated. According to former cosmonaut and Salyut designer Dr Konstantin Feoktistov, the wreckage of the booster burned for twenty hours.

Titov later related his memories of the event. He stated that he was not happy with the way the craft behaved in the seconds leading to the aborted launch and that he sensed a ripple of unusual vibrations (not experienced on his first flight of T-8, five months earlier). As the escape system boosted them suddenly into the night sky, they soon summised that events were not going as planned. 'There was no time to fear', Titov said, and he recalled his mixed emotions as he floated down to the emergency landing minutes after launch. Strongest of all was his bitterness at the wasted preparations for this and the previous failed T-8 flight. He recalled how both cosmonauts had remained passive during the whole incident.

It was later revealed by the Soviets that the escape system could be activated automatically by the cosmonaut crew or by manual command from the control bunker by the specialists responsible for the launch vehicle and spacecraft. It was also revealed that despite many unmanned system tests with dummies and animals, no tests had been conducted with human subjects onboard. This was either because of their reluctance to subject cosmonauts to excessive risk or because of lack of confidence in the system. It was felt that the chances of actual use in flight were small, based upon the reliability of the Soyuz launch vehicle. This was similar to the American tests with Mercury, Gemini and Apollo escape systems and Shuttle abort profiles. In any event, Titov and Strekalov were holders of the unenviable title of becoming the first space travellers to test a pad abort system in an actual mission situation. The launch abort was not immediately acknowledged by the Soviets, nor designated with a Soyuz number.

The next successful launch was Soyuz T-10 in February 1984. In the West, the abort was labelled T-10A and the 1984 flight T-10B. The initial reports leaked out of the Soviet Union claimed, falsely, that a crew of three, including a female cosmonaut, had been aboard the Soyuz. It was also estimated that the cost of repairing the launch pad facilities would be around $250–500 million.

Interestingly enough, the Soviets had publicly criticised the near-fatal accident of STS-8 when its SRB almost burnt through just weeks before. The failure to send the replacement cosmonauts to Salyut 7 pushed the duration of the T-9 crew to five months instead of three months. They also had to conduct the two EVA originally planned for Titov and Strekalov. Fatigue and lack of training for the flight, which was a replacement for the failed T-8 mission in April, began to show on the crew as the mission drew to a close.

Rumours of the Soyuz T-9 craft completing the flight well beyond its orbital design life and of a major engine problem and fuel leak on Salyut, furthered the drama of a 'stranded' crew. There were even Western media reports of a plan to send the Shuttle *Columbia* to rescue them, which although a theoretical possibility was never a viable solution, if only for the lack of a compatible docking system. However, the crew was able to achieve the majority of their mission objectives, and were recovered safely in November, after 150 days in space. The rumour of an engine problem proved to be true, and the major repairs on the Salyut fuel system were finally completed in 1984.

Rumours of a rocket launch failure continued to spread in the West. Finally, during the IAF conference in Budapest, Hungary, in mid-October, under pressure from the West, External Services of Radio Moscow noted 'another accident' in the Soyuz programme and the successful operation of the Soyuz safety system, with no harm coming to the two-man crew. The citizens of the Soviet Union did not learn of the mishap until after the landing of T-9 in November.

Following the *Challenger* accident in 1986, the Soviets issued a deluge of articles and speeches on space safety and launch accidents, including a detailed report of the T10A abort. It was, as one Soviet official put it, 'A very serious accident ... just six seconds from a Soviet *Challenger*!'

# The STS-25 (51-L) launch explosion, 1986

The year 1986 marked the 25th anniversary of the first manned spaceflight. As these had became more frequent and apparently routine, plans circulated for some time to send non-career astronauts into space for the promotion of spaceflight for ordinary citizens. The unique flying qualities of the Shuttle dispensed with all the crew having to pass rigorous medical examinations or be top test pilots. The advent of the Space Shuttle had seen the Americans fly a new breed of space traveller – the Payload Specialist, mostly scientists and engineers. In the Soviet Union, 'guest cosmonauts' from other countries had been selected to fly to Salyut space stations, although these had all been military officers.

By the early weeks of 1986, two US politicians (acting as observers) and a Saudi Prince had flown on the Space Shuttle. On the very next mission – number 25 – an American high-school teacher was to broadcast from orbit and introduce the space environment to schoolchildren in a fresh, new and yet familiar way. And shortly thereafter, a professional journalist, possibly veteran space reporter Walter Cronkite, would be selected to fly.

In trying to prove that Shuttle would provide America with all her launch needs, NASA set ever more ambitious targets for each successive year's operations. Problems with payloads, hardware and procedures stretched the four-orbiter fleet and the near non-existent spares capacity to breaking point. Engines were being removed from a fully loaded orbiter for transfer to another orbiter, so that it could fly on time. It would be only a matter of time before a serious setback befell the programme and threatened the whole existence of NASA and the space programme. That day finally arrived one cold morning in January 1986.

For the 25th mission of the Shuttle programme, NASA manifested the deployment of the long awaited and much delayed second Tracking and Data Relay Satellite (TDRS), which would improve communications with the Shuttle during phases of the orbit out of range of ground tracking stations. A series of mid-deck experiments would be conducted, as well as the teacher-in-space experiments. A free-flying satellite would be deployed and retrieved later in the mission, and the crew was to perform the usual series of on-orbit tests of the Shuttle systems and a range of TV broadcasts to Earth – all in all a nominal mission, resembling many earlier ones. There would be no planned spacewalks or dramatic repairs in space – just a straightforward mission.

Although it was to be the 25th launch of the Shuttle system, it would be the tenth launch of *Challenger* and the first from the Launch Complex Pad B. NASA originally envisaged regular launches of the Shuttle system from two pads at the Kennedy Space Center in Florida and one from the USAF Vandenberg AFB in California, for Department of Defense missions over the polar regions of Earth. To identify missions prior to launch, a new numbering system had been devised which used a three-digit code to identify the fiscal year of the mission (October–September), the location of launch site (either 1, KSC, or 2, VAFB) and the manifest designates (A–Z). This made for a theoretical launch rate of 26 missions each from the Cape and Vandenberg in a twelve-month period – 52 Shuttle launches a year! Therefore, the 25th mission was manifested as a Fiscal 1985 (5) mission launched from KSC (1) and carrying the twelfth (L) assigned manifest. It became more commonly known as mission 51-L.

The mission had originally been scheduled for a December 1985 launch, and could have been the first time American astronauts would be in space over Christmas since the Skylab 4 mission in 1973. Several delays and scrubs saw the launch slip into the new year and conflict with the previous mission, Shuttle 24 (61-C). These added to an already busy launch schedule and a very confusing numbering system. Work continued, however, on the thousands of steps necessary for every Shuttle mission as each was crossed off the pre-launch checklist.

Hardware preparations for the mission began in the Vehicle Assembly Building on 28 October 1985, with the vertical stacking of the twin solid rocket boosters on top of the second mobile launcher platform in the VAB's High Bay 1 area. The second booster began stacking on 4 December, followed by the mechanical attachment of the huge external tank on 10 December. The combination was checked and prepared to receive the orbiter assigned to the mission = OV-099, *Challenger.*

*Challenger* had long been known as the astronauts' favourite orbiter. The vehicle had returned from space on its ninth mission on 9 November 1985, carrying the long module Spacelab D1 payload. It landed at Edwards Air Force Base, California, and was later air-lifted by Boeing 747 Shuttle Carrier Aircraft to the Kennedy Space Center, Florida, to begin processing for its next mission, 51-L. It arrived at the Cape on 11 November, and was immediately offloaded and towed to the Orbiter Processing Facility (OPF). There, it would have the payload from the last mission removed, correction and replacements to all faulty equipment attended to, and payload for the next mission installed.

Inside the OPF, the Spacelab long module was removed, and the fitting designed to hold the inertial upper stage and its TDRS payload was installed, together with the support structure for the Spartan–Halley free-flyer and the installation of the remote manipulator system (RMS – the Canadian robotic arm). Inside the orbiter living compartments, the mid-deck area was reconfigured for carrying three astronauts instead of the four carried on the previous mission, and all lockers were offloaded and resupplied with new inserts carrying equipment and consumables needed by the next crew for their mission.

As preparations for the mission continued, the ground processing crews faced a continuing battle to find flightworthy spares for each mission, and for this reason the launch would be delayed until January.

On 10 December, *Challenger* was towed across to the VAB to be mated to the ET/SRB combination and have its mechanical and electrical connections checked and verified. Six days later the stack was rolled out to Launch Complex 39 Pad B for final preparations for launch, still scheduled for late December. *Challenger* was to remain on Pad B for the next 37 days – a situation that certainly contributed to the accident that followed. Pad 39B was a new launch site for Shuttle missions, and like most of the Shuttle launch facilities it was refurbished from the Apollo Moon programme.

The availability of a second pad for Shuttle flights made the 1986 launch schedule one of the busiest planned in recent years. Determined to complete a full Shuttle programme, NASA had some very important launches planned and a tight schedule arranged to meet them. But by late 1985 fate was already dealing its first blow to that important schedule. Severe weather, and hardware difficulties, had forced a postponement of mission 61-C (Shuttle 24) six times from December into early January, and the required time between launches in order to complete the necessary simulations on-pad forced the delay of the 51-L (Shuttle 25) mission to 23 January 1986. The unusual number of problems in launching mission 24 (known as 'Mission Impossible') saw the NASA Program Requirement Board issuing a decision to delay the launch of 51-L from 23 January to 26 January.

Delays to the launch allowed the potential use of the Casablanca Trans-Atlantic Launch Abort site. As the management teams reviewed the status of sites across the globe and the weather forecasts for the duration of the 26 January launch window, it became apparent that a launch on that day was not be feasible. A decision was made to delay launch for another 24 hours, and with this additional postponement most of the specially invited guests went home to view the eventual launch via TV links.

On 27 January it seemed that the launch would take place. Activities at the Cape centred on the fuelling of the huge external tank, which began at 00.30 hours EST. The seven-person crew had been awakened in their quarters at 05.07 am, and by 07.56 had been strapped into their seats and were going through their numerous checklists, awaiting the final seconds of the count and launch.

The crew for this mission was a mixture of American culture and background. In command of the week-long spaceflight was veteran astronaut Dick Scobee, a USAF Major. Pilot on this flight was Naval Commander Mike Smith. Between and behind them was Mission Specialist Judy Resnik, PhD, who would serve as 'Flight Engineer' during launch and entry, assisting Scobee and Smith in ploughing through the mammoth checklist procedure. Sitting on the flight deck with the other three astronauts was Air Force Captain Ellison Onizuka, a native of Hawaii. Below them, on the mid-deck area, were the other three members of the crew. Seated in the forward mid-deck seats were Payload Specialists Greg Jarvis, an engineer from the Hughes Aircraft Corporation, and Sharon Christa McAuliffe, at that time perhaps the most famous non-career astronaut ever selected. A history and social studies teacher, McAuliffe had been selected from hundreds of applicants to participate in the Teacher in Space programme. She was to conduct 'field trips' from orbit, teaching school children and students on the effects of weightlessness, and helping to explain the space programme as well as promote it for the general public. It was the inclusion of McAuliffe and her magnetic personality on which most of the media attention was focused. The flight of the teacher in space brought this mission to the forefront of media coverage perhaps more than any other flight since STS-1 five years earlier.

Rounding out the crew and sitting by the side hatch was physicist and career astronaut Ron McNair, an African-American who had played his saxophone on his previous mission. The crew represented a cross-section of America: white, African-American and Asian, male and female, military and civilian. There were pilots (Scobee and Smith) engineers (Resnik, Onizuka and Jarvis), scientists (McNair) and professionals (McAuliffe). There were also experienced space explorers (Scobee, Resnik, Onizuka and McNair) and rookies (Smith, Jarvis and McAuliffe), professional career astronauts and representatives from the aerospace industry and public service. This mixture of race, culture and background would become a media feature in the accounts of the fate of the seven astronauts.

At 09.10, as the closeout crew proceeded with their tasks, they reported a problem securing the exterior hatch handle after seating the crew and closing the entry hatch. It took 80 minutes to solve the problem, and as they completed it, reports of high cross-winds were being received at the Cape. This would seriously hamper any attempts to land the orbiter in the event of an emergency return to the Kennedy landing facility shortly after launch. At 12:35 the command to scrub the launch was given, and the crew was obliged to eventually leave the orbiter and return to their crew stations disappointed, but nonetheless hopeful for the next day.

As the crew settled down for another night on Earth in the warmth of the crew quarters, the temperature around *Challenger* dropped to well below freezing level as an extreme cold front swept over the Cape area. Investigating the launch constraints with this sudden drop in temperature, the mission management team evaluated the data to hand, and was told that the winds were expected to die down. The morning was predicted as clear but very cold, with the temperatures having reached the low 20s during the night.

One of the most important factors for launching in these conditions was the effect of the weather on the vehicle and structures. Ice on the pad was a very real threat, as it could damage launch equipment. As ice falls off the cryogenic-filled ET, it risks hitting and damaging orbiter tiles. This had happened on past missions, but this time the ice was much worse. The bare aluminium skin of the orbiter structure might be revealed and seriously damaged, especially during the re-entry sequence. At 01.35 am on 28 January, the first ice inspection team, consisting of engineers and technicians, was dispatched to the pad to evaluate the status of the area. They dutifully completed their inspection and reported their findings. The report was not as encouraging as hoped, but it was estimated that the majority of the ice could be cleared away or even melted well before the scheduled launch time of 08.30 am.

The crew were awakened at 06.18, and enjoyed a leisurely breakfast before a weather briefing and suiting for the trip to *Challenger* for the second attempt at launch. Following breakfast, the crew dressed in light blue flight suits. 'Space' pressure suits had never been worn for Shuttle launches. The first four missions saw

The STS 51-L crew at breakfast on launch day: (*left–right*) E. Onizuka, C. McAuliffe, M. Smith, R. Scobee, J. Resnik, R. McNair and G. Jarvis.

the two-man crews wear specially adapted high-altitude pressure garments from the SR-71 spyplane programme, so they could use the ejection seats in an emergency. From the fifth mission, the crew wore only a launch and entry helmet that provided oxygen and communications. The 51-L pilots then received their final weather briefing before leaving the crew quarters. Through the glare of photographers and cheers of well-wishers, they boarded the Astro Van for the three-mile trip to the pad and the waiting launch vehicle.

The weather report given to the crew included the news that the Casablanca site was a 'no-go' situation due to rain and low cloud levels. The primary site at Dakar was clear, which was fortunate, for if both sites were closed then the launch would have been postponed yet again. The ice and the extreme cold weather were reported to the crew. However, as it was later revealed, despite informing the crew of the conditions no apparent information concerning their effects on the Shuttle during launch was provided. They were also not informed of the concerns expressed by several Morton Thiokol engineers – who built the SRB – that the cold could have a hazardous effect on their integrity – concerns that had been raised since the beginning of the Shuttle programme.

The seven astronauts arrived at the base of LC-39B at 08.03 am, some 2 hrs 11 min before planned launch. They then ascended in the elevator to the crew access arm level, and were assisted into the vehicle and strapped to their seats. Astronaut Support Person (ASP) 'Sonny' Carter ensured that they were comfortable in their seats and that communications were established with the orbiter test conductor and test director in the launch control building. As they were being strapped in, they were informed that a T-38 NASA jet had taken off to determine the weather and winds higher in the atmosphere, and that a second ice inspection team had examined the pad area. The temperature inside the orbiter crew module was a comfortable 61° F. Just outside the windows, KSC was shivering in temperatures in the 20s.

Scobee was the first to enter the vehicle and take up his position in the forward

STS 51-L crew-members in their launch positions on the flight deck during training: (*left–right*) Mike Smith, Ellison Onizuka, Judy Resnik and Richard Scobee.

left-hand seat, followed by Smith, next to him in the right-hand seat. As he climbed into the seat he mentioned that the Sun was warm through his window. Carter replied that he should have been out there at 2 o'clock that morning. Scobee joked that they must have been having fun ice-skating on the mobile launch platform instead of working. Onizuka followed onto the flight deck, and Resnik then took up her position in the centre seat.

On the mid-deck, Carter paid a little more attention to strapping in the two Payload Specialists, McAuliffe and Jarvis. Finally, McNair clambered aboard. For the next two hours, all they could do was wait for clearance from the ice teams to launch. Resnik commented that her behind was numb. Several light-hearted conversations between the crew included an offer from Jarvis that he and Onizuka could always arrange a massage to solve the problem for her!

Onizuka thought it was snowing as ice blew off the supercold ET and was carried past the window. He wondered where they obtained all the water, dripping off fawcets on the launch tower. Resnik offered the suggestion that it was probably from Onizuka's tax dollars. Scobee added that it was a special grade of water. Jarvis joked that the hold was probably making extra money for the space programme by selling coffee and doughnuts in the viewing area during the hold. Scobee reflected that he should have picked up some coffee for his crew. McAuliffe, knowing her family would be watching the launch from the crew family area, reflected that it would be cold for them, standing out there at the viewing site. McNair remained quiet during this period.

The mid-deck area during training: (*left–right*) back-up payload teacher Barbara Morgan (who became a career NASA astronaut in 1998) Christa McAuliffe, Greg Jarvis and Ron McNair, in their launch positions.

Ice on the launch pad, with *Challenger*, the external tank and the left-hand SRB behind.

The ice team reported to the launch director following completion of their inspection at 08.44 am. As a result of their findings, the launch was delayed to enable the rising Sun to melt more of the ice cover on the pad. The team recorded a thickness of three inches on the Rotating Service Structure as they completed coverage of the 160 × 135-foot Mobile Launch Platform. The team also reported finding ice floating on the water troughs beneath the platform. The water was used to prevent rocket exhaust back-pressure inadvertently moving the orbiter's body flap and other aerodynamic surfaces. There were also rows of icicles hanging at the 120- and 220-foot levels on the service structure, measuring 6–12 inches long and ⅝ of an inch thick. As they reported their findings, the outside temperature around the vehicle rose from 26.1° F to 36.2° F.

At 10.30 am the ice team made a third trip to the pad to remove further ice from the platform and from the water troughs, using long-handled fishing nets! On the same visit to the pad, temperature readings of each solid rocket booster were taken. The left-hand SRB was recorded at 33° F, and the right-hand at 19° F. At the time there was no significant concern with this discrepancy, as it was thought to be a result of wind blowing off the huge ET (filled with cryogenic super-cold fuel) and onto the right-hand SRB.

On the flight deck, Scobee, Smith and Resnik were going through their pre-flight checklist and performing communications checks between the seven members of the crew and the ground. Down on the mid-deck the rest of the crew had nothing to do but wait for launch. When McAuliffe's turn had come to enter the orbiter she was handed an apple by the close-out crew. It was also discovered that one of the crew had left their gloves in the Astro Van, and these were quickly retrieved by one of the close-out crew-members.

At 1 hr 10 min before launch, Sonny Carter, the ASP, and the close-out crew had closed the hatch to a round of applause from flight controllers and, following pressure checks, the confirmation of a sealed hatch was given during an unscheduled hold. The close-out crew then retreated from the pad to a safe area.

KSC COM: 'T [launch] minus 70 minutes and holding. We are presently in an unscheduled hold [awaiting results from] an ongoing meeting on the ice situation. We will possibly be extending the hold 30 minutes. NASA TV now shows some of the ice around the Shuttle. Waiting to resume the count. The countdown clock continues to hold. Engineers are taking a look at the situation. Presently, temperature is expected to be 32 [degrees at launch]. Looks like we will be coming out of this hold on time, the launch team is being updated on weather. Looks good. We have picked up the count at the T–1 hour mark. It does appear we have satisfactory weather at this time. Out on the pad, the crew performed a cabin leak check. We'll do some cleaning of the ice around the pad. Winds have shifted around.'

A nine-person ice inspection team was once again dispatched to the pad to report on the ice situation and to clean off as much as possible. A 'GO' on the weather was given 24 minutes before planned launch, and five minutes later the ice inspection team returned and confirmed that the pad was clear of all personnel. They had checked and cleared what ice they could from the base of the vehicle.

*Challenger* was now in a planned hold pending resumption of the countdown from the T–9 min mark. As the clock reached 11.29 am, the countdown started again. The whole countdown procedure for 51-L, as with all Shuttles, was a mammoth checklist more than 2,000 pages long and taking up four volumes!

As the countdown clock in front of the viewing stands began to move again, cheers went up from the thousands of spectators, including family and friends of the seven astronauts and pupils from McAuliffe's school. The long wait for launch seemed to be almost over as the computers and flight crew performed their final actions before lift-off.

KSC Comm: 'T–7 minutes and counting'.

T–5 min: auxiliary power units were started on *Challenger*. Liquid oxygen replenish was terminated and LOX drain back was initiated to the ET.

T–4 min 30 sec: SRB and ET safe and arm devices were armed. SSME fuel valve heaters were also turned on at this point in preparation for engine start.

T–4 min: spacecraft test conductor (STC) reminded the flight crew to close the airtight visors on their launch and entry helmets, and at the same time the final purge of SSME below them began.

T–3 min 45 sec: the final orbiter aerodynamic surface test began, in which the flight-control surfaces were moved in a preprogrammed pattern verifying them for launch. The ground support equipment was turned off, and *Challenger* switched to internal power.

Inside *Challenger*, Smith noted that the gaseous oxygen vent arm covering the top of the ET (commonly called the 'beanie cap') was moved away from the top of the vehicle. Onizuka questioned if it had moved the wrong way, which brought a round of laughter from the crew. Smith also noted that helium pressures in the right SSME on *Challenger* were down slightly, and Scobee commented that they were the same the previous day prior to the abort. The countdown continued normally. At 2 min 22 sec the computer handling the caution and warning memory was routinely cleared for flight operations.

   The countdown reached the final stages and the PAO counted down the last ten seconds. At six seconds before lift-off the three SSME ignited to build up thrust pending ignition of the twin SRB and lift-off.

T–6 sec: Scobee: 'There they go, guys [the three main engines].'
Resnik: 'Alright!'
Scobee: 'Three [engines] at a hundred [percent of rated power].'

At T–0 (lift-off), the twin SRB each side of *Challenger*'s ET ignited on time, shaking the ground and spewing out billowing clouds of exhaust as the vehicle was finally released to begin its trip to space. *Challenger* lifted off to the cheers, shouts and screams of the crowd of onlookers, and began its tenth mission at 11:38:00:010 EST, 28 January 1986.

T–00:00: Resnik: 'Alright!'
T–0:01: Smith: 'Here we go!'
KSC PAO (Hugh Harris): 'Lift-off! Lift-off of the twenty-fifth Space Shuttle mission and it has cleared the tower! Houston is now controlling.'

   As the pride of the American space programme cleared the service structure, control of the mission was switched from the Cape to Mission Control, Houston, Texas, where it would remain until the vehicle made its planned landing a week later.

The launch of *Challenger* reveals a puff of black smoke from the lower right of the right-hand SRB (arrowed), as recorded by automatic camera.

The following are the comments of Challenger's last, short flight, taken from onboard audio tapes recovered from the crew module and released by NASA after they had been restored. The Public Affairs Officer at JSC was Steve Nesbitt, who informed the media of the flight's progress. The CapCom was astronaut Dick Covey, a highly experienced Shuttle pilot. He was the point of contact between Flight Controllers and the crew in space. The in-cabin crew chatter was not broadcast publicly, but was recorded by NASA.

MCC Covey: 'Watch your roll, *Challenger*', alerting the crew to the pre-programme roll of the vehicle.
T + 0:07: Scobee: 'Houston, *Challenger* roll programme.'

This first manoeuvre after leaving the ground was designed to rotate the vehicle to fly a 'heads down' (towards Earth) orientation to provide the most favourable configuration for flight through the Earth's atmosphere, where aerodynamic stresses on the vehicle are less severe.

A second view of *Challenger* shows the puff of black smoke (arrowed) that indicated the breach of the SRB 'O' ring.

Houston PAO: 'Roll programme confirmed.'

As *Challenger* climbed, the engines were throttled down to 94% thrust. During the period of maximum dynamic pressure on the vehicle (as it goes supersonic), the engines were to be throttled down to 65% and then throttled back up to continue the ascent as the atmosphere thinned. All telemetry received from the vehicle indicated that all three engines were performing normally. All three fuel cells, which provided electrical power to the orbiter, and all three APU, which powered the hydraulic flight control surfaces, were running well.

Houston PAO: 'Velocity 2,257 feet per second [1,538 mph], altitude 4.3 miles. Down-range 3 nautical miles. Engines throttling up. Three engines now at 104%.'
T + 0:11: Smith: 'Go, you mother!'
T + 0:14: Resnik reminds the flight crew about switch positions for local vertical, local horizontal (LVLH) to switch from a ground-based to a vehicle-based frame of reference.

T + 0:15: Resnik: 'Shit hot!' (Originally deleted by NASA; an expression meaning 'all seems to be going well.')

T + 0:16: Scobee: 'Okay!'

T + 0:19: Smith: 'Looks like we've got a lot of wind up here today'. As *Challenger* encountered its first wind sheer, the onboard computers evaluated the moves needed to keep the vehicle on its correct heading and continually instructed the nozzles of SRBs and main engines to gimball in order to compensate. These variations were recorded on the flight deck displays from the inertial guidance system.

T + 0:20: Scobee: 'Yeah. It's a little hard to see out my window here.'

T + 0:28: Smith: 'There's 10,000 feet and Mach point 5.'

T + 0:35: Scobee: 'Point nine.'

T + 0:40: Smith: 'There's Mach one,' (through the sound barrier).

T + 0:41: Scobee: 'Going through 19,000 feet.'

T + 0:43: Scobee: 'Okay, we're throttling down.'

The computers aboard *Challenger* now signalled the main three engines to throttle down from 94% thrust to 65% thrust as the vehicle passed though a 36-second period of maximum dynamic stress on the structure (Max Q), following which the engines would be automatically throttled back up to full power at 104% performance.

T + 0:57: Scobee: 'Throttling up.'

T + 0:58: Smith: 'Throttle up.'

T + 0:59: Scobee: 'Roger.'

T + 1:02: Smith: '35,000. Going through 1.5.'

T + 1:05: Scobee: 'Reading 486 on mine'. (The meaning has been disputed, but according to NASA it was an airspeed check.)

T + 1:07: Smith: 'Yep. That's what I've got too.'

CapCom: 'Go at throttle up.'

T + 1:10: Scobee: 'Roger, go at throttle up.'

By now a rapid sequence of events that had begun the moment that *Challenger* lifted off the ground was unfolding, unseen by the crew and most of the onlookers on the ground. It was only during replays of launch pad video and long-range cameras that the full extent of what happened next was revealed. As *Challenger* soared onwards at 2,900 fps at 50,000 feet altitude and seven miles from the Cape, disaster struck.

The orbiter was heading east (out to sea) and flying on its back following the roll manoeuvre. Long Range Camera 207 – one of several located around the launch complex – recorded a bright glow on the right-hand side of the external tank, which increased in size and brightness until the whole cloud engulfed *Challenger* and her crew. From the ground the twin SRBs could be seen still burning normally as Smith uttered the last comment from the cabin.

T + 1:13: Smith: 'Uh-oh!'

As though they had had imminent warning of the impending event, Scobee activated

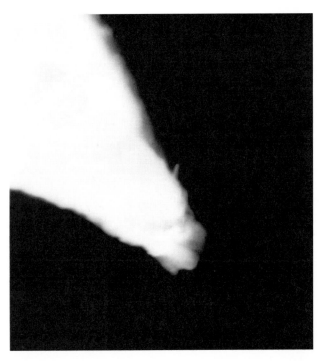

A flash from the region between the orbiter and the external tank at MET 73.200, taken by a long-range camera a split second before the explosion of the ET which led to the aerodynamic destruction of the orbiter and the death of the seven astronauts.

his voice microphone but had no time to speak. Houston was unaware of Smith's comment, and contact was lost.

From the ground, many of the onlookers were at first unaware of the tragedy that was unfolding before their eyes, erroneously assuming that they were seeing a staging. However, when the two SRB emerged from the cloud of flame and arced away each side, many began to realise that all was not well. As eyewitnesses saw hundreds of items of debris fall from the cloud and streaks of flame spread across the sky, the reality of what was happening finally sank in. Then, the Range Safety Office detonated both boosters by remote control. They exploded at 101,300 feet (19.13 miles), one of them being still filled with burning propellant at 5,600° F and apparently heading for the coastal community of New Smyrna Beach.

*Challenger* and the external tank had disappeared in a ball of flame as the external tank's LOX and LH had ignited. Thousands of items of debris, ranging in size from fractions of an inch to the crew module and one of the wings, rained down from the explosion.

PAO: 'Flight Controllers are looking very carefully at the situation. Obviously a major malfunction. We have no downlink. We have a report from the Flight Dynamics Officer that the vehicle has exploded. The Flight Director confirms that.' The Flight Director was Jay Greene, who became the first FD to lose a crew.

An SRB is still burning as the vehicle disintegrates just 73 seconds after launch.

KSC PAO: 'There are ships on their way to where the orbiter is. Apparently we were full throttle when it exploded. We had an apparent explosion [whereupon] data was lost with the vehicle. No reported problems with the SRBs.'

PAO Steve Nesbitt continued to explain that the recovery forces were being assembled to see what could be done and that all contingency procedures were in operation. Many observers, including some NASA employees, hoped that the crew had survived the explosion that tore their vehicle apart. Expectations of the orbiter appearing from the clouds limping towards a runway landing were dashed as the seconds passed. Some thought, in error, that they had seen parachutes, and hopes were raised that perhaps the crew had escaped from the orbiter and was heading for a landing in the ocean. However, there were no parachutes stored on the orbiter. To those knowledgeable about the Shuttle programme, it was clear that the crew was dead.

In his book *The Space Shuttle: Roles, Missions and Accomplishments* (Wiley–Praxis, 1998) author David Harland states that a call for an RTLS abort could have been initiated from the first indication of SRB pressure drop at T+60 sec. The decision for Flight Controllers in looking at their data was whether the change was

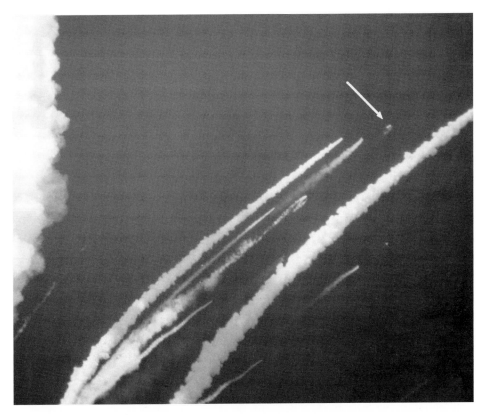

Debris including the pressurised crew module trailing cables (arrowed) emerges from the cloud and begins the long fall into the Atlantic Ocean.

temporary or a more serious problem. The SRB was underperforming, but this was not evident in the telemetry. It would have taken about three seconds to initiate an RTLS abort at this point, but by T + 66 seconds and the fracturing of the ET and venting of hydrogen the loss of supply for the SSME would have minimised the chances of *Challenger*'s manoeuvring for a safe landing. Harland observes that, despite being very small, a window of opportunity for an RTLS abort lasted for six seconds, but no call was made by the flight controllers, although the reason is unknown.

Telemetry received on the ground had indicated a fault in the SRB but not its cause. It was the lack of information on the cause of the problem in the SRB that prevented the call for an abort to the crew who, according to some interpretations of the in-cabin comments, were suddenly aware of a drop in SRB performance, but had no time to initialise an abort from the flight deck; and without confirmation from Mission Control they would probably not have initiated the abort themselves.

Harland also observes that the TV views of the ascent available to the Flight Director were not from a viewpoint from which the flame plume was observed by remotely operated cameras. With subsequent development in split-screen technology, the advantage of a selection of views of the launch from different camera angles at the same time was not available to Flight Director Greene at MCC, Houston; but even if it had been, could they have acted quickly enough to do anything about what they would have surely questioned they were seeing develop in the 13 seconds they had to react?

Unlike earlier programmes such as Mercury and Apollo – which employed escape towers to pull the crew module clear of an exploding launch vehicle – the Shuttle did not have such an escape system. Earlier in the programme, ejection seats for the two-man ALT and OFT programme missions were provided in (respectively) *Enterprise* and *Columbia*, but they were removed in 1983 when full crews began to fly. As a result, the Shuttle flew from November 1982 to January 1986 (24 missions) with no effective escape method for the crew during the SRB phase of the ascent. The complication of increased crew size, the added difficulty of split-level crew stations, and the belief in the safety of the design, preempted proposals for escape towers, ejection seats or a detachable and recoverable crew module early in the design phase.

On 28 January 1986 the tragic result of these decisions was the loss of seven astronauts and *Challenger*. The American media reported the tragic loss of seven people at the pinnacle of their lives, who at the time were riding a symbol of national achievement which was suddenly lost in a violent explosion

Pictures of the short flight of *Challenger* were received in real-time at Mission Control Center, Houston, where astronauts and flight controllers stared with disbelief at the events unfolding before them. All flight controllers were told to preserve data on their display consoles for the inevitable post-accident investigations. What had started as relaxed expectations for the 25th mission, carrying the first schoolteacher into space, ended with the American space programme in disarray and its most appalling disaster played out in full view of the world's media.

At the Cape, onlookers stood in shocked disbelief as the vehicle suddenly exploded before their eyes. Families of the crew were immediately protected by NASA protocol officers and were soon whisked away to cope with their private grief. Many schoolchildren – including those from Christa McAuliffe's own school – were at the Cape or were watching live coverage of the launch on TV. They too witnessed the last flight of *Challenger*, carrying a teacher with whom they were very familiar.

Almost immediately, messages of sympathy flooded into the United States from around the world. It was nineteen years and one day since America suffered its last worst-case space-related accident, in the flash fire inside the Apollo 1 capsule during a pre-launch training session.

As this book records, near-tragic events had occurred on other US missions. The Soviet Union also suffered in-flight accidents and losses of crew-members, and they too shared with the US the tragedy of lost space explorers. Some reports from the Soviet Union, though, took the opportunity of criticising the Americans for military uses of space, cited the loss as a warning against such operations and stated that the loss resulted from NASA's increased flight rate.

As the day wore on, Vice President George Bush assigned Congressional Senator Jake Garn, a former Shuttle Payload Specialist, and Senator John Glenn, a former NASA astronaut, to fly to Kennedy with him to meet the families at the Cape.

As the press wired their impressions of the disaster around the world and formed theories of the cause and effects on the American space programme, NASA called its first post-disaster news conference, which began at 4.30 at the Kennedy Space Center. It was received at other NASA centres, including Johnson in Houston, JPL in California – where staff had just celebrated the success of the Voyager 2 encounter with Uranus – and Marshall at Huntsville, Alabama. Outgoing Associate Administrator for Space Flight, Jesse Moore, gave the opening address to the assembled newsmen:

'It is with deep, heartfelt sorrow that I address you here this afternoon. At 11:30 am this morning, the space programme experienced a tragedy with the explosion of the Space Shuttle *Challenger* approximately a minute and a half after launch from here at the Kennedy Space Center. I regret that I have to report that, based on very preliminary searches of the ocean where *Challenger* impacted this morning, these searches have not revealed any evidence that the *Challenger* crew survived.'

Initial indications, according to the press conference, were that everything appeared normal for about the first minute of flight and that no flight controller reported any unusual reading or incident. A search team was dispatched to recover as much debris from the sea as possible, and evidence from all sources was impounded pending investigations. NASA also set up the Interim Mishap Review Board. It comprised astronauts Robert Overmyer (who flew on two Shuttle flights) and Deputy Chief of the Astronaut Office and astronaut Bob Crippen (who had flown, at that time, a record four Shuttle missions). They were joined by Richard Smith, Director of KSC; Arnold Aldrich, Shuttle Project Manager JSC; Walt Williams, NASA consultant and former executive on the Mercury and Gemini programmes; William Lucas, Director of Marshall Space Flight Center; James Harrington, Director Shuttle Programme Integration Office NASA HQ Washington; and members of the National Transportation Safety Board.

As NASA was questioned on the events of the day, President Reagan addressed the Nation on TV that evening. Using extracts from the famous poem 'High Flight' by Canadian John Gillespie Magee Jr, a 19-year old pilot killed in December 1941, he commented on how they had 'slipped the surly bonds of Earth and touched the face of God'. President Reagan also recognised the crew's admirable dedication to the peaceful exploration of space, and pledged that despite this tragedy the space programme and further Shuttle flights would continue, although nobody could say when. On Friday, 31 January, President Reagan headed the memorial service for the crew at the JSC, along with several astronauts. Personally meeting each of the families, the President reinforced the commitment to space as many from JSC, including former astronauts from past programmes, attended to give their personal support to the grieving families and the grounded space programme.

The official Presidential Commission was formed on 3 February, and was chaired by former Secretary of State in the Nixon Administration, William Rogers. Other members of the Rogers Commission included former astronaut and the first man on the Moon, Neil Armstrong; Chuck Yeager, the first person to break the sound barrier; active Shuttle Mission Specialist Sally Ride; Robert Hotz, former editor of the leading aerospace magazine *Aviation Week and Space Technology*; and Richard

Feynman, a Caltech Nobel Laureate in physics. The commission worked over the next four months, investigating the cause of the accident and the events of the fateful day that claimed the lives of the 'Challenger Seven', as the media was now calling them. Apart from recorded data, a substantial element of the review of the accident was the debris of the vehicle recovered from the sea – a recovery and reconstruction process that took about six months.

The evidence of what exactly caused the accident lay at the bottom of the ocean, and a mammoth operation to recover as much wreckage as possible of the vehicle was begun the day that the accident occurred. Following the explosion, debris rained from the sky for around 45 minutes. TV monitors recorded debris splashing into the sea, preventing search aircraft from entering the area for more than an hour, despite being airborne two minutes after the accident. Search teams immediately spotted floating wreckage. *Challenger* had exploded some nine miles above the ocean, and the force of the explosion had scattered the wreckage over a wide area twenty miles off shore.

By the afternoon of the 28 January, as the Gulf Stream drift took its effect on floating debris and the daylight hours ended, there were a dozen aircraft and around ten ships searching for debris. Over the next ten days the surface of the ocean was combed repeatedly, and much lightweight debris was picked up before it sank. The plan was to recover all surface debris before starting the much more difficult underwater search. The search initially covered the area south of the Cape to Cocoa Beach and north to Daytona Beach (12,959 sq km), and this was extended to 270,000 sq km of the Atlantic Ocean by the end of the surface search on 7 February. Within the first few days after the accident, several large elements of the spacecraft had been recovered and were returned to the Cape for later examination. They would try to reassemble as much of the debris as possible to reconstruct the effect of the explosion on the vehicle and SRB to help determine its exact cause.

Jagged portions of the right-hand wing and fuselage, including a large segment of the outer shell that surrounded the crew compartment, were among the first to be recovered. A large number of thermal protection system tiles were also recovered, as were parts of the body flap, vertical tail, the payload bay doors and elements of the top of the external tank. In early February, search teams found several mid-deck lockers and some of the crew's launch and entry helmets floating on the surface.

As well as the surface recovery, elements of the spacecraft and its boosters were beginning to be washed up along the shores surrounding the Cape. A tight security net was established, and every piece of debris recovered was photographed and its location plotted on a chart for post-recovery analysis. All debris was designated Federal Property, and citizens were urged to report finds.

Using a combination of video recorders, sonar, radar and visual sightings, a complete surface search area was covered. Not all the debris was retrieved – only those pieces thought to be significant to the post-flight review board and those linked to probable causes of the accident.

From the data on the accident, NASA examined each frame of available videotape and stored telemetry, in order to establish a probable cause. By 2 February the agency released a video that clearly revealed a close-up of the ascent of

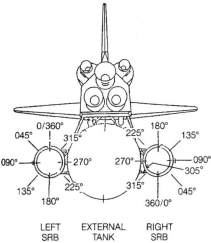

(*Left*) Part of the wreckage of mission 51-L recovered from the Atlantic Ocean, including a piece of the right-hand SRB casing, which clearly shows evidence of fracture and burn-through. (*Right*) The position of the breach at 305° on the right-hand SRB.

*Challenger*. The film also showed the presence of an unusual plume of flame extending from the right hand SRB. It appeared that the plume originated from a point on the aft field joint of the SRB, and it was clear that this initiated the sequence of events that resulted in the explosion of the external tank, and the consequential destruction of *Challenger* and loss of the crew.

The priorities for the salvage team were now the right-hand SRB, the crew compartment, the payload bay, the left-hand SRB and the external tank.

For the underwater search, a second impressive flotilla of ships and submersibles was assembled by NASA, the US Navy, the Coast Guard, the US Department of Defense, mining companies and commercial and private oil and exploration concerns. Heavy lifting gear and a range of manned and unmanned submersibles continued to search the ocean bed for the next few weeks, trying to locate the right-hand SRB and crew compartment.

By 13 April the recovery forces had found elements of the right-hand SRB. One showed a large hole of 0.61 feet on the edge of the 4,103-lb piece of metal, designated Sonar Contact 131, found at 550 feet depth some 46 miles north-east of LC 39B. The burn was centred on the 305° position of the segment which was the closest to the ET. No trace of the O-rings was seen. At last, the recovery forces had found an element in the debris that finally helped to resolve the sequence of events that claimed *Challenger*.

As the search continued, elements of the external tank, the left-hand wing, the engine compartment and elements of the IUS/TDRS payload of *Challenger* were gradually recovered and returned to the Cape. The search saw the largest salvage operation in maritime history. Fifteen surface ships, scores of USN divers and five submarines completed the underwater search. Divers likened the task before them to

View of the recovered wreckage of mission 51L from the Atlantic Ocean, showing part of the side fuselage of the orbiter with the charred remains of the name of the vehicle.

trying to find a coin in a murky pool the size of a football field, on their hands and knees!

By then the Presidential Commission had begun investigating the accident, and as the elements of the vehicle were recovered NASA allocated three buildings to house them at the Cape. As all the elements were being assembled, the search went on for the most important item of the spacecraft for the families and public – the crew compartment and the bodies of the seven astronauts. As the search continued into March, weather hampered operations and delayed the recovery. With most of the world waiting for news, the announcement that the crew compartment had finally been recovered and the crew's remains retrieved shook the public.

NASA did not release film and details of what was found on the sea bed to the general public or the media. Neither were the details revealed by the Presidential Commission. The grim details of the state of the *Challenger* cabin and its contents will probably, as with the Apollo 1 tragedy, remain classified as a mark of respect and dignity to the astronauts' families.

The painful discovery of the crew compartment was made on 7 March by two divers, working in 87 feet of water, 20 miles north-east of LC 39B, while following up a large sonar contact. They were investigating a piece of wreckage measuring 6–8 feet high, when in the murky surroundings they found one of the empty EVA suits with its feet floating towards the surface. It was known that these suits were stored inside *Challenger*'s airlock, in the mid-deck area of the crew compartment for launch, so it was reasonable to assume that the divers were close to the rest of the crew module.

On 8 March, positive identification of the crew module had been confirmed, lying on its left side. Over the next few weeks, repeated dives and searches by the recovery forces retrieved flight computers, onboard voice tapes and instrument recorders.

Elements of this information were presented to the Commission, and tapes of flight deck comments were released to the media.

One by one, the remains of the crew were discovered and returned to the surface vessels inside containers. Some of the dives on the wreckage were completed by astronauts, including Jim Bagian. Several times, recovery vessels docked at the Cape without running lights and away from the glare of the media and official photographers, to offload the crews' remains. On 15 April the remains of the seventh and final astronaut (PS Greg Jarvis), were recovered, and the bodies were taken to the Life Sciences Support Facility Hanger 'L' at the Cape, where preliminary forensic analysis and post-mortems were performed. Identifications completed, the seven bodies were moved by hearse, each accompanied by an astronaut representing the Astronaut Office, to the Shuttle Landing Facility. They were then put on board a USAF C-141 Starlifter transport aircraft, to be flown to Dover, Delaware, and then returned to their respective families.

The official burial ceremonies for the crew were conducted separately from private family memorials. Pilots Scobee and Smith were buried at Arlington National Cemetery, while the others were buried in or near their home towns. Mission Specialist Resnik was buried in Akron, Ohio, McNair in Lake City, South Carolina, and Onizuka in Honolulu, Hawaii. Payload Specialist Jarvis was buried in Mowhawk, New York, and McAuliffe in Concord, New Hampshire.

The remains of the seven *Challenger* astronauts receive an honour guard as they leave Cape Kennedy following recovery from the ocean.

The process adopted by the recovery forces for the return of the crew's remains was reported by Colonel Edward O'Conner Jnr, Director of Operations 6555th Aerospace Test Group, Patrick Air Force Base, Florida, Lead Officer for the recovery forces:

'It was decided that the most reasonable approach to [the recovery of the crew] was to say nothing until the process was completed. NASA astronauts were out there when we first discovered we had the crew compartment. Many engineers were involved in it, so the necessary information to assess crew safety was captured in real time.'

O'Conner went on to explain that there were several areas for concern in this operation. One was that if it were made visible in any way, there would certainly be a desire for media and onlookers to venture out to the site and witness operations first hand. So, for reasons of safety to the divers 'and our other assets out there', the area was simply closed off. All those not involved with the process of recovery – such as media, public and non-essential personnel – were excluded. Each of the crews' families was consulted on this phase of the operation, using the astronaut assigned to each family as a liaison between them and the recovery effort. It was agreed that instead of making the recovery a large event, it would be cleaner, simpler and more suitable to proceed with the job, issue one press release and then complete the operation with the necessary ceremonials.

By 30 April the deep-water recovery effort was complete, although the recovery forces continued to retrieve previously identified elements of debris through to the official ending of the salvage operation on 28 August 1986. It was estimated that the entire recovery process cost $100 million and resulted in the collection of 111 tons of debris comprising about 50% of the SRBs, 50% of the ET, and only 30–35% of *Challenger* itself. Significant elements of the TDRS and Spartan–Halley payload were also salvaged for engineering analysis. When completed, the *Challenger* Salvage Data package – comprising hundreds of photographs, sketches, videos, underwater photographs and films, schedules, logs and reports – weighed some 400 lbs. This detailed the recovery of *Challenger* and 42 significant pieces of booster. Following examination by NASA and the Presidential Commission, the recovered debris was prepared for long-term storage. On 3 October 1986, NASA revealed that it would place all of the recovered debris in two 31,000 cubic foot storage spaces formerly used as Minuteman test silos at Complex 31 and 32 at the Cape Canaveral Air Force Station, which was completed during the early months of 1987. As with the CM from the Apollo 1 fire, future public display of such hardware would be inappropriate.

Whilst the ocean search was running above and below the surface, the investigation panel set up by the President continued to piece together the events leading up to the tragic accident, as well as putting forward recommendations and possible remedies to begin the long road back to space.

The Commission worked on the information provided by NASA and other agencies in trying to determine the exact cause of the accident. They worked through photographic and taped data, examined recovered wreckage, and completed personal visits to the Cape and leading contractors and a series of televised and taped hearings with key personnel from NASA and shuttle contractors, as well as the recovery teams and observers. In addition, the Commission requested a series of tests

Members of the Presidential Commission investigating the *Challenger* accident examine propellent contained in one of the stored solid rocket booster segments at Kennedy Space Center ordnance facilities during February 1986.

and evaluations to assist them in pinpointing the exact cause of the failure. Early in the work of the Commission, it became very clear that the start of *Challenger*'s problems lay in the plume observed emitting from the side of the right-hand SRB. Much of the Commission's effort was directed to examination of this area in general and to the cause of the gaping hole that was seen on the recovered elements of the booster in particular.

One by one, elements of the Shuttle stack – including the ET, payload and the orbiter itself – were eliminated from the list of probable causes. One of the first theories was that the IUS had suddenly exploded or shifted inside the payload bay, but recovered debris of the payload soon proved this theory to be incorrect.

Adverse weather was a significant factor in the tragic sequence of events, as was the increased pressure on the ground crews to make the launch on time, and the lack of sufficient spares and ground procedures. As time wore on, it became evident to all those concerned that it was the right-hand SRB that had developed the fault in its O-ring sealing system, which resulted in the flame leak and the destruction of the vehicle.

As the available video of the last launch of *Challenger* was examined frame by frame, evidence from contractors and launch personnel, as well as data from past missions, indicated that failure and questionable structural integrity of the O-rings was not a new or unknown problem. It soon became clear that those who had a direct say in the launch of *Challenger* that morning were not fully aware of the fault or the implications of launching in such adverse temperatures. The effects of cold on the O-rings had been known to engineers to some degree, but their findings had not been passed on to the astronauts who put their lives on the line by flying each mission. Reaction was, as expected, one of surprise and shock.

The most publicised response to this new information were the memos from the Chief of the Astronaut Office, John Young, veteran of a record six space flights, including two Shuttle missions. The Young memos went further than criticising those who had confidence in the system despite knowledge of a potential disaster area. They also included comments on crew safety, pressure to launch, major systems such as the landing system tyres, the orbiter brakes, the main propulsion, reaction control and manoeuvring systems, and the RMS. Released during the time of the Commission, this soon became public – and the media scooped it up. At the time, the United States had also suffered launch failures in its unmanned programme with the loss of a Titan and a Delta vehicle within weeks of *Challenger*. The American space programme was firmly grounded, and it seemed likely to stay that way for a long time to come.

As the media picked up on elements of the investigation, several astronauts voiced their opinions of both the accident and other aspects of safety and procedures in a way that NASA had never seen before. Comparisons to the Apollo fire were difficult. In those days NASA retained a small astronaut group (around forty flight-status astronauts), who all had a significant input in the preparations of the spacecraft they were to fly. By 1986, many of the earlier astronauts had long gone. New processes were employed that took the responsibilities which many of the original astronauts had away from the Astronaut Office and on to others in industry and the administration. A visit to Houston, and a tour around corporate buildings springing up around JSC, clearly reveals where much of the space programme's money was heading. Corporate backing is big business, and the old pioneering, national pride in just doing the very best to get the job done was long gone in the bureaucratic halls of the American space programme of the 1980s.

During the summer months of 1986, NASA was in the spotlight for very different reasons from those that pushed the agency towards world acclaim in the 1960s and early 1970s. The phenomenal success of the Shuttle for the first five years of apparently trouble-free flight operations served only to increase the impression that NASA could achieve anything. Spectacular EVAs to rescue stranded satellites, and the 'regular' launches and flights of politicians and foreign nationals, all painted a rosy picture of success. The truth – as became apparent during 1986 – was that for many years a disaster of this magnitude had always been possible.

Partly as a result of the *Challenger* accident, many astronauts decided to leave the agency to pursue new careers. Some were critical of the agency and the accident, but many just recognised that the agency and the programme they had joined would never be the same again. Top administration positions were vacated and changed. Safety became a high priority, with new posts created. Former astronauts were given key positions in the chain of events to which every Shuttle flight has to be subjected prior to launch. Leading administration positions in NASA were also open for former astronauts to fill, using their flight experience in new roles as America looked forward to beginning the programme once again.

In industry too, changes were enforced, particularly in Morton Thiokol, contractors for the SRB that were pinpointed as the prime cause of the *Challenger* accident. Warnings from engineers on the safety aspects of the SRB, from flight data of past missions, had been ignored or not passed down the line. If astronauts had been monitoring the development of the hardware for their flights, perhaps some of the concerns expressed would have been noted in time. However, the complexity of the programme and the pressure on astronaut time was such that personal input towards all aspects of the hardware was just not possible. Unlike the Apollo days, when a crew followed the life of their one-mission spacecraft almost from the drawing board to the museum after the mission, Shuttle astronauts were just the next crew on a reusable vehicle. In effect they became a 'component' in the launch process that continued from mission to mission, as the vehicle was 'turned around'. Indeed, NASA saw this as as inevitable progression, as it tried to develop an operational system akin to that of an airline. But NASA was not a commercial operation, and the Shuttle could never be an airliner.

In the final analysis, the Commission interviewed 160 individuals during a series of more than 35 investigatory sessions. The transcripts from these meetings totalled more than 12,000 pages; and in addition, 6,300 documents and hundreds of photographs became part of the Commission's official record. As the results and recommendations of the Presidential enquiry were released and acted upon, and as the media and public looked to the next flight, the families of the lost astronauts sued NASA and the Morton Thiokol company for damages. This battle took more than three years – much longer the Return To Flight on Mission 26 in September 1988.

Meanwhile, the sequence of events that led to the loss of *Challenger* was released and, at last, the full chronology of events in the few seconds of the tenth and last flight of *Challenger* was known.

STS-25 (51-L): sequence of major launch events

| Time (EST) | Elapsed (sec) | Event |
|---|---|---|
| 11:37:53.444 | −6.566 | SSME No.3 ignition command |
| 37:53.564 | −6.446 | SSME No.2 ignition command |
| 37:53:684 | −6.326 | SSME No.1 ignition command |
| 38:00.010 | 0.000 | Solid rocket motor ignition command (T = 0) |
| 38:00.018 | 0.008 | Hold-down post 2 pyrotechnics firing |
| 38:00.260 | 0.250 | First continuous vertical motion |
| 38:00.688 | 0.678 | Confirmed smoke above field joint on RH SRM (E60 camera) |
| 38:04.349 | 4.339 | SSME 104% command |
| 38:05.684 | 5.674 | RH SRM pressure 11.8 psi above nominal |
| 38:07.734 | 7.724 | Roll manoeuvre initiated |
| 38:19.869 | 19.859 | SSME 94% command |
| 38:21.134 | 21.124 | Roll manoeuvre completed |
| 38:35.389 | 35.379 | SSME 65% command |
| 38:37.000 | 36.990 | Roll and yaw response to wind |
| 38:51.870 | 51.860 | SSME 104% command |
| 38:58.798 | 58.788 | First evidence of flame on RH SRM |
| 38:59.010 | 59.000 | Reconstructed Max Q (maximum dynamic pressure) 720 psf |
| 38:59.272 | 59.262 | Continuous well-defined plume on RH SRM |

| 39:00.014 | 60.004 | SRM pressure divergence (RH–LH) |
|---|---|---|
| 39:00.258 | 60.248 | First evidence of SRB plume attaching to ET ring frame |
| 39:04.670 | 64.660 | Change in anomalous plume shape |
| 39:04.715 | 64.705 | Bright sustained glow on sides of ET |
| 39:06.774 | 66.764 | Start ET LH$^2$ ullage pressure deviations |
| 39:12.574 | 72.564 | Start of H$^2$ tank pressure decrease with two flow control valves open |
| 39:12.974 | 72:964 | Start of sharp MPS LOX inlet pressure drop |
| 39:13.020 | 73.010 | Last full computer frame of TDRS (payload) data |
| 39:13.054 | 73.044 | Start of sharp MPS LH$^2$ inlet pressure drop |
| 39:13.134 | 73.124 | Circumfential white pattern on ET aft dome (LH$^2$ tank failure) |
| 39:13.134 | 73.124 | RH SRM pressure 19 psi lower than LH SRM |
| 39:13.147 | 73.137 | First hint of vapour at intertank |
| 39:13.153 | 73.143 | All engine systems start responding to loss of fuel and LOX inlet pressure |
| 39:13.172 | 73.162 | Sudden cloud along ET between intertank and aft dome |
| 39:13.201 | 73.191 | Flash between orbiter and LH$^2$ tank |
| 39:13.223 | 73.213 | Flash near SRB forward attach and brightening of flash between orbiter and ET |
| 39:13.292 | 73.282 | First indication by intense white flash at SRB forward attach point |
| 39:13.337 | 73.327 | Greatly increased intensity of white flash |
| 39:13.393 | 73:383 | All three main engines approaching temperature redline limits |
| 39:13.492 | 73.482 | SSME No.2 in shut-down due to temperature redline exceedance |
| 39:13.513 | 73.503 | SSME No.3 in shut-down due to temperature redline exceedence |
| 39:13.533 | 73.523 | SSME No.1 in shut-down due to temperature redline exceedance |
| 39:13.628 | 73.618 | Last validated orbiter telemetry measurement |
| 39:14.140 | 74.130 | Last radio frequency signal from orbiter |
| 39:14.597 | 74.587 | Bright flash in the vicinity of orbiter nose |
| 39:50.260 | 76.437 | RH SRB nose cap separation and parachute deployment |
| 39:50.260 | 110.250 | RH SRB destruct by range safety system |
| 39:50.262 | 110.252 | LH SRB RSS destruct |

Data from Report of the Presidential Commission on the Space Shuttle *Challenger* Accident (1986)

The Presidential Commission's investigation revealed that a puff of black smoke was first detected by cameras as early as 0.445 seconds after SRB ignition, and was seen to extend half-way across the right-hand SRB, centred on the 130° point on the booster by 2.147 seconds. There were indications of some visual identification of smoke through 'tower clear' and the roll manoeuvre and even as late as 12–13 seconds into the flight. From then until 58.774 seconds into the mission all appeared to proceed normally, but there was an indication of smoke from the side of the right-hand SRB forward of the aft ET attach ring immediately after lift-off, as recorded by pad cameras. (An independent analysis of the NASA evidence was completed by Ali Abu Taha, who stated that a photograph showed an 11-foot plume of flame emerging from the SRB at T + 20 sec, although this photograph has yet to be published. However, it did coincide with SRB internal pressure profiles.) The well-defined and intense plume finally appeared shortly after the 59-second mark. After 60 seconds, received telemetry data show a small divergence starting in chamber pressure between the left-hand and right-hand SRBs. A camera recorded a significant glow on the right-hand SRB after 66 seconds, which merged with the plume around the 67.75 seconds mark. At 73.175 seconds, a sudden cloud was observed along the side of the external tank, and a flash from between the orbiter's belly and the external tank's liquid hydrogen tank was recorded a fraction of a second later, followed by an explosion near the right-hand SRB forward attachment point.

Recovered telemetry also recorded the effects of the explosion, with the last data being received at 73.621 seconds. Following the explosion, the ground camera tracked the SRBs until the Range Safety Officer destroyed them. From these data, an accurate reconstruction of the events leading to the destruction of *Challenger* was compiled. As *Challenger* lifted off, the combination of the extreme overnight weather conditions and high winds contributed to the accident. The plume of smoke recorded at lift-off was an early breach of the SRB O-rings, which on a normal flight would 'melt' and seal the joint between SRB segments, thereby preventing the hot gases escaping from anywhere other than the thrust nozzle, as designed. As the vehicle ascended, the O-ring was breached and gases continued to escape, gradually burning through the metal side-wall of the SRB. This produced a lateral thrust that continued until the vehicle was destroyed. It burned through the strut, and then the booster rotated and impinged on the top of the huge and still almost fully fuelled external tank.

The flame colour and shape revealed that at 64.660 seconds the external tank was breached, and hydrogen fuel from the tank mixed with that from the leaking SRB. At 72.20 seconds the strut supporting the lower part of the right-hand SRB broke or came away from the attachment, due to sideways thrust caused by the leak. Frame-by-frame investigation of video of the sequence revealed, at 73.124 seconds, a bright vapour pattern at the base of the lower dome of the external tank, representing a massive structural failure and resulting in the whole aft dome dropping away from the tank. At the same time, the released hydrogen fuel ignited, pushing the structure into the inter-tank area of the external tank. The right-hand SRB swung around on its upper attachment point and impacted with the oxygen tank at the same time as the hydrogen tank ruptured. Flame suddenly shot up the side of the external tank, and an explosion occurred at the top of the structure adjacent to the crew cabin of *Challenger*.

With the combination of these dramatic events, the massive explosion of the released propellants pushed *Challenger* away from the ET/SRB as the tank exploded. As a result of high aerodynamic pressures on the surfaces of the orbiter, *Challenger* began to break up in mid-air, and its forward RCS propellants exploded. The orbiter split up into many large and small pieces. The pressurised crew module remained intact and, trailing wires, fell nine miles into the ocean below. The forces of the explosion pushed elements of debris a further 11 miles up out of the atmosphere. It was not the explosion that destroyed *Challenger*. Pushed out of its flight path, the vehicle could not withstand the huge aerodynamic pressures on its structure, and it simply broke up.

Recovery of onboard tapes and examination of the crew module revealed some interesting and startling facts. Most prominent was that some, if not all, of the crew survived the initial explosion and break-up of the vehicle. Evidence indicated that on the flight deck, at least three emergency oxygen packs were activated and, indeed, used. Smith's was found to be between three-quarters and seven-eighths depleted. Two others were also activated, but the crew-members were unidentified, and a fourth – Scobee's – was not activated. The fact that the crew was not wearing pressure suits would have meant that they would have lost consciousness 6–15 seconds after the explosion in the event of a cabin depressurisation. If the integrity of the module had not been impinged, however, they could have remained conscious much longer – perhaps even to ocean impact.

Dr Joseph Kerwin – Director of JSC Life Sciences and former Skylab astronaut – headed the investigation to determine the cause of death of the crew. It was soon determined that the stress of the explosion itself was too small to result in their death. In ten seconds, loads reached 12–20 g, then back to 4 g – survivable as past tests had indicated, with a low probability of serious injury.

Some windows were broken, but Kerwin's team were unable to determine the time or cause of this, to assist in accurate determination of the loss of cabin pressure. The exact cause of death of the crew could not be determined. However, it was evident that the crew was probably unconscious some seconds after orbiter break-up as a result of cabin pressure loss.

It was also determined that the crew module continued an upward trajectory which peaked at 65,000 feet around 25 seconds after break-up, and then descended towards the ocean surface, impacting at 204 mph, 2 min 45 sec after the explosion. This far exceeded the structural design limits of the compartment and the probable survivable limits of the crew.

When recovered, all crew were found harnessed in their seats, which indicated a rapid loss of pressure at around 48,000 feet. The cabin remained above that height for the next minute, and loss of cabin pressure would have resulted in rapid decompression and loss of consciousness, which could not have been regained before impacting with the ocean. Impact damage was so severe that no conclusive evidence for or against in-flight pressure loss could be determined.

The autopsy findings were never publicly released and, like other documents surrounding the final minutes aboard the last flight of *Challenger*, as well as details of the recovery of the crew module, this information remains classified.

In outlining the course of action needed to restore NASA and Shuttle flight operations, the Presidential Commission cited nine recommendations that were to guide the space agency and aerospace community over the next two years.

*Recommendation 1: SRB Design* Marshall Space Flight Center headed a redesign board to review the design of the current SRB and to instigate improvements to the design of the system for Mission 26 and beyond. These redesigns (and an extended test programme) took up the next two years, as several firings of redesigned SRB took place at Morton Thiokol facilities. Several setbacks were encountered that slipped the planned launch of mission 26 several months, but these new designs were finally proven successful with the launch of mission 26, *Discovery*, in late 1988. Many more launches were needed to finally dispel fears of a second SRB failure. In addition, new designs of liquid boosters were also investigated for future use, but it was agreed that solid boosters were the only way to launch the Shuttle for the foreseeable future, and so the design improvements had to work immediately.

*Recommendation 2: Shuttle Management Structure* Over the two years following the *Challenger* accident, several significant changes were implemented to place many former and current astronauts and NASA managers in prominent and essential positions in the 'turnaround' process that must be completed prior to any Shuttle launch.

*Recommendation 3: Critical Item Review and Hazard Analysis* NASA instigated a review programme that led to the establishment of new levels of Critical Items within the Shuttle programme. NASA had identified these as Criticality 1, 1R, 2 and 2R. In Criticality 1 there were certain elements of hardware in which there were no back-ups which included components such as the wings, landing gear and control surfaces, and if these failed the orbiter and its crew would be lost. The number of items listed as Criticality 1 totalled over 700! A new review of these levels and hardware certification for flight would be conducted by the agency and its contractors to identify potential risk.

*Recommendation 4: Safety* NASA established a new level of safety, reliability and quality assurance, and assigned former astronauts to work in this office to assist on crew recommendations and requirements.

*Recommendation 5: Improved Communications* NASA established guidelines to improve interdepartment and interagency communications and those between the agency and contractors. Lack of communication had been cited as a factor contributing to the accident, both in awareness of the O-ring problem and the effects of severe weather on hardware.

*Recommendation 6: Landing Safety* Not directly related to Shuttle 25, but (as Young had noted) for some time concern had been raised into the effectiveness of the Shuttle landing gear, brakes and landing sites. This was especially directed at the Cape, where bad weather frequently forced a landing at contingency and secondary sites. Several astronauts voiced concern in this area at the same time as the investigations of launch safety, and NASA took positive steps on the whole subject of landing a Shuttle. A step in this direction was completed in 1987, when *Enterprise*, the non-flightworthy vehicle used for ALT, completed ground arrest tests at Dulles Airport, Washington.

*Recommendation 7: Launch Abort and Crew Escape* On 7 April NASA initiated a Shuttle Crew Egress and Escape Review. This review of feasibility studies, tests, costing and implementations, evaluated adequate crew escape system for certain launch and landing situations, in addition to a review of launch and landing abort modes. After extensive evaluation, a proposed 'tractor rocket personnel escape' method was eliminated in favour of a slide-pole, which was made available for

Shuttle mission 26. As a result of the *Challenger* accident, the French Hermes shuttle design was subsequently configured for a crew escape module, but with severe launch/weight penalties. (Hermes was eventually cancelled.) This approach was not practical for the American design. A more comprehensive method of escape would have to wait for the next generation of American Shuttle vehicles.

*Recommendation 8: Flight Rate* NASA was often criticised for pushing to the limit the rate it launched Shuttles to meet launch demands and political/media comments. NASA always stated that it launched the Shuttle only when safe to do so, but the Commission revealed that sometimes launches were made near the limits of launch teams and supplies. Spares were minimal and flight rates were in excess of practicability. To alleviate this, NASA opted for a mixed fleet rate, and a new orbiter was ordered. Payloads were changed, and the role of the Shuttle as America's sole satellite launcher was abandoned. Indeed, the USAF reverted to launching its own satellites by expendable launch vehicle, and the promise of serving military officers flying on the Shuttle as Payload Specialists after flights resumed was actually abandoned just a few weeks before STS-26 flew.

*Recommendation 9: Maintain Safeguards* Turnaround activities, spares, maintenance of orbiters and other flight hardware and launch-related equipment was reviewed, and new standards were initiated to support the new flight programme.

In addition, several other activities were initiated, taking advantage of the enforced grounding of the fleet. Changes were made in main engine performance* by downrating them from a peak of 106% for a 109% target to approximately 104% maximum on future flights. Enhanced crew training in areas of safety, and escape methods from the orbiter, were incorporated in the mission preparations. Pre-launch preparation improvements saw the implementation of weather protection enclosures around the Shuttles while on the pad, as well as static escape procedures both at the pad and off the vehicles.

Work was also adjusted on the developing space station programme, and the incorporation of new elements of crew safety and rescue in light of the *Challenger* accident were all continued. Work also proceeded on designs for the next generation of the Shuttle, and on in-depth reports on the future of the US manned and unmanned space programmes.

In September 1988 the long awaited return to flight came with a textbook mission 26. A successful DoD STS-27 followed later that year, and by 1989 and the third anniversary of the *Challenger* accident, NASA was indeed again looking ahead to a promising future.

However, problems with flight preparation, launch constraints and budget restrictions remained. America had returned to space after *Challenger* – but it needed to do so many more times before the image and memory of that fateful January 1986

---

\*   When designed, Shuttle engines were rated at 100% thrust, but in actual performance they operated better than designed. Thus the figures of 104%, 106% and 109% were above the design qualification based on flight performance.

day would be put to the back of the minds of the media and general public, as had the Apollo flights after the Apollo 1 fire. However, like the loss of President Kennedy and the Apollo 1 astronauts, the *Challenger* tragedy would remain a very emotional recollection for those both inside and outside the space programme.

Unlike the untelevised accident of Apollo 1, the much-recorded flight of 51L was one of the lasting images of the 1980s, and very probably of the century. Like so many other shattering events in history, the loss of *Challenger* and her crew of seven will long be remembered – and rightly so – as a tragic demonstration of the price of the conquest of space.

# Summary

The first forty years of human access to space have seen the use of chemical rocket engine technology to enable us to progress from the surface of our planet to the surface of our Moon. The next forty years will see the continued use of this technology as other types of propulsion are developed, but it will be many decades – perhaps another century – before an alternative method of escaping the gravity of Earth, to routinely and safely travel to and from space, is developed.

Without doubt the most dangerous part of a flight into space lies with the type of rocket that is to take you there. The early days of the programme that saw the development of former missiles to launch the first humans into space, and the creation of huge rockets such as the Saturn family, have been replaced by the semi-reusable Shuttle system. However, throughout these years of spaceflight the Russian R7 launch system has continued to perform a key role in providing access to space.

All vehicles that have been developed for carrying a human crew into space have been remarkably reliable, with only one crew being lost in a launch accident in more than 220 launches by the end of 1999. During the early years of the American programme, the direct involvement of astronauts in visiting contractors involved in the construction of the launch vehicle they were actually going to ride into space gave an added sense of responsibility and purpose to every worker. This helped to ensure that the workers' personal input, no matter how small, would be their very best. At the time of the race to the Moon in the 1960s, although on a tight deadline – the factors of cost and weight were critical – the amount of dedication was unlimited. With this dedication came safety as each worker knew that the life of an astronaut rested, in part, with the quality of his own workmanship, and it soon became a matter of pride that a mission would not fail because of the lack of attention to detail. That desire to be part of the nations' space programme often involved working hours far exceeding the normal requirement.

The team required to launch a Saturn V numbered several hundred, but this dropped to around 100 for the Shuttle – a less complex vehicle, but still demanding in its own way. For every American launch there was a way out for the crew to escape before the vehicle left the pad, or during stages of the ascent to orbit. For Mercury and Apollo this was the escape tower, and for Gemini and flight tests of the

Shuttle it was ejection seats. From the fifth to the twenty-fifth Shuttle flight, escape in flight involved flying one of several launch abort modes or contingency flight profiles, one of which was demonstrated on STS 51F, the nineteenth Shuttle launch. The twenty-fifth launch of the Shuttle created the requirement to provide a further method of personal escape for each crew-member, and the selected system of a slide-pole exit has been available since STS-26.

The Shuttle has also demonstrated how a complex system can be prone to several unrelated system malfunctions and technical difficulties, and one that – no matter how precise the mission planning and pre-launch preparations – could be very frustrating for the crew, controllers and customers alike. However, these apparently too numerous launch holds and delays are also part of the in-built safety system that is aimed at protecting the crews' lives. Although run by computers and micro-circuitry, if the launch system believes the vehicle is not ready, it will not launch.

In Russia, on the other hand, the same basic vehicle has launched every cosmonaut into orbit since Gagarin (although a few have launched on the Shuttle from Florida since 1995). In more than 85 R7 human launches, only two have failed – one on the pad in 1983, and one in flight in 1975. In both instances the crews were successfully recovered. The reliability of the R7 has been the mainstay of the Soviet/Russian space programme, and the learning curve of operating these vehicles has allowed the Russians to predict with some accuracy the exact moment of launch weeks in advance.

It was during the two Voskhod missions in 1964 and 1965 that there was no emergency crew escape system during launch. Had anything gone wrong with the launch vehicles and the capsules could not have performed a normal parachute recovery sequence, the crews would have perished.

In contrast to the ultra-clean, vehicle preparation areas of American launch complexes, the Russian system was more basic during the days of the Soviet Union. Indeed, one Western observer who questioned how the layers of dust and soil deposits blown by the wind off the steppes of the Baikonur cosmodrome in Kazakhstan, onto the rocket stages, was dealt with, was simply told that it fell off when the vehicle was launched! Such was the confidence in the system and vehicle. This was relatively simple, reliable technology that worked – and it worked well. This also boosted the confidence in the cosmonaut crews that rode them into orbit. The fact that the launch escape systems had not only been fully qualified under normal test conditions and was actually used on the 1983 pad abort, and that a second crew had survived a launch abort in 1975, gave added confidence in the Soyuz launch system.

There are instances during a launch that rely on the experience and skills of the ground controllers or the flight crew. The 1983 Soyuz pad abort was triggered from a nearby command bunker and not by the crew, and the support of the launch teams and mission control teams during the Apollo 12 and 13 launches, and several Shuttle ascents, notably 51F, have shown that training and team work play an important role in any space mission from before the moment of launch.

Perhaps the coolest response to a launch mishap remains with Walter Schirra on the 1965 Gemini 6 launch attempt, when sensing that the vehicle had not left the pad after Titan ignition and immediate shutdown. In remaining calm and by electing not to initiate the crew ejection seat sequence which would have delayed the launch beyond the flight of Gemini 7, the intended rendezvous target, their mission was saved. That level of decision is no longer in the hands of the crew as they sit on the pad.

In training for a launch, the chance of an abort is in the minds of every member of the crew. Astronaut Mike Collins summed up the awareness of a crew in the desire for a successful launch first time, against the effects of an abort to safety in the event of a malfunction. He wrote that during launch preparations for Apollo 11 he noticed the pressure suit of Armstrong rubbing against the abort handle, and could imagine the following day's headlines, if it was inadvertently triggered, as 'Moonshot falls into the sea; Armstrong's reported last word was "Oops".'

# Survival in space

# Overview

The previous section described how the launch phase is the shortest but perhaps the most dangerous part of any spaceflight. But actually travelling in space brings with it another set of problems. The very environment of space is hostile to both humans and their machines. The near-vacuum conditions, extremes of heat and cold, radiation, microparticles of cosmic dust and meteoroids, and the ever-growing problem of space debris, are very real and potentially terminal dangers.

For the crew, this is when the mission really begins. Years of training and preparation are rewarded by the sights seen through the window and by the feeling of weightlessness as the fun of zero g pushes any thoughts of danger to the back of the mind. The effects of suddenly entering this new environment, however, can often be felt and displayed by the onset of space motion sickness as the crew gain their 'space legs'.

Once in space, an abort or emergency landing becomes much more difficult than during the launch or landing phase, because returning the vehicle back to Earth requires correct orientation and re-entry. For a crew living and working in the vacuum of space, as with all aspects of a mission, training and contingency procedures are available for most situations. But there are some scenarios where there are no options, and death is the inevitable outcome.

Had the ascent engine of the Apollo Lunar Module (LM) not ignited to leave the Moon's surface, the two astronauts would have been stranded until their oxygen was depleted, with no hope of rescue. Equally, a correct firing of the service propulsion system (SPS) engine on the Service Module mother ship was the only chance to bring the crew home from lunar orbit. Luckily, no such failures occurred during the Apollo lunar missions. Despite this exceptional run of luck, the near-disaster of Apollo 13, coupled with budget restrictions due to other pressures at home and from abroad, saw NASA cancel further lunar missions beyond Apollo 17 and abandon all hope of a human base on the Moon before the end of the twentieth century.

However, despite some close calls, in the first four decades of human exploration of space the dreaded headlines of 'LOST IN SPACE' have yet to be added to an epitaph of a space explorer. Of course, the ill-informed news media have repeatedly called upon banner headlines of similar nature to sensationalise some less life-threatening mission events.

In-flight mishaps fall into four categories:

- Medical problems or physical injury of crew-members.
- Equipment failures in which hardware does not work as programmed – or, at least, the telemetry indicates that it does not work.
- Collision with another space vehicle or space debris, causing a major systems failure, decompression or, at worst, total destruction of the spacecraft. This is most likely when spacecraft are brought together and linked – rendezvous and docking.
- Incidents during spacewalks (extravehicular activity – EVA) outside the protective cocoon of the spacecraft.

## MEDICAL INCIDENTS

Medical problems have affected the in-flight performance of several missions, both American and Russian, although not until 1985 had a flight been threatened with termination by the illness of a crew-member. One of the more common incidents of illness is the adaptation of the body to the environment of space itself. Space adaptation syndrome (SAS) – or, to give it its more common name, space sickness – has been likened to travel sickness, although there are significant differences. No two persons are alike, and there is no way to predict who will become sick in space. Astronauts known to suffer travel sickness or motion sickness on Earth have never reported SAS. Equally, some of the top pilots, who suffer no effects from the most rigorous jet flights, are regular sufferers. Astronaut Bill Pogue is a good example. A former USAF Thunderbird aerial display pilot, he never became sick when flying an aircraft, but often made his passengers queasy or just plain scared! When he flew on his only space mission, in 1973, SAS affected him more than it did the other two crew-members. Known as Iron Belly in the astronaut office, he was one of the least likely to suffer – or so it was thought.

Normally, SAS affects only a small percentage of spacefarers, and for only the first two or three days. By then, the inner ear and body accept the situation, and the person overcomes the symptoms of vomiting and a feeling of being 'full in the head'. On most Shuttle or space station missions nowadays, experience of past SAS cases results in light duties and no EVA for the first 24–48 hours of a mission. If such a requirement is needed or is important for mission success, then a crew who has proven immune to SAS is usually selected to fly (such as on STS-26).

Early spaceflights were very short. Crew compartments restricted movement, and crews were medically examined and trained to peak performance. Cosmonaut Gherman Titov suffered the first reported case of SAS, in August 1961, and several small incidents followed over the years. Studying the effects of spaceflight on the human body was one of the reasons for going into space in the first place, and there have been many case histories to investigate.

During the Apollo programme, astronauts were able to leave their seats and move around the capsule for the first time. This brought on some bouts of sickness, until

Jim Irwin, Apollo 15 Lunar Module Pilot, whose hard training schedule and strenuous work on Apollo 15 gave cause for concern. He had his first heart attack in 1973, and died in 1991 – exactly 20 years after the splashdown of Apollo 15.

they learned that slow movements and no sudden head twists helped alleviate the onset of SAS. The physical condition of astronauts on early long-duration flights was also a problem until exercise equipment was developed to help condition the crew during flights.

The Apollo 7 crew had such bad head-colds that they refused to wear their helmets during entry, and came home pinching their noses to equalise pressure. Frank Borman was ill in 1968 on the first flight to the Moon, even though he had been fine for two weeks, two years earlier on Gemini 7. Rusty Schweickart was so sick on Apollo 9 that his EVA was delayed, curtailed and almost cancelled. Fred Haise was affected on Apollo 13, and developed a slight urine infection that made him very sick at the end of the mission (see p. 277).

Strenuous work also puts a strains on the heart. In 1971, during one Apollo 15 lunar EVA, high heart rates were recorded on both Dave Scott and Jim Irwin. The astronauts were not informed of the irregularities during the mission, which annoyed them. Medical tests on Irwin after the flight actually revealed bigeminy rhythm, in which both sides of his heart contracted at once. This was put down to the stress of mission training, as both he and Scott worked long hours in full pressure suits during several warm months. To compensate for heat loss they drank quantities of electrolyte solution which extracted potassium from their body systems. By the time they left Earth they were actually suffering from potassium deficiency which, when added to by the stress of their mission tasks, causes heat irregularities. NASA doctors hoped the two astronauts would soon return to normal levels of fitness, which they apparently did.

Scott also broke several blood vessels in his fingers in trying to get the lunar drill to penetrate the surface of the Moon, and strained his shoulder trying to extract the core tube. Irwin's heart problems returned on 4 April 1973, when he had his first heart attack at Lowry Air Force Base during a game of handball. He later attributed it to God's way of telling him to slow down. He had heart problems for the rest of his life, and in August 1991 he died from a heart attack.

In 1987, after one particularly strenuous EVA outside the Mir space station, cosmonaut Alexandr Laviekin's heart rate developed a hitherto undetected irregularity. It was several days before it returned to normal, and follow-on EVAs were delayed. After the first of two planned EVAs, the irregularity returned, and medical controllers decided to replace him with another cosmonaut for the remainder of the flight. Bitterly disappointed, he returned to Earth with the next visiting crew, leaving his replacement onboard to finish his work. He was later pronounced fully fit and was returned to flight status. In 1997, during the stress of the accidents on Mir, cosmonaut Vasili Tsibliyev suffered heart irregularities and was unable to perform planned EVAs. He, too, recovered after rest and the return from a very strenuous mission. However, neither cosmonaut flew in space again.

As flights became more frequent, the media showed more interest in space adaptation. With the openness of the American programme, almost everything was known about the crew in orbit – even to when they went to the bathroom, and how much they used it! Biomedical sensors were attached throughout early flights for the duration of the mission. Urine and solid waste was collected for post-flight examination, as were bags of vomit! Sometimes this irritated the astronauts, and they either removed the sensors or did not disclose bouts of illness to the ground or even on the private medical channel. By 1983 and the Shuttle programme, a code of privacy cloaked individual adaptation to space flight and, apart from the NASA doctors and the astronauts, no one knew who actually suffered SAS. More serious illness would become known if mission safety was compromised.

Crew compatibility is very important on long-duration flights or in the close confines of a space complex. In his autobiography, published in 1994, astronaut Deke Slayton – who selected the crews for early American missions – indicated that crew compatibility was very important both to the success of the mission and to the harmony and safety of the crew. In this the Soviets have years of experience, and crew-members are usually compatible, but in 1982, during the first mission to Salyut 7, Valentin Lebedev reported that he and his Commander Anatoly Berezovoi 'got on each other's nerves' during their 211-day space marathon. At times, they would not speak to each other for days on end, and would work at opposite ends of the space station, which was a single compartment only a few metres long!

As flights increase in duration and complexity and venture further from Earth, the problems of space debris and solar radiation shielding will become considerations for the health and welfare of the flight crew. Soviet space station missions have long been planned around periods of maximum solar activity, and with the continuous occupancy of Mir, monitoring of solar winds and solar flares are vital to mission planning. The added problem of space debris (as witnessed onboard Mir when it was pummelled by Perseid meteors and the solar arrays were holed) and microscopic particle penetration of spacecraft hulls and pressure suits are also concerns for

The original Soyuz T-12 back-up crew of (*left–right*) Vasyutin, Savinykh and Pronina. Vasyutin flew his first and only spaceflight as Commander of Soyuz T-14 in 1985 and suffered 'a psychological trauma' in space. (Astro Info Service Collection.)

extended endurance and deep space exploration. The effects and experiences of a multiracial, mixed gender and culturally varied crew on board the International Space Station will be of great importance to future planned human trips to Mars, where compatibility will be essential.

**1985: Vasyutin**

In September 1985 Cosmonaut Vladimir Vasyutin was launched as commander of a three-man crew to dock with Salyut 7, already manned by two cosmonauts who had just completed the repair of the crippled space station. After a period of hand-over, Vasyutin remained on board with two cosmonauts to command one of the last missions to the ageing space station before the new, modular Mir station was launched in 1986. Although he had been a cosmonaut since 1976, this was his first flight.

At first all seemed well, but by a mission duration of more than 55 days, things started to go wrong. After a week of official silence on the progress of the mission, with the cosmonauts reporting to the ground with an apparently coded system, an unexpected announcement of the immediate termination of the mission was made on 15 November. This was the first in-flight illness of a crew-member to result in the early termination of the mission. In 1976 the Soyuz 21 crew evacuated the Salyut 5 space station, but this was due to an acrid odour that resulted in an illness that affected the whole crew. The Vasyutin incident was different.

According to 'official' reports, Commander Vasyutin was suffering from 'a psychological trauma'. However, fellow cosmonauts indicated that he had been a bag of nerves for some time in orbit. His colleagues had tried in vain to calm him down to save the mission, before giving up and consulting the ground, who gave immediate instructions to return to Earth, even though conditions for landing were not ideal.

One report stated that Vasyutin had developed an infectious inflammation which could not be successfully treated in space. The absence of the standard Soviet statement on the health of the crew – 'The cosmonauts feel well' – hinted that there was a problem. It was reported that for some weeks Vasyutin had endured pain, discomfort and a high fever, and was sometimes confined to his sleeping bag. His condition became so serious that he could no longer effectively command the mission, and during the return to Earth, control of the capsule was assigned to the mission flight engineer, assisted by the research engineer. Once back on Earth, Vasyutin spent a month in a medical institute in Moscow, while the crew spent their normal post-mission isolation period near the landing site in Kazakhstan.

Rumours of an appendix problem affecting the cosmonaut were circulated in the Western press, and stories of the cosmonaut cracking up in the confines of Salyut during the long-duration flight also arose. It was thought more likely that the cosmonaut caught a more common Earthly viral infection prior to flight – one that the usually stringent medical examinations did not detect prior to launch. One source passed information to the Americans, who were interested in why such an emergency return was necessary, as they were planning their own long-duration space station missions on what was then called 'Freedom', a forerunner of the ISS. This source stated that Vasyutin had suffered 'a serious bout of prostates' which could not be treated on the station successfully with the available medical equipment. After his recovery, Vasyutin left the cosmonaut team to return to Air Force duties. His illness did not apparently affect his military career, as he rose through the ranks to become a Lieutenant General.

## EQUIPMENT ANOMALIES

A second element that threatens the lives of the crew in space is the failure of equipment or hardware – or at least the indication of failure. Several Shuttle missions have experienced small problems that inhibited closing of the payload bay doors or restraining equipment to allow safe entry, but these were all overcome. On the tethered satellite flights of STS-46 and 75, the fear of both mission planners and the crew was a snapped tether wrapping miles of wire around the Shuttle, necessitating an emergency EVA to free it.

Over the years, several incidents have highlighted the risks that the crews take in trusting the hundreds of items of equipment and circuitry that not only helps them to complete their flight but also helps to maintain their lives.

The effects of spaceflight on hardware can be seen in this view of the Apollo 16 LM ascent stage preparing to dock with the CM after lift-off from the Moon. Note the buckled thermal panels on the rear of the vehicle.

### 1962: Friendship 7

On 20 February 1962, astronaut John Glenn became the first American to orbit the Earth, inside the Mercury 6 spacecraft Friendship 7. Although only a short, three-orbit mission, it was to be an historic one for the Americans. But a potentially serious problem was detected before he completed his first orbit. One of the engineers at Mercury Control, KSC, reported a signal in Segment 51, which meant that the landing bag between the capsule and the heat shield had somehow inflated itself. Designed to be used after splashdown, it would deploy and fill with seawater to provide stability whilst the astronaut awaited recovery. If the reading was real, then the heat shield must be loose and the spacecraft, with Glenn inside it, would burn up on re-entry.

A contingency plan was available to solve this – at least on paper. The retro-rocket pack was held to the shield and the spacecraft by three straps. By leaving this during re-entry, it could last long enough to hold the shield in place, even if it had became dislodged. Concern was expressed that if only two of the three retros fired, then the third would be like a potential bomb waiting to explode. Then again, if only two retros fired, Glenn would not re-enter properly anyway. So the argument was soon settled: they decided to leave the pack on.

There was no point in worrying the astronaut unnecessarily, as there would be nothing he could do about it. During the third and final orbit, the CapCom advised Glenn of a possible landing bag deployment and recommended that he not jettison the retro pack – at least, not until he had passed over Texas during re-entry. As he passed over Hawaii for the last time, Glenn was asked to put the switch into the auto position to see if he obtained a light in the spacecraft. He did not, which indicated that the spacecraft systems showed the bag to be correctly stowed. As he fired the retros, Mercury Control monitored the signal light from the bag to determine whether the ignition shut off the light. It stayed on. There was nothing to do but wait. Adding to his tasks, Glenn had to manually override the automatic programme to fly the spacecraft through the re-entry phase. His final instructions were radioed up before the plasma ball created by the spacecraft entering the upper atmosphere blocked all radio contact. The ground waited... and waited... and waited.

CapCom (Alan Shepard): 'Friendship 7, how do you read?'
Glenn: 'Loud and clear. How me?'

A few minutes later he splashed down successfully in the Atlantic Ocean. Post-flight analysis revealed that the problem was nothing more than a faulty loose switch, and that the bag had not been loose.

Glenn later wrote of his experiences during the flight, and he highlighted the landing-bag incident. As he passed over the tracking stations around the world, he was asked a few innocent questions about the status of the heat shield. The CapCom, on the tracking ship in the Indian Ocean, gave him the first indication of suspected problems with the heat shield during his second orbit, when he was told to leave his landing bag switch in the OFF position. He thought ground control must have been obtaining some peculiar readings. A few minutes later, CapCom in Australia (astronaut Gordon Cooper) again asked him to confirm the position of the switch, and asked whether he had experienced any banging noises. This prompted him to think that ground was concerned, or they would not be asking such questions. Glenn

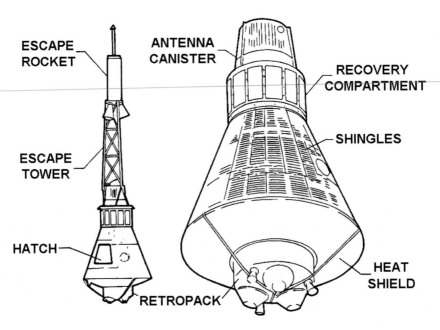

The Mercury spacecraft, showing the position of the retro-rocket pack strapped to the capsule.

himself was not too worried. He felt sure that if the shield was loose and shaking behind him, or banging against the hull, he would be feeling the effects inside Friendship 7 – and he was not. There was also no drifting back and forth to correct any such movement. However, there still remained some concern for re-entry and the safe functioning of the heat shield. 'This was the only thing that stood between me and disaster as we came through the atmosphere. If it was not tightly in place, we would be in real trouble.'

Towards the end of his flight, passing over Hawaii as he stowed equipment, the CapCom there asked him another question about the heat shield. He was informed of the Segment 51 signal and that the ground thought it was erroneous. He was asked to check the onboard switch to see if he obtained a 'deployed' signal. Glenn wrote, 'I thought this over for a second. This was a tricky situation and I was a little reluctant to try it. What if the bag had not deployed, but decided to do so when I activated the switch? We [may] have jumped from the frying pan into the fire. So I rapidly switched on and off again. The light did not come on – a pretty good indication to me that we were in good shape.'

Five minutes after retro-rocket firing, Glenn hoped he would be asked to jettison the pack. It was unknown what effect the pack may have on even heat distribution over the shield. It may have resulted in hot-spots that could penetrate the protective ablative coatings as well as upsetting other automatic sequences. He was therefore concerned when the Texas CapCom informed him to leave the pack on during the entire re-entry sequence. When he asked why, he was told that it was the judgement of Cape Flight Director who would tell him later. He finally learned for certain what the problem was from Shepard at the Cape station. It was now too late to worry about it, because the spacecraft was heading towards re-entry.

As he re-entered, Glenn heard a considerable thump from the capsule behind him, and he was sure the retro was breaking away. He reported this to Shepard, but the ionised layer around the capsule had built up and he was not heard. Shepard had also told Glenn to jettison his pack as g force built up, but the astronaut failed to hear this message. 'This was normal and I expected it to happen, but it left me more or less alone with my little problem.' A few seconds later he saw one of the restraining straps of the pack go flaming past his window as the bright orange glow of re-entry built up. He saw big flaming chunks of debris 6–8 inches across flash by, bumping against the capsule behind him before they took off, and he thought that perhaps the shield was breaking up after all. He knew there was nothing he could do but try to keep the capsule under control and sweat it out. 'This was a bad moment. I knew that if the worst was really happening it would be all over shortly. I would feel the heat pulse first at my back, and I waited for it. Pieces of flaming material [later identified as the pack] were still flying past my window [and although] it lasted for only about a minute, those few moments ticked off inside the capsule like days on a calendar, and I still waited for the heat.' It never came. The temperature diminished, and a safe recovery was achieved. The event, and Glenn's coolness, helped forge the image of the hero astronaut and the legend of John Glenn for all those who followed him.

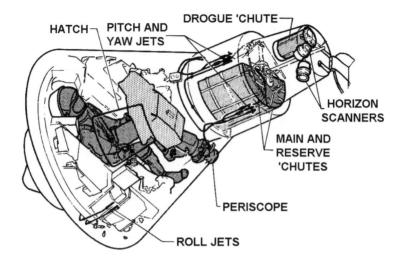

Cutaway of the Mercury spacecraft showing the astronaut's position and associated equipment.

### 1969: Apollo 10

In May 1969 the crew of Apollo 10 – Tom Stafford, John Young and Gene Cernan and – flew the second Apollo mission to the Moon. Their primary objective was a test-flight of the Lunar Module to within nine miles of the lunar surface, and a simulated abort profile lift-off, paving the way for the first lunar landing mission – Apollo 11 – the following July.

The crew of Apollo 10 – Stafford, Young and Cernan – proudly display their mission emblem showing the LM ascent stage swooping over the lunar surface to rendezvous with the Command Module. The LM mascot, Snoopy, sits in front of them.

On 22 May, after three hours of troubleshooting small niggling problems in the lunar module (code-named 'Snoopy'), astronauts Stafford and Cernan were ready to undock from Young alone in the command module ('Charlie Brown'). As the docked spacecraft slipped behind the limb of the Moon for the twelfth time, the astronauts received the 'go' for undocking over the far side. In the LM, Stafford was at his station on the left of the cramped cabin, ready to control the descent engine and attitude control system to fly Snoopy, while Cernan, on the right, would monitor and call out altitude and velocity readings as they descended. Both men were wearing full pressure suits and 'standing' facing their instruments, suspended in microgravity by a spring-loaded tether harness.

Over the far side of the Moon, Young undocked the command module, letting Snoopy begin its six hours of independent flight. An hour after the undocking the descent engine of Snoopy was fired to put the LM in an elliptical orbit (descent orbit), taking them on a long coasting flight from the 69-mile orbit where Charlie Brown would wait for their return to the 9-mile (47,500 feet) point, where on Apollo 11 the descent engine would be re-ignited to begin the final approach to the surface. On Apollo 10, however, at the nine-mile point they would begin to climb to a new orbital high point of 215 miles before beginning a second descent over the approach – Apollo Landing Site 2 on the Sea of Tranquillity.

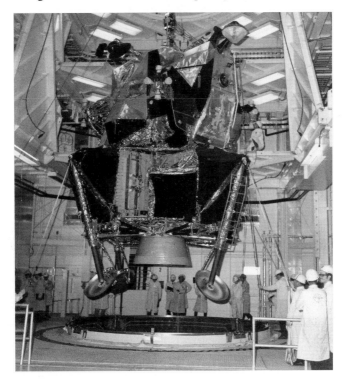

Final preparations for loading the Apollo 10 LM (flight vehicle 4) Snoopy aboard the Saturn V at Kennedy Space Center.

An artist's impression of the undocked LM – containing Stafford and Cernan, and monitored by Young in the CM – beginning its descent across the surface of the Moon.

During the next three hours the two astronauts put the LM through its paces as it exchanged altitude for speed and swept over the surface features at almost five times the speed of sound – 3,700 mph. For Stafford and Cernan, both former military jet pilots, this was flying at its best.

At the end of the second pass over Landing Site 2, Stafford and Cernan prepared to use the LM descent stage as a launch platform, separating the two stages of the lander and firing the small ascent engine to take them in a fast ride to rejoin Young in Charlie Brown, then about 300 miles ahead and 50 miles higher. The two astronauts planned to use the abort guidance system (AGS – 'Aggs') which was an auxiliary navigation to be used by future LM crews as they neared the surface. It was also the back-up system to be used in the event of an abort during a landing approach when the main computer handling the primary navigation guidance system (PNGS – 'Pings') failed. Stafford and Cernan had successfully tested the 'Pings' system, and it was time to test the 'Aggs' back-up system.

Both men were fully expecting that the ignition of the ascent engine would give them a firm 'kick in the pants'. During hundreds of training simulations they had run through the setting-up of the control switches for the ignition so often that they no longer looked up to check which switch they had operated. This 'second nature' approach would come back at them in a very sudden and violent way. Cernan had reached up and commanded the guidance and navigation system from 'Pings' to 'Aggs', as planned. Seconds later, and without looking, Stafford reached for the same control and inadvertently reversed the command, thinking it had not yet been changed from the primary to the abort mode. Cernan had not noticed that his command action had been reversed, and with both men totally unaware of the error they initiated the ignition of the ascent engine and triggered four explosive bolts to separate the two stages.

Cernan later observed that as they separated, 'all hell broke loose'. Snoopy went completely wild on them, trying to move in three different axes at once and pushing the gyroscopes which sense past their designed limits and beyond their automatic correction capability. Called 'gimbal lock', the navigation system suddenly became useless until it was reset. Travelling over the lunar surface at 47,000 feet and 3,000 mph, Snoopy's smaller and more agile ascent stage was spinning, bouncing and gyrating wildly as both astronauts fought to regain control. Over an open microphone, Cernan inadvertently yelled: 'Son of a bitch, what the hell happened!?'

Stafford immediately thought that 'Aggs' was at fault, and flipped the switch, he thought, to the 'Pings' system – but he actually placed it back into 'Aggs' where it should have been. This put the system into even more confusion and made the situation worse. Unable to lock onto the command module, the radar had searched for another target on which to lock on. It selected the Moon! As the blackness of

The ascent stage of Snoopy finally rendezvous with the CM after the astronauts regain control.

space alternated with the approaching lunar surface as Snoopy spun around, Stafford urgently informed the ground that they were 'in trouble' and were approaching gimball lock. Thinking they had a stuck thruster (similar to the Gemini 8 incident in March 1966 – see p. 261), Stafford overrode the computer and took manual control, which almost immediately cancelled the guidance programme confusion and, as suddenly as it had started, the problem disappeared as the astronauts regained control.

The whole incident had taken less than 15 seconds, during which Snoopy had made eight complete rolls. It was later estimated that had the spinning continued for only two more seconds, they could not have recovered control before heading for an impact with the surface of the Moon, from which there would have been no escape.

Stafford informed Houston that they still retained 'all our marbles' as they headed back to an uneventful docking with Young in the command module. The objective of these early Apollo flights had been to 'shake-down' the machines, procedures and men before commitment to a landing attempt. With Apollo 10 this 'shakedown' was taken to the limit, and the path for Apollo 11 had been cleared in a very dramatic way. In future more care would have to be taken to ensure that commands inserted into the controls by one astronaut were not overridden by another.

### 1972: Apollo 16

On 20 April 1972, astronaut Ken Mattingly was flying the Apollo 16 Command and Service Module (CSM) Casper, alone in lunar orbit. Having just undocked, he was monitoring the departing Lunar Module (LM) Orion carrying John Young and Charlie Duke towards a powered descent for the fifth lunar landing, in at the Descartes highlands. Mattingly was preparing to fire the Service Propulsion System (SPS) engine at the back of the Service Module (SM) that would circularise his orbit to 69 miles for the duration of the landing and surface operations. He had taken the docked combination down to an elliptical descent orbit of 69 x 9 miles for the LM separation manoeuvre. He had already routed electrical power to the engine and activated the gyros, and was checking the secondary control mode. But when he moved the yaw gimbal thumb-wheel which controlled the engine gimbal motors, he felt his spacecraft shake around him. He noticed the artificial horizon instrument (the 8-ball – a black and white 'cue-ball' type display that revolves around as a spacecraft manoeuvres about its axis) moving back and forth, but when he removed his hands the shaking stopped. After resetting some switches he tried again, and the shaking returned. After a third attempt he knew he had a sick spacecraft, and he reported the situation to his commander in the LM.

Young – then one of the most experienced people to fly in space – had no answer to the problem. Flying a thousand feet away, he was unable to see the engine and was not experiencing the shaking. Suddenly, Mattingly felt very alone. He was a CSM specialist who had been removed just 72 hours before the Apollo 13 mission, two years earlier, due to a measles scare (which was a false alarm) and then worked tirelessly in the CSM simulator helping to return the crew after the explosion. Despite long hours in training, this was a failure with which he was not immediately familiar. With no SPS engine control, mission rules were clear. He had to rendezvous with the LM, and the landing had to be delayed until Houston had reviewed the situation.

For several hours the two spacecraft flew in formation, using precious fuel and time, while Mission Control reviewed telemetry and consulted with NAA-Rockwell engineers at Downey and specialists at MIT. The crew feared their landing was over before it had begun. Ironically, Mattingly, having had the opportunity to review his previous experiences in CSM systems while alone in the orbiting CM, thought he knew what the problem might be. He remembered hearing about a test of the engine years before in which the cable that carried signals to and from the gimbal motors was a little short. It resulted in pins being pulled from electrical connectors when the cable pulled taught. If this was what had happened to the secondary system, how could they ensure that it would not happen to the primary one, as all control cables were routed the same way?

On the ground the investigation of data received from Casper revealed that despite the shaking, the steering signals were in fact reaching the engine. Rockwell engineers had fed the data into a test unit of the SPS, and concluded that if the crew had to use the secondary system the engine may well shake, but it would be controllable. During the Apollo 9 mission in 1969, the crew had also caused the SPS to vibrate when fired in one of many test objectives of the Earth orbit shakedown flight, and it had still performed well. The management decision was that it was safe for the mission to proceed.

The crew was jubilant on hearing the news. The burn proceeded safely, the CSM reached its new orbit and the LM crew landed safely on the Moon. With both control lines in the same cable, Mattingly had been sure the mission would be cancelled and expressed surprise when they proceeded. He was later teased that he was the only one who knew about the control lines being routed this way, and that if the ground had noticed then they would not have proceeded.

Post-flight analysis of the problem pointed to a cable-flexing problem, as Mattingly had thought. Routing of the cable harness to the actuator assembly was changed to prevent strain and flexing in the remaining Apollo missions and in the CSM assigned to Skylab and the Apollo–Soyuz Test Project. The primary system showed no indication of failure, and if the mission had been aborted it would have been because a secondary system revealed a probability of failure, losing the level of redundancy and infringing the crew safety rules

**1973: Skylab 3**

On 28 July 1973, during the approach of the Skylab 3 crew (Alan Bean, Jack Lousma and Owen Garriott) to the Skylab Orbital Workshop, at the start of a 59-day mission, one of four manoeuvring sets of thrusters (termed a 'quad', as there are four sets of four thruster packs around the circumference of the Apollo Service module) began to leak oxidiser on the Apollo Service Module and had to be shut down completely. Though presenting little danger to the crew, as they had twelve working thrusters left, when a similar problem appeared in a second quad on 2 August, with the CSM now attached to the Skylab OWS, it prompted NASA to plan a possible rescue mission to retrieve the three Skylab astronauts. Although two sets were

The Apollo Command and Service module docked to the Skylab space station during the Skylab 3 mission. One of the quads of thrusters began leaking oxidiser and had to be closed down. (Astro Info Service Collection.)

adequate for positioning the CSM for re-entry at the end of the mission, the reliability of the entire system came into question.

The spacecraft planned for Skylab 4 was moved to the pad early, with five 'seats' inside instead of the usual two. Two astronauts (Vance Brand and Don Lind) would have been launched to dock with Skylab's second docking port and bring back the three resident astronauts. This would have become the first rescue mission in spaceflight history, the two astronauts flying a stripped-down spacecraft to take into account the additional mass of the extra three astronauts and returned experiment results. However, analysis of the problem thrusters rendered this action unnecessary, and Skylab 3 continued for a record mission duration and the safe recovery of the crew. NASA decided that although faced with two identical leaks, they were unrelated and unlikely to spread to the remaining two quads. In addition, ground simulations showed that eight thrusters would be enough to bring the crew home.

### 1976: Soyuz 21
In 1976, during the first occupation of the military space station Salyut 5, cosmonauts Boris Volynov and Vitaly Zholobov were progressing though their planned 66-day mission when, after 48 days, Radio Moscow suddenly announced their immediate return to Earth. This was in sharp contrast to normal announce-

ments of a crew return around six days prior to the event. The crew began medical preparations for their return, packing up experiments and preparing to mothball the Salyut station. They landed at night (a first for the Soviets), just two hours after the announcement, and at their post-flight press conference the next day they reported that their mission had been 'Interesting but complicated' – which was interpreted as meaning that they had had problems. It was later reported that the two men suffered 'sensory deprivation', but with a low-key return the whole mission was played down in the media. A hint of what the cosmonauts experienced during orbit was offered many years later. They had suffered a combination of acrid odours in the spacecraft atmosphere. Zholobov fell ill, and both men suffered fatigue and tension.

International Space Station first elements linked together in orbit. It was during the STS-96 flight in the summer of 1999 that the crew encountered an unexpected odour inside the Russian Zarya Control Module (*right*) docked to the US Unity Node (*left*).

### 1999: STS-96/ISS

Twenty-three years later, in the summer of 1999, during the second Shuttle–International Space Station mission, a further incident of 'bad air' on a space station was to affect the activities of the crew. Launched on 27 May 1999, STS-96 docked with the ISS (Zarya/Unity combination) two days later, and on 30 May the seven-person crew entered the inside of the ISS for the first time.

For more than three days the crew had access to the inside of the ISS through the open hatchways and docking tunnel on the Shuttle. Although they worked on the ISS, their living quarters remained on the Shuttle mid-deck. After 79 hours and 30 minutes inside the station, and 5 days 18 hours docked to it, the Shuttle undocked on 4 June to complete what appeared to be a highly successful and incident-free mission with a landing on 6 June.

It was during the scheduled stowage and transfer post-flight crew debrief activities that a serious crew health issue came to light. During flight days 6, 7, and 8 (June 1, 2, and 3) several of the crew had experienced symptoms of headaches, burning or

Canadian astronaut Julie Payette works in the Zarya module. Her smile disguises the
fact that the crew were suffering from effects of 'bad air' inside the new space station.

itching eyes, a flushed face, nausea, and in one case actual vomiting, which had been
linked to the degraded quality of the air inside the station during the times they were
inside the ISS. These symptoms, were apparently not reported by the crew at the
time, but two weeks later during the post-flight debriefs.

If there were to be a serious degrading of the quality of the atmosphere on the ISS,
then this would need to be resolved before sending the next crew (STS-101) up to the
station. Space Station Operations Manager astronaut Frank Culbertson established
a six-person investigation team from staff at the Johnson space Center, Houston,
and headed by Dave Herbek of the ISS Program Office (ISSPO) of Mission
Integration. Other team members were Shuttle Flight Director Phil Engelauf of the
Mission Operations Directorate, Flight Surgeon Dr Phil Stepaniak of the Office of
Space and Life Sciences, Larry Gana from the Safety and Mission Assurance Branch
of the ISS Program Office, and Tony Sang from the Vehicle Office, Environmental
Control and Life Support Branch (ISSPO). Astronaut Kalpana Chawla represented
the Astronaut Office (Code CB).

The team conducted a 1½-hour special debrief of the crew, who expressed some
concern with the violation of their medical privacy rights and Freedom of
Information Act (FOIA) implications from the release of findings from the teams.
The Team Flight Surgeon (Stepaniak) reviewed each crew-member's medical data on
an individual basis, and provided only a relevant and generic overview to the rest of
the team. A FOIA exception waiver was prepared by the JSC Legal Office, but was
not used, as it was determined by Culbertson and Herbek that the crew had not
supplied confidential medical information in their report of the incident.

The investigation team also reviewed all post-flight reports and analysed all
available data, including air-to-ground commentary and onboard video, in an

To escape from the effects of the atmosphere inside the space station, the STS-96 crew sought refuge inside the crew compartments of the Shuttle by traversing the connecting tunnel between the Shuttle and the International Space Station.

attempt to determine the cause and source of the discomfort. The summary was completed in July 1999, but NASA did not initially release the completed report. However, information emerged from various sources on what was found and recommended as a corrective measure as a result of the investigation.

The crew was affected while working on the activation and stowage activities in the Zarya Control Module (in Russian, FGB) or to a lesser extend in the Unity docking node. Over several days (FD6–8) it was generally observed that the symptoms occurred when two or more astronauts were working either in close proximity for several hours or behind opened panel doors in enclosed spaces in the FGB, and usually towards the end of the working day.

The crew had found that the symptoms appeared after they had opened the FGB panels or had remained in close proximity in the node for one hour. Relief was gained by going back into the orbiter, where the symptoms took 10–15 minutes to a couple of hours to allow the crew member to 'feel better'. Medical treatment from onboard medical kits controlled the vomiting, and the crew set up Russian ventilator fans when working behind the FGB doors, to help circulate the air.

The crew reported that the onset of the condition did not seem to begin during any particular event or task, and described a musty or solvent odour but were unable to find its exact source. They did state that when working under the floor of the FGB they did not encounter any discomfort. They suspected an unexpected outgassing of onboard materials, noted an extensive amount of Velcro in the FGB, and voiced concern about the levels of adhesive used.

The investigation team found that there was a real problem with the quality of the atmosphere of the ISS at the time that the STS-96 crew were docked to the station. The main problem in accurately pinpointing the source was in the delay of reporting the event to the ground. Had immediate information been available, then samples of the atmosphere could have been taken for analysis back on Earth to pinpoint the source of the pollutant.

Samples that were taken during initial ingress and final egress showed no significant increase in carbon dioxide levels, and previous experience from some of the crew members indicated that either the problem was a build-up of stagnant air where carbon dioxide is not dispersed away by the air circulation system but is left to accumulate near the astronaut, or a disruption to the air exchange and $CO_2$ scrubbing by the orbiter system. It was found that the equipment on board the orbiter worked normally throughout the flight.

The STS-88 crew – the first to work in the Zarya/Unity ISS configuration – reported no such symptoms, which pointed to a different airflow pattern on STS-96. The Space Hab cargo module – carried on STS-96 but not on STS-88 – resulted in a lesser flow of air by volume from the orbiter into the ISS. During these initial missions, ventilation in the ISS depends on mixing air in the node to effect the change of air between the ISS and the orbiter. Any disruption in air flow in the node area would prevent this exchange of air. It was also known that by design the volume behind the closed panels in the FGB provides a controlled path of air around the module. By opening the wall panel doors in the FGB, the flow of air through the module is disrupted.

As there was no permanent resident crew aboard the ISS there were no carbon dioxide scrubbers or air circulation systems installed. These tasks are accomplished by the docked Shuttle. When the next element, the Russian Service Module, is docked to the station, this will contain the air quality equipment needed to maintain a safe and healthy environment for the resident crew.

The team recommended that provisions be made to allow the crew to monitor air quality with personal monitors or local monitoring instruments. A re-evaluation of the air circulation system of the FGB and the SM should be conducted before the first increment (long-duration or resident) crew of Bill Shepherd, Yuri Gidzenko and Sergei Krikalev take up residence on the ISS, and who would not have an orbiter docked for a safe haven from a repeat of the STS-96 symptoms. The review also suggested that further investigation into the performance and design should also determine whether restrictions in duration or quantity of panel door opening should be controlled.

It was also recommended that several 'quick-fix' solutions be prepared for the next docking mission (STS-101). Other recommendations for later missions included a fixing of air ducts from the orbiter into the Service Module to push air to the rear of the stack with return through the station to enforce air circulation. This was used during Shuttle–Mir dockings. STS-101 would also carry additional air sampling canisters and carry out an evaluation of the installation and activation of a self-contained Vozdukh carbon dioxide scrubber which was used on Mir, replacing the lithium hydroxide units, which required constant canister change-outs.

The report also suggested that more free exchange of information between the ground and station crews would help identify problems as they occurred in order to act upon them more swiftly.

The Russians indicated that this outgassing was normal on new space station modules, and was not of great concern. Anticipated by the Russians for some time, it was reported to the Stafford–Utkin panel when they investigated the Mir docking and fire incidents of 1997, and was passed on to NASA. The STS-96 crew actually reported the incidents to the Russian flight controllers during the mission, but apparently the information did not permeate to NASA officials at JSC. In addition, one of the reasons that Boris Morukov – a doctor–cosmonaut from the Institute of Medical and Biological Problems (IMBP) – was pushed by the Russians for an early assignment to an ISS–Shuttle mission was that he could analysis the air aboard the module to assess the level of contamination before the resident crew took up residence. This was originally on STS-88, but (at the time of writing) is scheduled for STS-106 in the autumn of 2000.

With the addition of further habitation modules to the ISS over the next few years, the establishment of a healthy and safe working environment for the crew, early in the life of the station, is essential to the success of permanent habitation. Maintaining it will be more of a challenge.

## DOCKING INCIDENTS

One of the most important elements of spaceflight and space exploration is the ability to link one or more spacecraft together to form larger vehicles, or return elements from separate mission objectives. This is commonly called rendezvous and docking. It also includes manoeuvres in the vicinity of one or more spacecraft, called proximity operations (Prox Ops) – one of the trickiest elements of any mission. It relies on techniques of orbital physics and dynamics, the skill of the crew and the reliability of sometimes automated or inaccessible equipment. The techniques were developed out in theory many years before spaceflight became possible. The first rendezvous attempts during the US Gemini programme led to a series of dockings with unmanned vehicles in 1966, then with manned spacecraft in 1969. Since 1971, spacecraft have docked regularly with space stations, but as we have seen with all elements of spaceflight, events do not always proceed as planned, and the very real dangers of collision and damage make this technique very risky.

In bringing spacecraft together, care has to be taken to ensure that the force at the point of contact is enough to allow a capture of docking devices, but not too hard to bounce off the other spacecraft and perhaps cause damage.

Docking is usually achieved in four stages. The two vehicles first complete a rendezvous at a safe distance and keep station (remain in their relevant positions as they move in space). At this point – Prox Ops – the final alignment is confirmed, and checks on both vehicles are completed. The go-ahead for the next stage is then given.

The vehicles now complete the terminal approach. One vehicle (the target) acts as a passive object, remaining still as the other spacecraft (the active element) moves in for contact. In lining up the docking targets on both vehicles, the astronaut or cosmonaut (or auto-system) knows that the docking equipment on the spacecraft is in direct line with the receptive equipment on the other vehicle, and gingerly moves in for contact.

Initial contact triggers small latches that hold the two vehicles together (soft dock) so that they can dampen out any unwanted movement prior to initiating final docking. Once all is well the docking equipment is retracted, drawing the vehicles together, whereupon latches around the circumference of the docking hatch fire to achieve a rigid, air-tight seal between the two vehicles – hard dock.

### 1971: Apollo 14

Stu Roosa was known as a careful driver. In the nineteen months after being named as CMP for Apollo 14 he simulated more rendezvous manoeuvres to extract the LM from the top of the Saturn third stage than he cared to remember. On 31 January 1971 he was in space for the first time and was completing the procedure for real on his way to the Moon. He slowly moved the CSM Kitty Hawk about 100 feet away from the spent third stage, turned 180° and gently moved back towards the roof of the LM Antares, which was nestling on the top of the Saturn stage.

This was not a new procedure. It was one of several docking techniques that had been devised for the Gemini programme more than five years earlier, and it had been carried out on five previous Apollo missions. But being a space rookie, Roosa knew that the focus of attention would be on his untried skills, and that flying with him as Mission Commander was Alan Shepard, the first American in space a decade earlier, known as either 'Smilin' Al' or the 'Icy Commander,' depending on his mood.

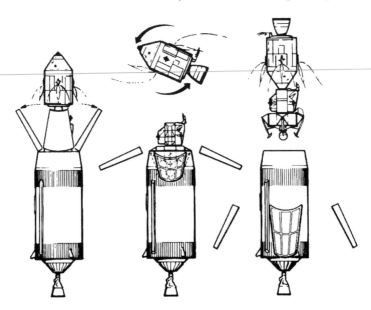

Apollo transposition and docking manoeuvre to extract the lunar Module from the top of the third stage of the Saturn V.

Roosa must have hoped that Shepard's long-awaited second flight might allow more of the former to shine through, whatever might happen next.

Roosa planned to break the record for economic fuel consumption on this flight, as he gently guided the probe on the nose of his spacecraft into the drogue on the top of the LM. Expecting the two spacecraft to lock together, he was a little surprised to feel them drift away again. He thought he must have approached too gently to activate the docking capture latches. Knowing the record was lost, he fired the jets to back off and try a little harder, but this also failed. Lunar Module Pilot Ed Mitchell confirmed that he was hitting the drogue dead centre, and Shepard gave him confidence by commenting that he was doing just fine and should try again. For the next 90 minutes, Roosa conferred with Mission Control and tried twice more, but without success.

Without a docking, the mission would be over before it had started. As the crew waited for advice from the ground and for someone to find the replica docking hardware for flight controllers to examine in Mission Control, they quietly contemplated alternative missions if a landing was not possible. The thought of not landing on the Moon was not an option that Shepard considered. There was no way such a problem was going to deprive him of his chance to walk on the Moon.

On the ground, controllers discussed various ideas to solve the problem. After a couple more attempts by Roosa, Shepard – concerned about depleting fuel reserves – called Houston in his typical brusque fashion and said that enough was enough. They could perform an EVA through the forward hatch of the CM. Shepard planned to open up the tunnel, extract the probe assembly, and then reach through the tunnel to manually steer the latches on the two vehicles for alignment. This was a very risky and totally impracticable idea. Room for mobility in the docking tunnel was very restricted without wearing a spacesuit, and a fully pressurised suit would restrict his mobility and his visibility; and if he happened to snag his suit glove in the latches then it would in all probability puncture, and he would die very quickly. Shepard apparently wanted to consider anything to save the landing!

An added problem was that in a few days – after the return from the Moon – the latches would have to work again to bring the crew home. Any problems then would result in an EVA transfer, and with a tired lunar crew this was not advisable. Mission rules would in any case prevent their getting this far.

The ground, having briefly discussed this option (which had never been attempted, simulated or even trained for), realised it was too dangerous and would not work. They came up with a new idea. As Roosa nudged the two craft together he was to fire the rear-facing quad thrusters on the Service Module, which would push the CSM against the LM. At the same time, Commander Shepard would retract the faulty probe out of the way and, if they were aligned correctly, the twelve latches would trigger, securing the two spacecraft as planned.

This had never been tried before, except briefly in a simulator by the back-up crew, just before the procedure was called up to the crew. The crew noticed scratches in the LM drogue, confirming that they were on target as they approached. Shepard told Roosa to forget messing around in trying to be gentle and about trying to conserve fuel: 'This time, JUICE IT!' The danger in this was that they could damage

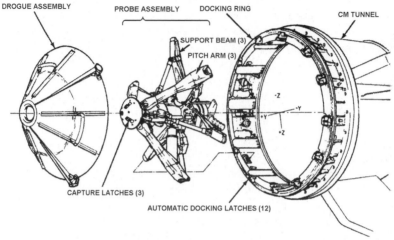

The Apollo docking mechanism (drogue on the Lunar Module at left, probe on the Command Module at right).

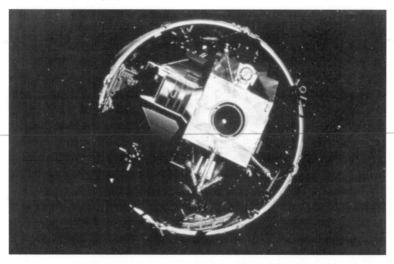

The view from the Command Module window, showing the Lunar Module housed in the top of the Saturn third-stage launch vehicle. Also visible is the roof of the LM containing the circular docking drogue where the CM probe enters.

the very hardware that they were to rely on for the return from the Moon. Another possibility was that a severe hard dock could jam the two spacecraft together and present even more problems when trying to undock for the descent to the Moon or in jettisoning the LM for re-entry into Earth's atmosphere if they had to abort the mission.

This photograph of a docking test of Apollo hardware clearly shows scratches in the docking cone. Similar marks were typical of actual docking manoeuvres during missions.

On their sixth attempt there was no immediate reaction from outside, which made their hearts sink. Then they heard the tell-tale 'ripple-bang' of the latches securing and saw the signal of hard dock on the flight control console. They had the LM safely in tow. The Icy Commander turned immediately into Smilin' Al. He was going to the Moon after all.

They went on to have a successful docking after the Moon landing, and they brought home the docking probe for analysis. Early theories were that ice particles had jammed the capture latches. It was determined that a foreign material interfered with the operation of the capture latch mechanism. Corrective action was put into force to ensure that this did not recur, including the provision of a cover over the problem mechanism prior to launch, a modification to the mechanism of the latch assembly, and added pre-flight tests.

### 1971: Soyuz 10

For more than three decades the Soviet/Russian space programme has repeatedly demonstrated its skill in performing routine docking of spacecraft to its space stations. However, problems have plagued the programme since its inception in the early 1970s. Some of these problems posed no threat to the crews, but others were definitely risky incidents and highlighted the dangers of bringing complex machines together in space.

Early attempts at docking Soyuz spacecraft together were frustrated by failure.

Soyuz 2 was cancelled before launch for a docking with Soyuz 1 in 1967 due to the problem encountered with the Soyuz 1 in orbit (see p. 369). The following year, Soyuz 3 was unable to dock with the unmanned Soyuz 2 because cosmonaut Georgi Beregovoi attempted to dock with both docking mechanisms totally missaligned. In effect he tried to dock 'upside down' in relationship to Soyuz 2, despite several attempts. In 1969, Soyuz 8 could not dock to Soyuz 7 due to several minutes of 'uncontrolable' actions aboard Soyuz 8.

On 19 April 1971 the Soviets launched their first space station, Salyut 1, followed on 23 April by the three cosmonauts (Vladimir Shatalov, Alexei Yeliseyev and Nikolai Rukavishnikov) aboard Soyuz 10, who intended to occupy the station for three to four weeks. Commander Shatalov completed a manual docking with Salyut on 24 April. However, no crew was transferred, and after 5½ hours they undocked, and landed several hours later. Initially it was stated that entry into the station was not part of the plan and that they had completed their mission of simply testing a docking unit successfully. In truth, however, the gear mechanism in the new docking system had failed, and the electrical connectors had also been unable to link up with their counterparts on the Salyut. Without these, the crew had been unable to open the hatch to enter the Salyut. There was no other access between the two spacecraft in the docked configuration, and EVA was neither trained for (although Yeliseyev was an experienced walker) nor practicable on the first mission. If the hatches were forced open and the docking alignment was not precise, the seal between the spacecraft could have ruptured, causing a rapid decompression and almost instant death for the crew. Even if the seals remained intact, forcing the mechanism might have prevented the closing of the hatches at the end of the mission. It was safer to leave the equipment alone, return early, fix the fault and try again on the next mission. It was a bad start for the Salyut programme, and the first of a series of docking problems. Soyuz 10 was commanded to undock from the Salyut and perform a flyaround inspection, allowing the crew to photograph the docking mechanism and the condition of the Salyut's exterior. Instead of a month in orbit, they found themselves returning to Earth after only two days. Having experienced the disappointment of not entering the Salyut, the crew also experienced a risky landing, just missing a lake by 50 yards and almost becoming the first Soviet crew to achieve a splashdown in water instead of the normal 'dustdown' on the steppes of Kazakhstan!

Two months later the Soyuz 11 cosmonauts successfully docked to the Salyut station and completed a three-week stay on board. Unfortunately a tragic accident during the recovery sequence lead to their deaths because they were not wearing pressure suits (see p. 389), and all Soyuz missions were grounded for two years while the accident was investigated.

When the next Soyuz carried a cosmonaut crew in a two-day test flight in September 1973 the changes resulting from the events on Soyuz 10 and 11 saw a two-person crew wearing full pressure garments for all critical phases of the mission such as launch, docking and recovery. In addition, changes in the Soyuz spacecraft utilised batteries instead of solar arrays for power, as spacecraft would be powered down apart from essential systems when docked to the Salyut station. Independent flight of Soyuz to and from a Salyut would be no more than two days. However, if a Soyuz failed to dock to a Salyut there would not be sufficient power margins for repeated attempts at docking, and the crew would be forced to terminate their flight almost immediately. This design change from solar panels to two-day batteries led to some of the frustrating missions of the Soyuz series during the next decade.

### 1974: Soyuz 15

In August 1974, Lev Dyomin and Gennedy Sarafanov – flying the new version of the Soyuz ferry spacecraft with a two-day battery supply – were evaluating a new automatic docking device being developed for the unmanned Progress cargo vehicle. Using the onboard computers, data from Soyuz and Salyut were automatically reviewed, and the manoeuvring engines fired to align for docking. However, on Soyuz 15 the computer crashed with a logic error. The crew tried again, but once more the automatic system malfunctioned. Each time, the computer pushed the Soyuz out of control with excessive engine burns just 100–165 feet from the unmanned Salyut 3 space station. Flight Engineer Dyomin remarked that they were shooting past the side of the Salyut. Commander Sarafanov took over manual control but it suffered a short circuit, and he also noticed that he was running low on fuel before he could dock to the station. Ground control told the cosmonauts to land as soon as possible. The flight ended after two days.

A 1975 ASTP version of the Soyuz spacecraft in orbit. These reliable spacecraft were also used to ferry crews to and from Soviet/Russian space stations for almost 30 years. This version carried solar panels for power supply, but the battery-powered two-day ferry versions were limited in orbital duration and in the event of a docking failure had to return immediately.

In their post-flight report, both cosmonauts indicated their frustration about the total reliance on the computer programme to dock. The State Commission, however, found both cosmonauts were over-critical of the computer, and it actually blamed the crew for the loss of the mission. The full report has never been seen outside of NPO Energiya. Sarafanov trained for a further mission on the subsequently cancelled Transport Logistics Spacecraft (in Russian, TKS) military space station with an attached three-person crew, similar to the cancelled USAF MOL missions. However, despite prolonged efforts by fellow cosmonauts to help clear their names, neither man flew in space again. Soviet explanations for the short flight and the non-docking to the Salyut varied. They cited testing new docking systems and procedures for unmanned tanker spacecraft, testing the psychological compatibility of the two cosmonauts (the CDR was 32 and the FE was 48 – both on their first flights) in space (on a two-day flight!) and the testing of emergency landings at night!

**1976: Soyuz 23**
Launched on 14 October 1976, Soyuz 23 attempted to dock with the unmanned military Salyut 5 space station the following day. The crew of Vyacheslav Zudov and Valeri Rozhdestvensky (both on their first mission) were trained for a 14-day stay aboard the space station. The fully automatic docking approach system malfunctioned before the spacecraft had reached 328 feet from the Salyut, where the crew normally takes over and flies a manual docking. However, such was the faith in the automatic system on this flight that the cosmonauts were not trained in manual docking. After trying to override the automatic system and running low on fuel, they were ordered to land at the earliest opportunity, resulting in yet another disappointing two-day aborted mission. The landing turned out to be perhaps the most challenging part of their mission (see p. 364).

**1977: Soyuz 25**
Twelve months later, a new Soyuz crew left the launch pad. The crew was the first to try to reach the new Salyut 6 second-generation space station, which was to herald a new era in manned spaceflight, with almost continuous crewing of the station, as it had two docking ports. At 394 feet, Commander Vladimir Kovalyonok took manual control and tried to hard dock with the forward port of the station. The Soyuz contacted the station's drogue, but did not hard dock. A further three attempts were performed by the crew before they reported to the ground that the approach force did not seem sufficient to activate the docking latches. Each time they had soft docked with the probe latches, but either the docking ring latches would not engage or the probe retraction sensor would not work.

After consultation with Mission Control, Kovalyonok undocked and tried again, but still without success. With low fuel levels (using the ferry-type Soyuz) and no margin to fly around and try to dock with the aft port, they had to return home. It was a frustrating start to the Salyut 6 programme. In December, Soyuz 26 successfully docked to the rear port of the Salyut, and later that month cosmonauts Georgi Grechko and Yuri Romanenko performed an EVA to manually inspect the forward docking area. They confirmed that there was no damage to the area, although signs of Soyuz 25's attempt were evident. It was cleared for use, indicating that it was the Soyuz 25 probe that had been at fault. In January 1978, Soyuz 27 achieved a successful docking with the forward port, without incident.

### 1979: Soyuz 33

In April 1979 the Soviets were preparing a two-man visiting (Bulgarian) Interkosmos mission, which was to spend eight days at Salyut 6. Two cosmonauts (Vladimir Lyakhov and Valeri Ryumin) were already occupying the space station. Soyuz 33 was to be commanded by civilian cosmonaut Nikolai Rukavishnikov, and the first Bulgarian cosmonaut, Georgy Ivanov.

Launched on 10 April 1979, the early stages of the mission went normally. Twenty-four hours after launch – during their seventeenth orbit in what was now a routine operation in the Soviet programme – the two men prepared for their docking approach to the Salyut. At 21.54 Moscow Time, the final approach command was relayed to the cosmonauts. The spacecraft was some 3 km from the Salyut's aft docking port, and the crew prepared to fire the engines to slow their approach. Suddenly the crew and ground control noticed 'deviations in the regular operating mode of the approach-correcting propulsion unit of the Soyuz 33 engine'; or, as Rukavishnikov later understated, 'We noticed something wrong with the engine's function.' Commanded to burn for six seconds, the Soyuz 33 unit fired erratically for only three seconds and promptly shut down, to the accompaniment of abnormal vibrations that both cosmonauts felt clearly in the descent module. Despite urging

Soyuz 33 cosmonauts Nikolai Rukavishnikov (foreground) and Bulgarian Georgy Ivanov (background) during training.

from the crew for the approach to continue, strict mission rules were applied. The docking was aborted, and the crew was ordered to return to Earth.

The Soyuz 33 spacecraft was to replace the Soyuz 32 ferry, which had been docked to Salyut 6 for two months, and it remained in orbit for the Soyuz 32 crew to rely on in an emergency. With a faulty primary engine recorded on Soyuz 33, this was clearly an impossible objective. With no way of adequately diagnosing the fault from within the spacecraft, due to the lack of appropriate instrumentation, the reserve engine's successful operation was imperative to ensure a safe return of the flight crew. The crew's desire to use the reserve engine and hand controllers to achieve a manual docking was overruled by ground control, and so the two unhappy Soyuz 33 cosmonauts resigned themselves to a short spaceflight. For Rukavishnikov, it was doubly disappointing, as he had been unable to enter the Salyut 1 space station following the first Salyut docking by Soyuz 10 in 1971. In 1973, having witnessed the loss of a further unmanned Salyut (Cosmos 557) shortly after launch, he was reassigned to the joint US/USSR Apollo–Soyuz Test Project. A Salyut specialist, he was frustrated at not being able to board a station he had helped to design for a second time. He was also the first civilian commander of a Soviet mission in eighteen years of launches – a great honour, but one which would have been greater if he could have completed his mission.

The Salyut crew followed the approach with interest as they, like all crews, were pleased to receive visitors in their lonely outpost. In his diary, Ryumin wrote that he and Lyakhov watched the Soyuz 33 approach blunt end forward, waiting for the final braking burn that never came. They saw the engine ignite and flicker, and then the exhaust changed colour and died. When they saw the little spacecraft speed past their station instead of slowing down, they knew a serious problem had affected their colleagues. That night they tried to determine how to use their own spacecraft to rescue the Soyuz 33 crew in the event that they became stranded in orbit. They then began worrying about their own engine on the back of Soyuz 32, and the fact that their ferry was nearing the end of its service life.

Lyakhov and Ryumin were disappointed in not receiving their guests. They knew it would be some weeks before a Hungarian visiting mission would be launched. In the event, this was also cancelled, and an automated Soyuz 34 was instead launched to replace Soyuz 32. So, instead of hosting two visiting crews, they would receive no visiting missions. It was to be a very lonely six-month mission in orbit.

On board Soyuz 33 the crew had to overcome the disappointment of not being able to dock, as they were preoccupied with preparing for an emergency landing. With batteries designed for only 2½ days of independent flight, their entry was planned for the next day – 12 April, Cosmonautics Day. Prior to their last orbit, Director of Cosmonaut Training Vladimir Shatalov (who had flown with Rukavishnikov on Soyuz 10 eight years earlier) spoke to the crew, boosting their confidence and expressing his faith in the reliability of the re-entry engine.

Because of the desire to land at the first opportunity and the lack of reliability with the primary engine, it was decided to fly a ballistic entry, which would mean a rougher ride for the crew than the lifting entry normally flown. A manual, timed burn of 188 seconds was planned. Anything less than a 90-second burn would leave the two men stranded in orbit.

When Rukavishnikov ignited the reserve engine, it overburned for some 25 seconds past the planned cut-off time, to 213 seconds, forcing him to manually cut off the burn. The increased burn-time meant that the capsule would enter the atmosphere at a steeper angle than normal, and therefore impart a higher g load on the two cosmonauts. Soyuz spacecraft normally completed the return from the Salyut stations in a two-step process. After an initial undocking the Soyuz crew would lower the orbit of their spacecraft before initiating the re-entry burn and beginning the entry sequence. Soyuz 33 followed a one-step profile from the Salyut orbit to a steep entry profile with their reserve engine, in order to fire it once, rather than twice.

Other factors to consider were the available fuel after the aborted docking attempt, for controlling the vehicle during re-entry, and exactly where they were to land. A desire to return the crew to Earth as soon as possible in the primary recovery area, covered by full search and rescue support, was the deciding factor in the decision to fly the ballistic entry profile. This entailed performing visual confirmation of their using a daylight horizon with the Earth's surface below them for reference. From their couches in the Soyuz DM, cosmonauts are unable to see directly forward due to the location of the large OM. They use a periscope extended outside to view past the OM for forward views, and TV cameras mounted in the forward OM compartment. The only way to build in all these factors for the crew's safety was to go for a ballistic entry as soon as correct orientation had been confirmed.

For 530 seconds of their descent the cosmonauts experienced 8–10 g, which was well above the usual 3–4 g of a normal entry by Soyuz. Breathing was difficult, although the crew were able to talk to each other. The temperature outside the capsule rose to 3,000° C, and Rukavishnikov later stated that it was like flying 'inside the flame of a blowtorch' as they sped earthwards. Deputy Flight Director Viktor Blagov later described it as 'the most complicated flight we have ever had.'

A successful re-entry and landing was accomplished in darkness, and so the recovery parachute was not spotted until the spacecraft dropped gently to a dust-down on the steppes of Kazakhstan. It was only the second ballistic re-entry flown by a Soyuz vehicle, the first being just four years earlier when the Soyuz 18-A launcher failed late in the ascent and forced an emergency landing after only a sub-orbital flight (see p. 155). Normal Soyuz entries follow an aerodynamic lifting profile, in which the density of the atmosphere is used to minimise the g loads on the crew before the vehicle is allowed to plunge through the atmosphere again to complete a parachute landing.

Examination of recorded telemetry indicated that the primary engine's combustion chamber pressures were lower than normal during the final approach burn programme. Unfortunately the hardware was not recoverable for post-flight examination. The engine unit itself was lost in the atmospheric burn-up of the Service Module.

One of the problems in these early Soyuz spacecraft was the lack of instrumentation available to the crew. For engine burns, the only monitoring instrument was a stopwatch! The crew timed the length of the burn, and then waited for ground control to inform them if they had been successful. The single-velocity accelerometer could measure only changes of velocity in the forward or aft axis, and was unable to determine the precise burn duration.

Rukavishnikov later commented that although the flight had lasted only two days, it had felt like a month. He also revealed that during the orbital ordeal the cosmonauts broke open one of the packages intended for the Salyut crew: 'We fortified ourselves. I had very little. Georgi [Ivanov] took a good drink.' The contents of the drink were not identified, however. Both cosmonauts received high honours for the courage they displayed.

The true nature of the final orbits of Soyuz 33 was not revealed for four years. In 1983 it was reported that both the cosmonauts in space and the controllers on the ground believed that the crew of Soyuz 33 could easily have been marooned in space, with no hope of rescue. Calculations made at the time of the emergency indicated that natural decay of the Soyuz 33 capsule orbit would take ten days. There was less than five days breathable oxygen aboard the spacecraft, and only one day's electrical power.

Details of the accident also revealed the cause of the shut-down after only 3 seconds of the engine's seventh burn of the mission. The engine was fired again, but shut off immediately. This was when the Salyut crew noticed the glow at the rear of the spacecraft from a distance of some 3 km. This alarmed ground controllers, who realised a lateral plume was abnormal on such a burn. As they already knew the thrust main engine was at fault, it was feared that this glow could have resulted in serious damage to the back-up engine, which was located immediately next to the primary unit.

After a long night of evaluation and discussion with the crew, the re-entry was ordered, with a 188-second burn of the back-up engine – the only option they had. It was during this burn that it, too, malfunctioned. This time the engine did not automatically shut down, and it continued to fire 25 seconds longer than planned. It had to be manually turned off by Rukavishnikov, with a crew compartment switch. The over-burn forced the high-stress ballistic return profile of 8–10 g, and an undershoot of the landing point by several hundred miles.

The crew had already prepared for their landing when Flight Director Vladimir Shatalov added a rather unnecessary comment: 'We are waiting for you with impatience', reminding Bulgarian cosmonaut Ivanov to watch out for his moustache when closing the faceplate of his spacesuit. One had to be careful on the important matters of spaceflight! The two men looked at each other and breathed a sigh of relief when the engine lit, but the tension grew when it failed to shut off automatically.

As the two men began their fiery re-entry and the fireball grew outside their windows, the veteran Rukavishnikov told the rookie Ivanov, 'Now the fun will begin'. As they emerged from the re-entry, ground control repeatedly called for statements of their condition. Rukavishnikov told the ground controllers to be quiet for a while. They would answer later, because they were pinned into their seats by the heavy g loads. Suddenly the buffeting and flames ceased, the parachute was deployed and they descended gently to Earth.

**1983: Soyuz T-8**

In April 1983 the Soyuz T-8 crew was making their final approach to the Salyut 7 station, but unbeknown to them the launch shroud had torn off the all-important rendezvous radar antenna during launch. As they tried to activate the system they received no signals from the antenna, forcing a postponement of the docking while ground evaluated the situation. The next day the crew was ordered to try a manual docking. The normal mission rules were overlooked, as this was a highly trained crew and an important mission for the next stage of the Salyut 7 programme. They needed to get on the station. Mission Commander Vladimir Titov had never trained for manual approach and docking before the flight, and he had to use optical devices instead of radar. Such an attempt had never previously been performed on a Soviet mission. Seeing the station in his periscope, he had to determine the approximate diameter of the Salyut to estimate the range, and then mission control could compute the closing velocity so that the necessary burn could be made to close in on the station.

As the 1,000-foot point was reached, the Soyuz passed out of range of the ground communications network, and the crew lost their source of information on relative velocity. He continued in, but upon entering darkness Titov was no longer sure of the distances as he approached the Salyut, illuminated by its running lights and Soyuz's floodlight. He closed to within 575 feet, but feared a collision. He therefore decided to fire the Soyuz engine, and flew past the Salyut, aborting the approach and thus the mission. They were so low on fuel that there was barely enough for retro fire, and so the Soyuz was placed in the spin-stabilised mode used to save fuel, and the attitude control system was switched off.

They landed safely, but it was a close call, and for Vladimir Titov and Gennedy Strekalov it was a taste of things to come with their next launch attempt in September (see p. 161). The lack of training of a cosmonaut crew in new docking techniques and concerns on closing distances was to haunt the Russian space station programme fourteen years later in the 1997 Mir docking accident with an unmanned Progress vehicle.

**1987: Kvant**

On 31 March 1987, the Soviet Union launched the Kvant astrophysical science module to a planned docking with the Mir Core module, which had been occupied by Yuri Romanenko and Alexandr Laveikin since February. By 5 April the Kvant was aligned for an early morning docking with the aft port of the Mir station. It was a scheduled rest-day for the resident cosmonaut crew. The docking systems on the Kvant had been checked by ground crews receiving a telemetry down-link from the spacecraft during the flight, and since this was an automatic docking the crew had nothing to do but merely 'watch' as interested spectators.

The Kvant was large, weighing over 20 tons, and was a two-part vehicle consisting of a pressurised laboratory section and the unpressurised telescope and science equipment section. For the rendezvous and docking, a separate engine block – the Functional Auxiliary Block (FAB) – was also attached to the Kvant hardware. Forerunners of these modules had only been docked to unmanned space stations between changes of the cosmonaut teams. With Mir it was planned that a succession of permanent crews would be aboard during the station's lifetime, and that large scientific research modules would be docked to the main core of the station over the years. Kvant was only the first (and smallest) of these modules.

The Kvant radar located Mir at a distance of 10.6 miles, with Mir acting as the passive element and the new module as the active element in the docking. Closing speed was reported to be 6.6 fps, and when the module was only 164 feet from the station, the two cosmonauts on Mir transferred to their Soyuz TM-2 spacecraft docked to the forward Mir port and closed all internal hatches. They wore their pressure suits as a precaution in the event of a docking malfunction. This was routine practice for the Soviets in their docking manoeuvres since 1978, to allow quick escape from the station if something went wrong.

Flight Control Centre data received from Kvant indicated correct deployment of the docking equipment, and TV from the module displayed the approaching Mir docking port and closing approach data. At 656 feet from Mir, former cosmonaut and Mir Mission Director Valeri Ryumin informed the cosmonauts that data had been received that Kvant had failed to 'lock-on' to Mir, and that the module appeared to be drifting off course. Ryumin requested that Romanenko re-enter the Mir and visually observe the path of the module from the airlock: '[The Kvant] is drifting away. The module is behind and below us, outlined against the Earth. It is slightly rotating and drifting away slowly,' he reported.

He continued to state that the Kvant had passed just a few metres away from the hull of the station itself. From this, it appeared that the Soviets came very close to a collision between the uncontrollable Kvant and the manned Mir station, which could have resulted in serious damage to the station and an enforced emergency landing by the two cosmonauts. Once again, this was a near miss that would be reflected upon a few years later when a spacecraft struck the hull of one of the Mir modules.

Though subsequent Soviet news reports failed to mention this, Ryumin's reply indicated ground control's concern and lack of understanding of the problem now confronting them. 'Stay calm Yuri, everything is alright. We need a couple of hours to check the telemetry and then we will give you the result.'

The subsequent official report of the mishap indicated that it was due to a malfunction of the control system on board Kvant. Ryumin explained that it was not very often that the Soviets had tried to dock spacecraft the size of Kvant and Mir; 'That is why we were especially cautious and, perhaps, overdid it.' The FAB carried reserves for several docking approaches, and Ryumin added that the steering margins for the module were very narrow. 'We have somewhat overdone it by setting too high requirements. We understand what happened,' he continued, stating that it would take a few days to solve the problem, set up another approach and be ready to try to dock the module to the station.

During the night of 8–9 April, a second attempt resulted in a soft docking to the Mir. Unfortunately an unidentified piece of debris was preventing the Kvant docking capture latches from securing a hard dock to the station, frustrating the attempt to secure the first element of the Mir complex to the core module. Future plans to link up to four much larger scientific modules to the station seemed to be in jeopardy. Docking specialists soon identified the cause of the problem, but external video was unable to reveal the presence of a foreign body or its location inside the docking apparatus. The cosmonauts were also unable to visually see the problem, so an impromptu EVA was ordered to investigate the area from the outside the station.

On 12 April, the 3 hr 40 min EVA by the two cosmonauts resulted in Laveykin discovering a small white cloth and removing it from the docking area. With the cosmonauts observing from a safe distance on the hull of Mir, Kvant was then commanded to hard dock to the station – this time successfully. The foreign object was probably either a pre-flight cover left on from ground preparations, or an item of debris from station operations, such as a waste collection bag from the previous Progress loading.

After the drama of the automatic approach, the successful conclusion of this EVA clearly demonstrated the value of having a human crew onboard and the opportunity of real-time planning which allowed the cosmonauts to perform an unscheduled EVA they had not trained for and visually inspect an area out of sight of automatic TV cameras. If this had been a totally automated operation, then the docking would not have occurred and the Kvant mission would have ended in failure.

## SPACE SHUTTLE SUB-SYSTEM ANOMALIES

### 1981: STS-2
The original plan for STS-2 – the second of four planned orbital flight tests of the Space Shuttle in November 1981 – was for a five-day mission. It would carry a scientific instrument package in the payload bay and perform the first extensive operations with the Canadian-built robotic Remote Manipulator System. Early on the first day, however, one of the three fuel cells on *Columbia* developed problems. In the fuel cells, hydrogen reacts with oxygen to produce electrical power for the orbiter and drinking water for the flight crew, and they are therefore an important sub-system on the orbiter. Mission rules insisted that if one fuel cell failed then the mission would have to be reduced to a 'minimum mission' (lasting 54 hours, or 36 orbits). The problems were first reported just over 2 hrs 30 min into the mission, when the crew informed the CapCom that they had 'two or three things to tell you about', one of which was a high reading on fuel cell 2. Over the next couple of orbits, the astronauts, acting upon suggestions from Mission Control, attempted manual gas purges of each of the fuel cells in turn. All looked fine at that time, and the crew was given the 'GO' to remain in orbit. This allowed them to remove their launch and entry flight suits.

Five hours into the mission, the problems with the fuel cell resurfaced and forced a shut-down of fuel cell 1. An evaluation of data on the ground resulted in a decision to burn off the reactants and shut down the fuel cell permanently. With one fuel cell off-line, this brought into play the minimum mission rule. The flight plan was adjusted to move forward the highest priority mission objectives within the 54-hour duration, and it was also decided that if the two remaining fuel cells remained healthy then the flight duration could be extended. However, during the night an evaluation of the situation by the mission management team resulted in the decision to bring the crew home early, as most of the mission objectives would have been accomplished. This was a disappointment for the crew of Joe Engle and Dick Truly.

CapCom (Sally Ride): 'First the bad news. Our plan is that we're running a minimum mission and you'll be coming in tomorrow.'
*Columbia*: 'Oh boy. I'll tell you what. You're garbled and unreadable there, Sally.'
CapCom: 'Want me to say it again? OK, you get to hear the bad news one more time then. We're running a minimum mission and you'll be coming in tomorrow.'
*Columbia*: 'Oh, OK, that's not so good.'
CapCom: 'Think of it that you got all of the good OSTA [science package] data and all of the RMS data and you just did a good job. We're going to bring you in early.'
*Columbia*: 'OK, understood.'

It was therefore with some reluctance that the crew ended the STS-2 mission after a flight of 2 days 6 hrs 13 min. Although their mission was cut by half, they managed to cram in extra work in the available time, and achieved more than 90% of mission objectives. Engle later commented, during the post-flight press conference, that if there was a mistake made, it was in calling the mission a 'minimum mission' instead of a 'maximum accomplishment mission', due to the amount that they achieved in such a short space of time.

### 1983: STS-9

In December 1983 the American Space Shuttle *Columbia* was in orbit for the sixth time – this time carrying an international crew of six astronauts and the first Spacelab scientific module. Towards the end of the mission the crew faced an orbital malfunction which – had it not been for the redundancy of Shuttle systems and the skill of six-time space veteran John Young – could have ended in disaster.

STS-9 was launched on 28 November 1983, and for much of the ten-day mission, while the scientists were busy in the lab, Young spent his time on the flight deck with his Pilot, Brewster Shaw, manoeuvring *Columbia* to accomplish the various scientific objectives of the Spacelab mission. The two astronauts completed 182 manoeuvres and 212 attitude changes in support of the science programme and to compensate for faulty equipment encountered in flight. As the crew prepared for their return on 8 December, a sequence of events took them by surprise.

At 05.15 EST, with the crew strapped to their seats, Young fired the nose-mounted reaction control jets in order to perform alignment manoeuvres for re-entry. Suddenly there was a loud bang, and the whole spacecraft shook. 'My stomach churned and my legs turned to jelly,' Young later remarked.

At the same time as the RCS firing, one of the four onboard computers failed. As the crew fired the RCS again, a second computer dropped off-line. The effect of what was termed a 'hard RCS burn,' – which results either from impurities in the fuel supply or is brought on by mechanical failures – was to force Mission Control to cancel the landing attempt while the problem was being investigated. The Shuttle orbiter computer system relies on four general purpose computers with an overseer

STS-9 Commander John Young (here at the Command station on STS-1 in 1981) had to struggle with the controls of *Columbia* during computer failures prior to landing the first Spacelab mission in 1983.

fifth GPC acting as a back-up link which needs to be in line with the main four GPCs to provide a redundant system in the event of a failure

The re-entry was delayed for four orbits as the crew began a nine-hour battle to control the vehicle without the use of the nose jets. Almost immediately, one of three inertial measurement units (IMU) failed, adding to the already serious problems. On the ground, neither the team of Flight Controllers, nor the support astronauts working in the simulators, could reproduce the events that temporarily crippled *Columbia*.

It was believed that shockwaves generated by the excessive firing of the RCS during the mission caused the initial problem. With one of the failed computers temporarily up and running, the crew finally manoeuvred *Columbia* into a safe attitude for entry, and a 156-second firing of the twin OMS engines brought *Columbia* on a safe heading for landing at Edwards AFB, California. As the nose-wheel of the vehicle hit the ground, the second GPC failed again, giving credence to the theory that shockwaves had knocked out the systems in the first place.

It was also revealed that for two minutes prior to the landing, and shortly afterwards, hydrazine fuel from two of the three APU leaked and caught fire, without initially being seen. It caused considerable damage to the area behind the aft bulkhead, burning wire and electrical equipment at the rear of the vehicle. The crew was extremely lucky. Had the leak and fire occurred during the re-entry sequence it would have inevitably affected the third and final APU. *Columbia* could have been without hydraulic power for the control of aerodynamic surfaces (flaps and speed brake), and would almost certainly have lived up to its nickname of a 'flying brick' and crashed into the desert.

Following the mission the orbiter was transferred to the Rockwell facility in California for a pre-planned two-year refurbishment and upgrading programme, finally returning to space in January 1986 on mission 24 (61-C). While at Downey, investigations into the in-flight and re-entry problems were also completed. It was determined that the computers had failed in flight due to microscopic debris inside integrated circuits, which had not been picked up during pre-flight checks. The APU problem was traced to the failure of two rubber O-rings inside the hydrazine fuel lines, which had perished, allowing the fuel to leak and catch fire around the extremely hot APU hardware.

**1991: STS-44**
In November 1991, during the seventh flight day of STS-44, flying a DoD unclassified mission, inertial measurement unit 2 failed. Attempts to recycle the unit also failed, and invoked the mission rule to fly a minimum duration flight for the loss of one IMU. A little over a day later, three days sooner than planned, *Atlantis* swooped to a landing at Edwards AFB, California. Ironically it had been just over ten years since the failure of the fuel cell on STS-2 had shortened the *Columbia* mission. STS-44 was only the second Shuttle mission to come home early, although many of the flight objectives had already been accomplished, and so the impact on pre-flight goals was minimal.

**1997: STS-83**
Shortly after lift-off of STS-83 on 4 April 1997, fuel cell 2 displayed a degradation of voltage in one of the two sets of cells, within one of three sub-stacks of electrolyte chambers. As a result, the crew was asked to purge the fuel cell to remove any contaminants. The fuel cell had showed differential voltage during pre-launch start-up operations and had remained above the operations, and maintenance limit of 150 mV, for a considerable time. However, after a reactant purge and start-up sequence, the levels dropped significantly to below 50 mV, allowing a waiver to pass the unit for flight. Following the insertion into orbit, the fuel cell voltage levels continued to move upwards and, continued to climb despite a two-minute purge. Levels increased towards the 150 mV mark, despite a further ten-minute purge of the fuel cell during the next flight day.

While evaluation of the problem continued on the ground, the flight crew continued to operate onboard experiments on the Spacelab Materials Science Laboratory payload. Two days into the flight, following a Mission Management Team (MMT) meeting to review the situation, the decision to shut down the faulty cell was passed up to the crew, in order to prevent crossover of problems to the other two functioning fuel cells.

CapCom: Canadian astronaut Chris Hadfield: 'The MMT had all the players in on

the meeting right through from the factory. The consensus is they just do not understand the behaviour in fuel cell 2. Even though your efforts have done a good job towards stabilising the problem, it's significantly out [of line]. So, we'll shorten the mission.'
Columbia: CDR Jim Halsell: 'That's certainly a disappointment, but we know you guys put your best effort forward and you're doing the right thing. We appreciate all the work that's gone into that.'

Four days after launch, *Columbia* was back on Earth after an early termination to a 16-day, two-shift science mission. Upon return to the orbiter processing facility at the Cape, the fuel cell was removed from *Columbia* to undergo further tests and evaluation. The results were inconclusive, but revealed that 'an undetermined and isolated incident caused a slight change in the voltage of about one-fourth of the 96 cells that make up each fuel cell. To ensure the health of fuel cells pre-launch, the power plants will be started earlier than usual to allow for additional monitoring before lift-off.' Installation of new monitors that would reveal the health of each cell, rather than a single monitor for each of the 32-cell sub-stacks, would be implemented.

Two weeks after *Columbia* landed, a decision was made to re-fly the MSL payload on *Columbia* in July. The second attempt to complete the mission was an outstanding success, and reflected the hard work required by the ground processing teams to turn the vehicle around quickly. There were no significant fuel cell problems on the new mission, which was redesignated STS-94. This was the first re-flight of a complete vehicle, payload and crew.

## EXTRAVEHICULAR ACTIVITY INCIDENTS

The dangers of flying a complex manned spacecraft in the environment of space are challenging enough, but for true manned space exploration a space explorer must occasionally leave the protective environment of his spacecraft and venture outside into the void. To do this safely requires a pressurised suit, and here, too, events have highlighted the fine line every EVA astronaut walks in opening the door and stepping outside.

### 1965: Voskhod 2
On 18 March 1965, Soviet cosmonaut Alexei Leonov became the first person to leave his spacecraft in orbit and perform an EVA, or spacewalk. Protected by a suit pressurised at 5.87 psi, he found difficulty in bending his arms and legs, due to the ballooning of the suit. At the end of the EVA, when he tried to re-enter the air lock chamber, he was unable to bend his legs to scramble back inside. Several minutes of struggling resulted only in his approaching heat stroke and sweating profusely. He solved the problem by lowering his suit pressure to 3.67 psi (a dangerous move) to reduce the ballooning effect. He was still unable to enter feet first, and had to return to the safety of the airlock head first. He spent several harrowing minutes turning

In March 1965, Alexei Leonov became the first man to walk in space. Due to the stiffness of his suit, caused by the internal pressure, he experienced some difficulty when re-entering the Voskhod 2 spacecraft. (Courtesy Novosti Press Agency via Astro Info Service Collection.)

around inside the airlock to close the outer hatch, allowing repressurisation, then again in opening the inner hatch to enter the Voskhod 2 main cabin. Due to his efforts he lost 12 lbs body weight!

### 1966: Gemini

Leonov's short journey outside was followed in June 1965 by Ed White spending a happy 20 minutes outside Gemini 4 and making EVA look easy. He was physically fit, but still ended up exhausted when he finally closed the hatch – a fact not picked up by the mission planners and post-flight evaluation teams. After Gemini 4, extensive EVA experiments for Gemini were re-evaluated, and ambitious EVA objectives, including working with advanced manoeuvring backpacks, were assigned to Gemini 8–12. Unfortunately, Gemini 8's short-lived mission pre-empted the EVA attempt, but NASA did not adjust the next EVA programme to compensate for this loss. On Gemini 9, Gene Cernan would try to strap himself into the astronaut manoeuvring unit (AMU) at the back of the Adapter Module and then release it to fly (tethered) away from the spacecraft.

Cernan later wrote that when he opened the hatch on Gemini 9 one year and two days after Ed White, NASA knew 'diddly-squat' about EVA. Gemini was a logical step-by-step programme to build experience and knowledge after the Mercury flights, ready for the Apollo lunar missions. And Apollo itself was also to be flown in a five- or six-step method before attempting the first landing. To go from a relatively simple 20-minute excursion outside a spacecraft to a two-hour test flight of a rocket-

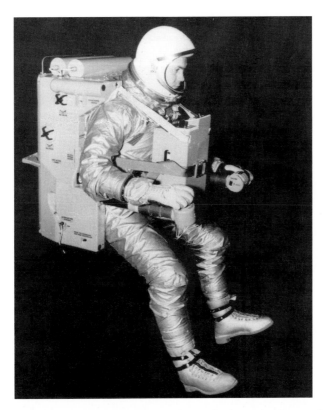

A demonstration of the Astronaut Manoeuvring Unit planned to be test flown by Gene Cernan on Gemini 9 in June 1966.

propelled backpack and a host of experiments, was ambitious to say the least. It was part of what Cernan called 'Go fever' – the pressure to achieve overly ambitious goals, which put the astronauts into new, uncharted and potentially dangerous situations.

At the very start of his EVA, when he pressurised his reinforced suit, Cernan found that it became very stiff and did not bend easily. Once outside he found that he had no good hand- or footholds, and any effort to restrain his body sapped his strength. As he worked away at the rear of the spacecraft, his faceplate fogged restricted his vision. No amount of rest alleviated the problem, and trying to hold himself stationary to perform a two-handed task with one hand only added to his problems. As he tried to mentally follow his checklists, he began to question the safety element. He could not fly the AMU if he could not see, and although he wished to complete his mission, when Stafford realised he was having difficulties with vision, communications and suit heating, it was decided to terminate the EVA.

Unfortunately, Cernan had ripped the outer rear seams of his seven-layer suit due to his physical exertion early in the EVA, and this gave him quite a sunburn on his

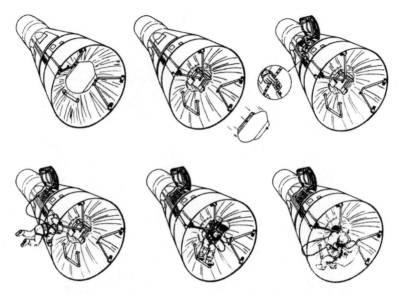

Gemini 9 EVA, showing how Cernan had to traverse to the rear of the spacecraft to put on the AMU pack.

lower back as the solar heat penetrated the weakened EVA suit. The heat also strained the suit's environmental system. He also experienced great problems in trying to squeeze back into the cramped compartment of the Gemini at the end of his EVA, cramping down while Stafford helped him push his weightless legs deep into the spacecraft.

Another factor not noticed in early Gemini missions was the razor-sharp serrated edge of the Gemini aft skirt where it had separated from the Titan booster, which Cernan had to pass to get to the AMU workstation, and which he feared might snag his suit. With assistance from Stafford inside Gemini 9, he returned to the pilot hatch and scrambled back inside. Even with faceplates touching, Stafford was unable to see Cernan's face due to fogging.

They eventually closed and sealed the hatch, and, breaking flight rules, Stafford squirted a stream of water at Cernan's face once he removed his helmet. Cernan's glove locking rings had cut into the flesh of his wrists as his hand swelled due to exertion.

Cernan's inability to complete the EVA tasks was a result of the lack of understanding on how much effort was required by the EVA astronaut to perform the most simple tasks without the aid of foot and body restraints. The contrast between White's and Cernan's EVAs presented the planners with much to think about for the remaining Gemini EVAs, including the provision of an adequate restraint device and new training programmes. These would not be ready in time for Gemini 10 or 11, but were being developed for the last Gemini mission, flight 12.

In July 1966, during Gemini 10's first EVA – in which Mike Collins 'stood up' in the open hatch – John Young found that they were unable to complete their tasks,

The rear of the GT9 adapter section, showing the location of the AMU and the hand and foot bars that Cernan would use to put on the unit in space. (Courtesy McDonnell Aircraft Corporation.)

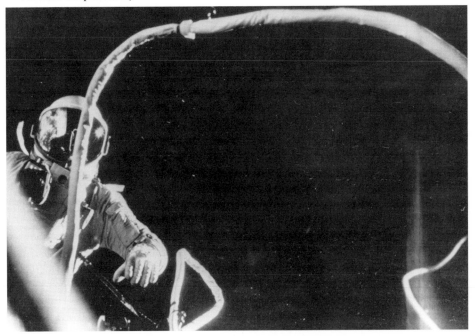

Gene Cernan returns to the hatch after encountering difficulties while performing his tasks at the rear of the spacecraft during the Gemini 9 flight.

due to their eyes filling with tears. At first it was thought to be a result of the new anti-fogging compound inside their visors, but they also commented on a strange odour. They thought it might have been the lithium hydroxide used in the ECS that caused their smarting eyes when both suit fans were turned on at the same time. Collins also experienced restraint difficulties.

On Gemini 11, one of Dick Gordon's tasks was to attach a tether from the Gemini to the docked Agena for a later experiment demonstrating the dynamics of two tethered vehicles in orbit. As with previous Gemini astronauts on GT9 and 10, restraint was a major problem, with no hand- or foot-holds, forcing Gordon to straddle the spacecraft to gain a firm foot-hold, and prompting Pete Conrad to quip 'Ride 'em, cowboy!' Gordon had some trouble with his helmet visor before opening the hatch, and had already overloaded his environmental system before going outside. Hot and sweaty, he struggled for six minutes, and soon tired. With sweat stinging his eyes as he groped his way back to the hatch, he again needed the help of his Commander, Conrad, to verbally guide him back towards the hatch.

The problems encountered on Gemini 9–11 resulted in a major review of EVA procedures for Gemini 12. Buzz Aldrin – assisted by an array of restraint tethers, foot-holds and hand-holds – set a record 5 hrs 30 min total EVA time – a grand finale to the programme. Problems of restraint and vision pointed to potentially dangerous situations, and with Apollo lunar missions ahead, EVA astronauts would not wish to experience them on the Moon.

### 1969–1972: Apollo

Some of the most spectacular shots of astronauts performing EVA on the Moon include several TV pictures of them falling over as they went about their surface tasks. The reduced gravity of ⅙ that of Earth made it appear like a slow-motion event, but just one snag of the suit on a rock or sharp instrument would have meant instant death. With no way of bringing a deceased colleague back into the LM, the sole remaining Moon-walker would have to abandon his companion on the Moon and return to orbit alone. Luckily, during the Apollo programme such circumstances never arose.

The problems of a space-sick astronaut wearing a pressure garment were highlighted on Apollo 9 in Earth orbit in March 1969. During the second day in space, after eating breakfast astronaut Rusty Schweickart suddenly vomited while donning his pressure suit for a planned EVA, but he managed to keep his mouth shut until he reached a bag. Feeling slightly disorientated, he moved to the LM. An hour or so later he vomited again, leading to a hurried conference with medical teams on Earth. Concerns for the safety of the EVA astronaut should he vomit again in a full pressure suit and helmet outside the spacecraft, caused his planned EVA to be first postponed and then curtailed. On the day, he was able to perform most of his EVA tasks, but there remained the fear of a vomiting astronaut on EVA being unable to use a bag and choking on his own vomit in the confines of the helmet.

During the Apollo 14 mission in February 1971, Alan Shepard and Ed Mitchell's long walk to Cone Crater took its toll. A stiff climb up a ridge on their second EVA saw Shepard's heartbeat rise to 150 in his efforts to find the rim of the crater. The

John Young leaps into the void above the lunar surface for Charlie Duke in taking a salute on the Moon in April 1972. When Duke tried to accomplish a similar leap he toppled over and landed on his back.

disorientating effect of the local lunar surface added to the frustration and exhaustion. Future lunar EVA astronauts would have the benefit of a Lunar Roving Vehicle, which was less strenuous and had the capability to navigate inertially.

From Apollo 14 the lunar surface astronauts did have a back-up system – in the event of a failure in one of the Portable Life Support Systems (PLSS – 'Pliss') – called the Buddy Secondary life Support System (BSLSS), which was similar to a scuba-diving system. It was designed to support two astronauts for 40–75 minutes. By using a length of connecting hose, one astronaut's system could support the second astronaut's system by supplying oxygen from the oxygen purge system (OPS), but not coolant water. It was estimated that on a short walk back to the Lunar Module OPS oxygen flow could still remove body heat and maintain an even temperature. Longer distances could have resulted in the depletion of oxygen before return to the LM. Fortunately the system was never called upon to be used during an Apollo mission.

Despite several Apollo EVA astronauts falling down (notably Dave Scott at Station 9A on the rim of Hadley Rille, and Jack Schmitt at Station 3 on Apollo 17), no serious effects were encountered, although the dangers of such action were later recalled by Charlie Duke, in his autobiography, of his fall on Apollo 16 in April 1972. In a demonstration of low lunar gravity he jumped straight up about four feet – in a 200-lb suit!). However, the weight of his backpack pitched him over backwards, and he came down heavily on his back in the lunar dust. Duke thought he had killed himself, as the full force of the impact hit the pack, which had not been designed with a 4-foot fall in mind. Had he split the pack, rapid decompression – or, as a friend of his called it, 'high-altitude hiss-out' – would have killed him instantly. Fortunately all was well, apart from the typical comment from Commander John Young: 'Charlie! That ain't any fun, is it? That ain't very smart.'

### 1973: Skylab

During the second EVA on Mission 1, to extract the stuck solar wing from the orbital workshop, astronaut Pete Conrad was nearly catapulted into space on the end of his umbilical tether. Skylab had been launched unmanned, and one of its solar wings had been ripped off during the ascent. The second one was stuck, so when the first crew arrived their first task was to lever the stuck wing out to allow the folded solar panels to deploy and power the station. During an EVA on 1 June 1973, Conrad and Joe Kerwin were outside on the side of the station, cutting the jammed wing free of debris and attaching lanyards to lever the wing out from its launch configuration. As they hauled on the rope, Conrad later commented, 'I gave a mighty heave, whereupon everything went black and I shot up into the air [*sic*]. By the time I settled back down, the wing had come out and fully deployed.' If he had not been secured by a safety tether, he would have been thrown into space, twisting away from the safety of his spacecraft.

Skylab space station, showing the astronaut-deployed solar shield and the absence of the solar array (which would have been on the left).

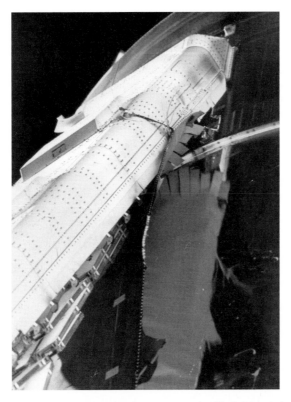

The damaged and partially deployed solar array on the Skylab workshop, taken by the astronauts prior to cutting the cable that was restricting full deployment, on 7 June 1973.

## 1977–1999: Salyut and Mir

EVA from Salyut and Mir stations have become a routine operation in the twenty or so years since the first inspection of the Salyut 6 docking hatch in 1977. Before this, no Russian had walked in space since 1969. Over the years, minor problems have plagued EVA, but a successful demonstration of routine EVA operations has generally been the case. One of the significant factors of Soviet/Russian EVA operations from space stations has been the regularity of excursions outside. With the construction of the International Space Station requiring hundreds of hours of EVA, the knowledge brought to manual operations in orbit will be especially valuable.

However, not all EVAs go exactly to plan, Unlike the one-flight Apollo suits and the short-duration Shuttle missions flying EVA hardware that is serviced and refurbished after each mission, the Salyut–Mir EVA equipment remained onboard the station for months or even years, being repeatedly used by successive cosmonauts. This provided for the necessity of space-borne maintenance and systems housekeeping, but it also led to in-flight failures, usually when the cosmonaut was wearing the suit outside.

Astronauts Conrad (background) and Kerwin (foreground) during the EVA to deploy Skylab's stuck solar array, on 7 June 1973.

## 1978: Soyuz 26

During the first Soviet EVA for nearly nine years, cosmonauts Georgi Grechko and Yuri Romanenko were to inspect the forward port of the new Salyut 6 space station. The previous Soyuz (25) failed to hard dock with the station (see p.234) and had been forced to return to Earth after only two days. Soyuz 26 carried Grechko and Romanenko to the second, rear port, and was successful. On 20 December, with Romanenko monitoring the EVA from the inside the airlock, Grechko exited the forward hatch as far as his head and shoulders to visually inspect the forward docking port. He found no serious damage from the contact with Soyuz 25's probe, and all associated equipment appeared to be in working order. The unit was cleared for future use, and the cosmonauts were instructed to complete their EVA by ground control before they went out of communication range. It was later that Grechko added a life-saving 'space rescue' story that kept Soviet space-watchers fascinated for years as they tried to unravel the true story

According to Grechko's early account, Romanenko wanted to take a look outside, and after poking his head out the hatch he gently pushed off, believing his safety tether would restrain him. It did not – simply because he had forgotten to attach it! He had seconds to attach his dangling tether to any part of the space station before floating off as a human satellite.

Grechko reportedly asked his commander where he thought he was going as he drifted past. Grechko was tethered to the hull of the station and, after double-checking, he pushed off and reached out to grab his commander before he floated out of reach. What was forgotten was that Romanenko was also attached to the Salyut systems by umbilicals, but these could have been pulled loose. As late as 1995

Grechko made much of his efforts in saving his commander from the jaws of death! Romanenko, however, had always denied the incident, and later Grechko himself made light of the 'little joke' and finally admitted that there was never any danger of losing Romanenko overboard. This was a clear case of how a good story can sometimes evolve into dramatic events and confuse true accounts of incidents.

By the end of their EVA both men were sweating heavily and breathing hard when they finally closed the hatch. But they were then faced with a more serious situation. When they activated the control to close the valve that had depressurised the airlock, it did not work. Data displayed that the valve was jammed open. They could not repressurise the airlock without losing the valuable air out through the valve, although their suits were drawing oxygen from the station supply. They did not have an individual air supplies from their back-packs, which prevented them from making a long traverse across the length of the Salyut to the docked Soyuz 26 at the other port, entering through the small hatch in the Orbital Module and then into the Salyut. Their main problem was that if they could not repress the airlock they would be unable to enter the main module.

When they re-established contact with mission control, they discussed what they could do next. Flight controllers were shocked to hear of the problem with the valve, and the fact that the cosmonauts were not yet back in Salyut as they should have been. There was only one option – to try and repressurise the compartment again, hope that the instrument reading was wrong. It was. The pressure held, and the two men finally floated out of their suits and headed for their sleeping bags for a well earned ten-hour sleep.

### 1990: Soyuz TM-9

In February 1990, Soyuz TM-9 blasted off from the Baikonour Cosmodrome to take Anatoly Solovyov and Alexandr Balandin to Mir as the next resident crew. Shortly after reaching low Earth orbit, the crew reported that an object was stuck outside the hull of their spacecraft, partially blocking the view from their Vzor periscope. They docked successfully. When TM-8 undocked to take home the previous resident crew, they performed a flyaround of the station in an attempt to determine the nature of the mysterious object which appeared to be stuck to the Soyuz.

They found that some of the blankets of the thermal shielding had 'peeled' away from the hull of the Soyuz Descent Module, like petals on a flower. Apparently broken free by the explosive separation of the aerodynamic launch shroud during the ascent, eight blankets were unfastened at the base. The blankets were open around the connection between the Orbital Module and the Descent Module. These were used for thermal protection during orbital flight, and were not essential for the protection of the crew during re-entry. Although a second Soyuz could be launched to replace the damaged one if required, photographs of Soyuz TM-8 revealed that it could be possible to fix the blankets back securely by EVA, or cut them away to prevent blocking of the horizon sensors. Special EVA equipment would not be ready for a while, so it was important to ensure that the Soyuz would not be alternately exposed to 130° C in direct sunlight and then –130° C in shadow. The attitude of the space complex would prevent the skin of the spacecraft from roasting and then

From 1977, Soviet/Russian cosmonauts conducted several dozen EVAs from Salyut and Mir Space stations, wearing upgraded versions of the Orlan EVA suit which was developed from the original Soviet lunar surface EVA suit.

freezing, thereby preventing condensation from forming inside, which would pose a further risk to onboard electronics.

On 17 July they were finally ready to begin the repairs to their spacecraft, and had moved the Soyuz to the forward port of Mir to allow easier access to the damaged areas. They exited to station via the Kvant 2 airlock, and made their way along the hull of the module towards the Soyuz. This was not the shortest route, but the use of a ladder prevented passage across the delicate antennae and other instruments on the external surfaces of the ferry craft. There was so much work for them to carry out in setting up the work-site and accomplishing the repair that they inadvertantly opened the hatch before the airlock was fully decompressed. When the catch was released it allowed the hatch to swing on its hinges in the rush of the last of the air from the airlock. It passed its designed opening arc and surprised the cosmonauts with a flood of sunlight earlier than expected. However, they carried on with their tasks, unaware that this action had damaged the hinge and would hinder their attempts to close the hatch at the end of the EVA.

The new Mir Orlan EVA suits had independent life support systems, which had a design working life of eight hours. No umbilicals were used during Mir EVA unless a fault developed in one of the suits, whereupon the cosmonaut would remain near the hatch area, supporting his colleague's activities while he continued with the main tasks of the EVA. The EVA was much harder to accomplish due to the difficult location in which they were working and the very strenuous task of reattaching the thermal blankets to the orbital module. They achieved two of the three planned repairs before beginning a hurried trip back to the airlock, almost at the limits of their consumables after six hours.

Although the trip back was quicker, as they left most of their equipment at the work-site, when they tried to close the hatch it just would not close that last few fractions of an inch to allow it to be locked and so repressurise the airlock. They therefore turned to emergency procedures. Fortunately, the central compartment of Kvant 2 was designed as an emergency airlock for such situations. Leaving the outer hatch slightly ajar, they closed the inner hatch and finally exited their suits after 7 hrs 30 min – just 30 minutes short of the duration of supplies in the suit. The supply of oxygen was not the problem, as they could have easily plugged into the station systems while awaiting repressurisation, but the carbon dioxide expelled every time they breathed out passed through the lithium hydroxide to be cleansed. This had reached saturation point, and carbon dioxide would soon have risen to a dangerous level. The cosmonauts would soon feel stuffy and have difficulty breathing and staying conscious. This would eventually lead to death. Fortunately, the emergency procedure worked, and nine days later a second EVA was completed.

During this trip outside, the cosmonauts tried in vain to repair the hatch. The next resident crew also tried to effect repairs, but it was not until January 1991 that a third crew completed the work, this time armed with the correct tool to effect the repair.

**Orlan suit problems**
Since 1977 the Russians have used the Orlan (Bald Eagle) design of EVA suit, which was developed from the cancelled Soviet Krechet (Falcon) lunar surface EVA suit. Originally used with umbilicals on Salyut EVAs from Mir, it had operated with independent life support systems in the improved back-pack unit. This has proved very reliable over the past two decades, but on a couple of occasions suit systems have presented wearers with cause for concern. Suit technicians and designers have always promoted the reliability of the garment and its systems, and acknowledge that systems can sometimes fail, as with any item of technology. But then, it is not the designer wearing it in open space when failure might occur.

*27 July 1991* EVA cosmonaut Anatoly Artsebarski's helmet visor fogged up due to his suit heat exchanger running out of water. Fellow cosmonaut Sergei Krikalev had to guide his commander back to the hatch entrance.

*20 February 1992* Cosmonaut Alexandr Volkov's suit exchanger clogged at the very beginning of his EVA. This forced him to remain close to the EVA hatch, attached to the spacecraft cooling system. The fogging was caused by the length of time that the suit had been stored on Mir (several months).

*28 September 1993* Cosmonaut Vasili Tsibliyev's suit cooling system failed, and again he had to remain near the hatch, performing only support tasks for the second

EVA cosmonaut. On the very next EVA, Alexandr Serebrov's suit suffered a problem in its oxygen flow system. The suit had been used on thirteen different occasions by different cosmonauts, and had exceeded its design lifetime.

*19 July 1995* Cosmonaut Anatoly Solovyov's suit cooling system failed.

None of these incidents were what may be termed 'life threatening', but they could have been. The fogging of visors seriously impairs vision, and suit cooling is extremely important not only for the comfort of the wearer but also the maintenance of the working components inside that keep the wearer alive. Fortunately, being close to the main hatch when these systems fail allows one to plug into the main station systems. However, if you are further away or separated from the main supply the need to return to the onboard system support becomes very urgent.

**Shuttle EVAs, 1982–1999**
One of the major headline events of Shuttle missions has been the spacewalks. None have been life threatening, but some smaller events illustrate that walking in space is never free from risk. And instant death is just the other side of the suit you are wearing!

On STS 41-C in 1984, astronaut George 'Pinky' Nelson experienced what was termed a 'minor urine contamination problem'. In other words, it leaked! The liquid

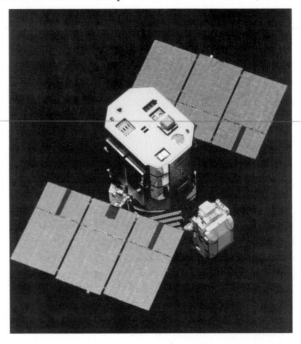

George 'Pinky' Nelson tries to slow the rotation of the Solar Maximum Mission satellite by hand during the recovery and repair EVA carried during the STS 41-C mission. During the EVA his suit urine collection system failed.

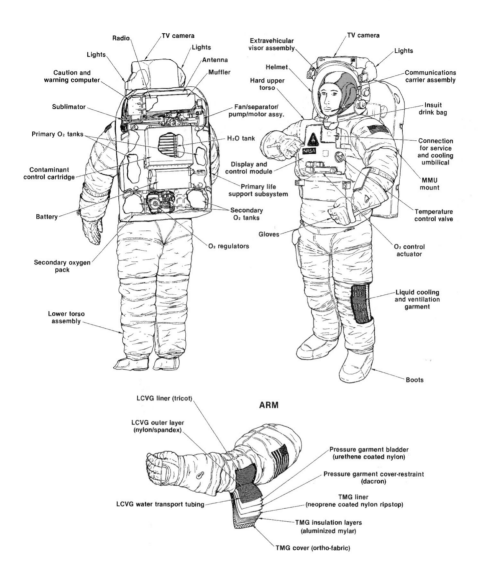

Radio

TV camera

Lights

Lights

Antenna

Caution and
warning computer

Muffler

Sublimator

Primary O₂ tanks

Contaminant
control cartridge

Battery

Secondary oxygen
pack

Lower torso
assembly

TV camera

Extravehicular
visor assembly

Helmet

Lights

Hard upper
torso

Communications
carrier assembly

Fan/separator/
pump/motor assy.

Insuit
drink bag

H₂O tank

Connection
for service
and cooling
umbilical

Display and
control module

MMU
mount

Primary life
support subsystem

Secondary
O₂ tanks

Temperature
control valve

Gloves

O₂ regulators

O₂ control
actuator

Liquid cooling
and ventilation
garment

Boots

LCVG liner (tricot)

**ARM**

LCVG outer layer
(nylon/spandex)

Pressure garment bladder
(urethene coated nylon)

Pressure garment cover-restraint
(dacron)

TMG liner
(neoprene coated nylon ripstop)

LCVG water transport tubing

TMG insulation layers
(aluminized mylar)

TMG cover (ortho-fabric)

Cutaway showing the construction of the Shuttle EVA suit and thermal protection and pressure garment layers which protect astronauts during spacewalks. (Courtesy Hamilton Standard.)

cooling garment he was wearing acted as a sponge and soaked up most of the moisture. Some helmet fogging was noticed, and post-flight inspection revealed that fortunately no urine had migrated to his helmet. The fogging was a result of turning down the liquid cooling ventilation garment after he felt cold as the urine soaked into his clothing. The danger of a floating globule of liquid in a helmet is that if inhaled, the astronaut could drown in little more than a teaspoon of liquid – and a globule of urine could make this even more uncomfortable! Once back inside the Shuttle crew compartment, the smell from the leak, combined with sweat from over six hours in the suit, made the odour similar 'to the inside of a toilet that had not been cleaned.' Apparently it was so bad that the rest of Nelson's crew threatened to throw him outside for the remainder of the mission!

One of the tasks evaluated for space station construction techniques was the ability of the EMU suits to withstand extreme levels of low temperatures. Cold fingers had been reported on STS 51-I, STS-63 and several other missions, where prolonged exposure to the sunlight heated the suits, then plunged them into freezing temperatures the moment the spacecraft moved into shadow. The environmental control systems of the suit can handle most of this variation of temperature, and some of the experiments were planned to evaluate future spacesuit design requirements. But the problems of heat and cold could affect a space-walker if a suit system failed, or if he was overexerted and unable to return to a safer location.

The planned construction of the International Space Station is envisaged to log in excess of 1,500 hours in 75 EVAs by succeeding teams of crew-members during Shuttle construction missions and by the ISS main resident crews. During these extensive and vital EVAs, the reliability and safety of the Shuttle, EMU and Orlan suits will be fully tested during the construction period. Incidents will inevitably occur, but with the development of the latest state-of-the-art technology, in-flight experience and adequate training, these should be kept to a minimum.

# Gemini 8 in-flight abort, 1966

Between April 1961 and December 1965, the Soviet Union and the United States successfully completed nineteen manned space missions, of which two were sub-orbital hops and seventeen were orbital flights. In addition, the X-15 rocket research aircraft completed nine flights of over 50 miles high during short 'sub-orbital' type flights, qualifying for the designation of 'astro-flights' just beyond the recognised boundary of 'air' and 'space'. In all of these missions, most of the pre-flight objectives were met, and no mission had to be curtailed early because of serious malfunction.

By the begining of 1966, NASA was completing final preparations for Gemini 8, which was planned to perform the world's first docking with an unmanned target vehicle, Agena. This would be an important milestone for human spaceflight, as well as for the forthcoming Apollo lunar programme. NASA and McDonnell Douglas had conceived Gemini as a logical step between the one-man Mercury spacecraft and the three-man Apollo vehicle. Gemini would have a number of key objectives: that a flight of up to two weeks was achievable, that activities outside of the spacecraft by an adequately protected astronaut would be possible, and that rendezvous and eventual docking between vehicles in space – crucial for Apollo – was feasible. As the Gemini 7 spacecraft hit the ocean following a two-week mission shortly before Christmas 1965, all of these objectives, except docking with the Agena, had been achieved.

It had been intended to do the same on Gemini 6, but it had lost its target vehicle in a launch mishap and had been reassigned to rendezvous with Gemini 7 in space during its 14-day marathon, completing another major step on the road to the Moon. It fell to Gemini 8, in March 1966, to hopefully achieve the long-awaited docking and lead the way to similar activities during the remaining four flights of the programme before Apollo's first flight. In command of the mission was Neil A. Armstrong. His pilot was Major David R. Scott, USAF. Not only were they scheduled to perform the first docking operation in space, but it was also planned that Dave Scott would perform a two-hour EVA during the three-day mission, using improved equipment following the brief experience of Ed White on Gemini 4.

At 7.00 am on 16 March, Armstrong and Scott were awakened in the crew

Armstrong and Scott enter the White Room on Complex 19 at Kennedy Space Center on the morning of launch, 16 March 1966. Both men are smiling in anticipation of their first spaceflight.

quarters to undergo the usual pre-flight medical and breakfast routines as they learned the status of their spacecraft and target vehicle. There were some small problems, but none were serious enough to threaten the two launches. Both astronauts put on their suits and were taken to the waiting spacecraft, sliding into their seats in Gemini 8, some 14 minutes before the Agena vehicle lifted off from pad 14 at 10.00 am. The target vehicle was boosted into an almost circular orbit of 184 × 188 nautical miles. After a few tense moments on the ground – especially as the previous target vehicle of Gemini 6 had been lost – Armstrong confirmed their desire to 'take that one'.

At 10:40:59 am, Gemini 8 lifted off, 3 seconds late, to begin its chase of Agena 8, orbiting high above them. Armstrong reported, 'It looks good from up here', as he and Scott rode the Titan II rocket to an orbit of 100 × 169 nautical miles. During the planned six-hour chase, coming in from behind and below, they would fly 105,000 miles to catch Agena. On entry into orbit, the spacecraft were 1,209 miles apart, with Agena in front. Armstrong was to use a combination of the orbital manoeuvring system thrusters and intricate orbital mechanics to catch up with the target vehicle. A series of five major manoeuvres was programmed for Gemini 8, and Armstrong recalled the hours of training he had completed in classrooms and simulators, practising for this event. He had learned that by firing the thrusters with the direction of flight, the craft moved into a higher and thus slower orbit. An engine fired against the direction of flight dropped the vehicle closer to Earth into a lower but faster orbit. Using a combination of both, the astronauts could gradually catch up with the target vehicle and then match their orbit to achieve rendezvous.

On Gemini's design, attitude control was achieved from small thruster units (discharging 25, 85 or 100 lb thrust) and was fuelled by self-igniting monomethyl hydrazine and nitrogen tetroxide. Located in the retrograde section were six orbital attitude and manoeuvring system (OAMS) thrusters. Four allowed for movement up, down, left and right, and two had nozzles that faced towards the front of the vehicle, allowing for translating rearwards. A set of ten OAMS thrusters in the equipment section afforded the astronauts control in pitch, yaw, roll and forward translation. On the re-entry module, two rings of eight thrusters were provided for attitude control manoeuvres. The system worked by infrared sensors scanning spacecraft movements in three axes and sensing the movements against the known location in relationship to the Earth's horizon. This provided attitude information for the computer which, in automatic mode, triggered the respective thrusters, depending on what was programmed. By selecting the manual mode the command astronaut could, from input of the three-axis hand-controller, orientate the spacecraft using the thrusters as desired.

With Dave Scott busily feeding commands to the onboard computer, Armstrong controlled the spacecraft as they settled down for their chase around the Earth. As Gemini 8 flew over Texas at GET 1 hr 34 min, Armstrong fired thrusters for five seconds in a retrograde burn to slightly lower the apogee, but in doing so he noticed a problem when cutting off the thrusters. There was a slight residual thrust, resulting in a variation of computer readings and making it difficult to determine the exact deceleration the burn had achieved. At 2 hrs 18 min 25 sec, Armstrong again initiated a burn, this time posigrade, to add 15 m/sec to their speed. Again, tail-off of the thrust led to difficulties in acquiring accurate readings.

As the spacecraft flew over the Pacific Ocean, some 25 minutes before they completed their second orbit (GET 02:45:50), Armstrong yawed Gemini 8's nose 90° south of their flight path. He also hit the aft thruster command button to impart a velocity change of 8 m/sec horizontally to effect a plane change, into that of the Agena. This was to compensate for lifting off three seconds late. Upon update from ground control, in the final few seconds before LOS with the tracking ship, in a hasty burn, he also added a further 0.6 m/sec to the speed. The next task was to test the rendezvous radar, which they did, achieving a lock-on at 332 km from the Agena.

At GET 3 hrs 48 min 10 sec after launch, Armstrong again nosed down the vehicle 20° and applied aft thrusters for a velocity change of 18 m/sec, resulting in a near-circular orbit just 28 km below the Agena. Gemini 8 was then ready for the terminal phase of its rendezvous programme.

Both astronauts saw the Agena as a bright shining object at a distance of 140 km, and as they approached, Scott transferred the computer from catch-up mode to rendezvous. The target slipped into darkness as the vehicles moved through the sunset, and Armstrong aligned the inertial platform for the translation manoeuvre.

With Agena just in front and above them, Armstrong pitched the nose of Gemini upwards and to the left, and at 5 hrs 14 min 56 sec he fired the aft thrusters once more. Some 28 minutes later, just before sunrise, he again fired the thruster to brake the Gemini. Using his eyes for guidance, he took information from radar and range rates as called to him by Scott. At a distance of only 46 m he had reduced the relative velocity between the two vehicles to zero. With rendezvous achieved, they were almost ready for the docking attempt.

Armstrong controlled the attitude of his spacecraft with delicate hand-controller manoeuvres for the next 36 minutes as Scott visually inspected the Agena with which they were about to link up.

The Agena Target Docking Adapter (TDA) was mounted on the forward end of the Agena stage, and was supported on seven dampers filled with hydraulic fluid which absorbed the energy of impact. The smooth nose of the Gemini was guided into a cone on the front of the target craft, and latches secured the craft. An electric motor then drove three gearboxes to retract the cone to form a firm contact with the main structure of the TDA. A display unit was mounted on the top of the cone at the front of the TDA, allowing both pilots to view Agena data. Scott could then control the engine on the back of the Agena via an encoder, with which he could input three-digit control codes.

Once all appeared safe to dock and the general stability of the target was confirmed, Armstrong was given the go-ahead to dock at GET 06:32:42. At a rate of only 8 fps he gently nudged the nose of the Gemini into the docking receptacle of the Agena.

Gemini 8 approaches the USAF Agena upper stage, modified for NASA Gemini docking missions. (Courtesy USAF.)

The view from Dave Scott's window of the final moments prior to docking GT8 with Agena.

Armstrong: 'Flight, we are docked! It's a real smoothy, no noticeable oscillations at all.'

MCC: 'Hey, congratulations. Real good.' Once their achievement was obvious, Mission Control erupted with loud cheers, handshakes and backslapping.

Armstrong: 'You couldn't have the thrill down there that we have up here. We've got a real winner here!'

The crew was now scheduled to conduct a series of post-docking chores. When they experienced difficulty in verifying the up link data of the Agena, Mission Control uplinked some new information to the crew.

MCC: 'Roger 8, reading you loud and clear. I have some information for you. First of all it's about the yaw manoeuvre. If you run into trouble and the attitude control system of Agena goes wild, just send in [encoder] command 400 to turn it off and take control of the spacecraft.'

As the combination passed out of range of the Tananarive station in Madagascar, the two astronauts busied themselves with their docking checklist. Scott commanded Agena's attitude control system to turn the combination 90° to the right and then dialled the next order on the encoder. As he did so, he noticed that the control panel in front of him was giving the wrong information. Gemini 8 should be in level flight, but the 'ball' indicator showed they were in a 30° roll. With the spacecraft out of range of ground tracking stations and on the night-side pass, visual confirmation with ground control was not possible, so Scott reported to Armstrong, who confirmed that his indicator was showing the same attitude.

Armstrong used the thruster to correct the roll, but when it started again both men thought Agena was being non-cooperative. Scott dutifully turned off the Agena attitude control system, as the CapCom (Jim Lovell) had instructed, just prior to loss of contact. For the next four minutes the combination stabilised, and it appeared that the problem had cleared up.

As Armstrong manoeuvred the combination using the Gemini thrusters to attain the correct horizontal orientation, the roll suddenly occurred again, this time much faster. As they considered the problem, they thought that the test objective to determine the stress tolerance between the two docked vehicles was somewhat academic. Armstrong struggled with the controls as Scott snatched photographs of the interaction from his window. Armstrong noted that OAMS propellant had reached the 30% level, which indicated to the astronauts that they might have a

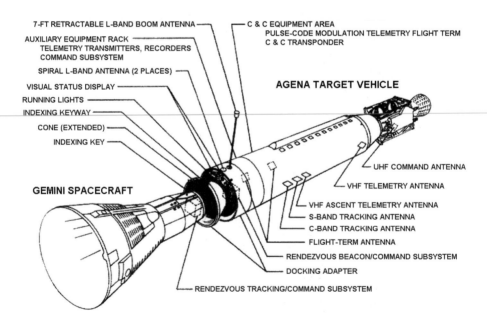

7-FT RETRACTABLE L-BAND BOOM ANTENNA
C & C EQUIPMENT AREA
PULSE-CODE MODULATION TELEMETRY FLIGHT TERM
AUXILIARY EQUIPMENT RACK
TELEMETRY TRANSMITTERS, RECORDERS
COMMAND SUBSYSTEM
C & C TRANSPONDER

SPIRAL L-BAND ANTENNA (2 PLACES)

**AGENA TARGET VEHICLE**

VISUAL STATUS DISPLAY
RUNNING LIGHTS
INDEXING KEYWAY
CONE (EXTENDED)
INDEXING KEY

UHF COMMAND ANTENNA

VHF TELEMETRY ANTENNA

**GEMINI SPACECRAFT**

VHF ASCENT TELEMETRY ANTENNA
S-BAND TRACKING ANTENNA
C-BAND TRACKING ANTENNA
FLIGHT-TERM ANTENNA
RENDEZVOUS BEACON/COMMAND SUBSYSTEM
DOCKING ADAPTER
RENDEZVOUS TRACKING/COMMAND SUBSYSTEM

The configuration of a Gemini–Agena docking mode.

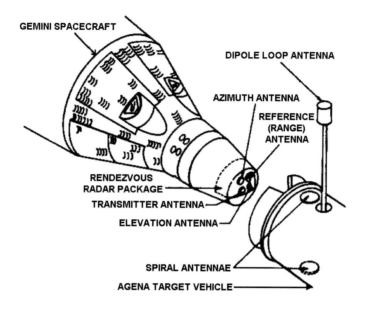

GEMINI SPACECRAFT

DIPOLE LOOP ANTENNA

AZIMUTH ANTENNA

REFERENCE
(RANGE)
ANTENNA

RENDEZVOUS
RADAR PACKAGE

TRANSMITTER ANTENNA

ELEVATION ANTENNA

SPIRAL ANTENNAE

AGENA TARGET VEHICLE

The combined Gemini–Agena rendezvous system.

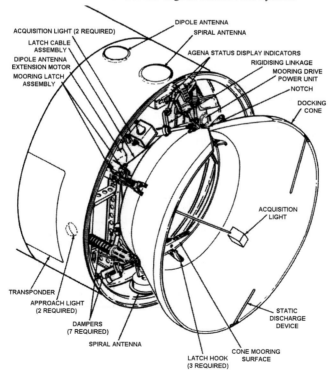

ACQUISITION LIGHT (2 REQUIRED)

LATCH CABLE
ASSEMBLY

DIPOLE ANTENNA
EXTENSION MOTOR

MOORING LATCH
ASSEMBLY

DIPOLE ANTENNA

SPIRAL ANTENNA

AGENA STATUS DISPLAY INDICATORS

RIGIDISING LINKAGE

MOORING DRIVE
POWER UNIT

NOTCH

DOCKING
CONE

ACQUISITION
LIGHT

TRANSPONDER

APPROACH LIGHT
(2 REQUIRED)

DAMPERS
(7 REQUIRED)

SPIRAL ANTENNA

LATCH HOOK
(3 REQUIRED)

CONE MOORING
SURFACE

STATIC
DISCHARGE
DEVICE

The Agena docking cone.

stuck thruster on Gemini. As Scott cycled the Agena switch off, on and off again, Armstrong tried the Gemini switches to determine if they could isolate the problem, but nothing seemed to solve it.

With sixteen thrusters and no time for a methodical diagnostic procedure, the two astronauts realised that the only way to stop the wild ride they were experiencing was to undock. By this time, the problem had become more complicated, because the roll rate had coupled with a yaw excursion, making it difficult to diagnose. As Armstrong tried to steady the spacecraft, Scott switched the control of the target to ground stations. (It had been inhibited to prevent rogue signals, but this action certainly saved the Agena for future experiments on a later mission.) Armstrong gave the command, Scott hit the undocking button and the two spacecraft parted, with Gemini pulling straight back as Armstrong gave the thrusters a long burst to clear the vicinity of the tumbling Agena. Almost immediately, their suspicions of a stuck thruster were confirmed as their yaw/roll rate increased alarmingly as they came into range of the next tracking station, the ship *Coastal Sentry Quebec* (CSQ).

MCC: 'How does it look?'
CSQ: 'We're indicating spacecraft free. I'm going to call the crew now. We're showing the spacecraft free – indicating they are not docked. Let me check with the crew. Gemini 8, CSQ CapCom, how do you read?'
GT8: 'We consider this problem serious. We're tumbling end over end but we are disengaged from the Agena.'
CSQ: 'We get your spacecraft free indication here. What seems to be the problem?'
GT8: 'It's a roll and we can't turn anything off. Continuously increasing in a left roll.'
MCC (to CSQ): 'Did he say he could not turn the Agena off?'
CSQ: 'No, he says he has separated from the Agena and he is in a roll and can't stop it. His reg. pressure is down to zero. His OAMS regulating pressure, that is.'

As the crew backed away from the Agena, their roll increase rate spun at the dizzy rate of one complete revolution of the spacecraft per second. Armstrong suspected that the fuel in the manoeuvring thruster was almost depleted. As they were fast approaching dizziness and blackout – which would have fatal consequences – Armstrong and Scott responded with the only chance left to them. They isolated the faulty OAMS system and initiated the reaction control system housed in the nose of the spacecraft. Armstrong tried the hand-controller again after checking the switch panel, and this time it responded. After they had steadied the vehicle, one ring of the RCS system was switched off to conserve precious fuel, and they reactivated each OAMS thruster in turn. As they did so they discovered that thruster No. 8 had failed 'on' and had indeed stuck in the open position, causing their current problem.

GT8: 'We're in a violent left roll here at the present time. Our RCS [reaction control system] is armed and we can't fire it. We apparently have a roll of a stuck hand-controller.'
CSQ: 'Flight, they seem to have a stuck thruster. We cannot seem to get any valid

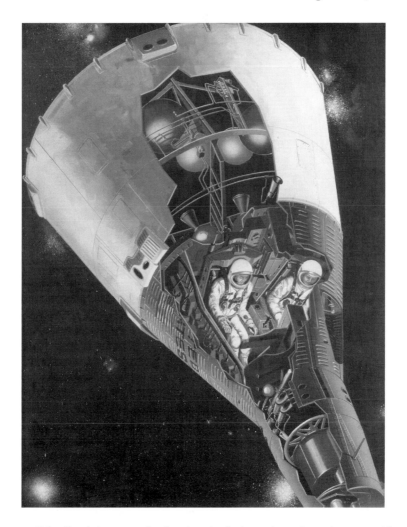

Cutaway of the Gemini spacecraft, showing the fuel supply tanks and crew positions.

data here. He seems to be in a pretty violent tumble rate. We're showing attitude control system on the Agena off and we've lost considerable gas pressure.'

GT8: 'OK, we're regaining control of the spacecraft slowly in RCS direct. We're pulsing the RCS pretty slowly here so we don't entirely roll right. We're trying to kill our roll.'

CSQ: 'We're showing pretty violent oscillations in roll attitude ... The Agena is also tumbling violently at this time.'

Flight: 'Find out how much RCS fuel he has used, and if he is just on one ring.'

CSQ: 'How much RCS have you used, and are you just on one ring?'

GT8: 'That's right. We're on one ring, trying to save the other ring; we started with two rings but now we are on one ring ... we are down to 1,700 pounds right now on

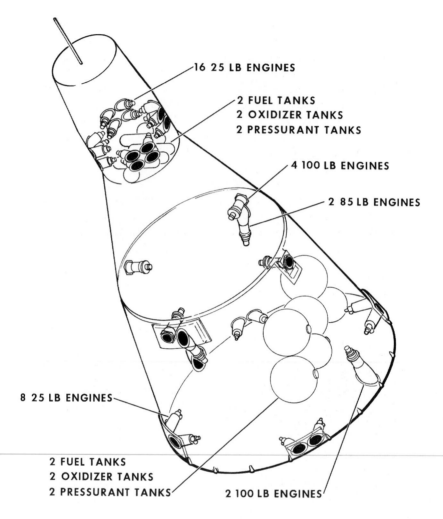

16 25 LB ENGINES

2 FUEL TANKS
2 OXIDIZER TANKS
2 PRESSURANT TANKS

4 100 LB ENGINES

2 85 LB ENGINES

8 25 LB ENGINES

2 FUEL TANKS
2 OXIDIZER TANKS
2 PRESSURANT TANKS

2 100 LB ENGINES

The location and composition of onboard Gemini propulsion systems. (Courtesy Rocketdyne.)

RCS and we have about 2,350 on [ring] A.'
CSQ: '8, CSQ. How are you doing?'
GT8: 'We're working on it'
CSQ: 'OK relax, everything is OK.'
GT8: The spacecraft–Agena combination took off in yaw and roll, and we've had ACS off. This happened at 7 hours ... we turned the spacecraft system on, and tried to stabilise the thing, and in doing so we may have burned out our OAMS roll-left thrusters.'
CQS: 'OK, I copy. Do you have visual sighting of the Agena right now?'
GT8: 'No. We haven't seen it since we undocked – a little while ago.'

By this time the use of the RCS thrusters impacted directly on mission rules, which stated that emergency use of RCS required return to Earth as soon as possible so that any leak in the system could not hamper or hinder re-entry. After barely ten hours in space, Armstrong and Scott would be coming home.

The effective training and past flight experience of the ground controllers now played their part as they quickly sized-up the new situation in real time and came to the aid of the crew in space. The contingency landing site had to be selected soon, or a 24-hour wait – which no one wanted – would have to be endured. It was decided that recovery in the Pacific Ocean would take place during the seventh revolution of Gemini 8.

Most of the public and NASA officials who had witnessed the Gemini launch were making their way home when news of the termination and pending recovery of the crew was received. McDonnell Douglas engineers were flying to Houston from the Cape at the time, and were 'caught with their pants down'. Never again would McDonnell Douglas engineers be in transit during a flight. In future, teams would remain both at Houston and the Cape for the duration of the mission, so that in the event of a similar malfunction, experts would be immediately on hand to support any remedial actions. This was carried on into the Apollo missions, and paid dividends during the Apollo 13 incident in 1970.

Back in space, Armstrong and Scott were preparing for re-entry and landing. They reported their experiences and actions, as well as the status of Agena, to

Mission Operations Control Room, Houston, during the recovery of Gemini 8 following the early termination of the mission. The view graph on the wall display at centre right shows the prime recovery area 500 miles east of Okinawa in the western Pacific Ocean.

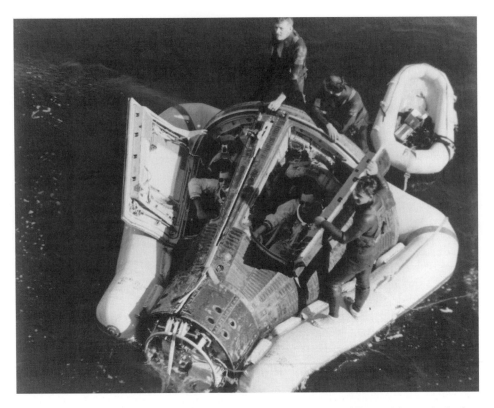

Scott (*left*) and Armstrong (*right*) sit with their hatches open while awaiting arrival of the prime recovery ship, the destroyer *Leonard F. Mason*.

ground control, and were advised to take a seasickness pill prior to re-entry. A more detailed account of the events would have to wait until the astronauts returned to Houston.

As Gemini 8 flew over the southern latitudes of the globe, *Leonard F. Mason* steamed as fast as it could for the expected landing site, 800 km east of Okinawa and 1,000 km south of Yokosuka, Japan. As the spacecraft passed over the Congo in darkness, the command for retrofire was given, and the crew punched the button to ignite the retrorocket package against the direction of flight. The two astronauts began their descent to an ocean recovery.

Some 30 minutes later, a C-54 recovery aircraft saw the Gemini gently swinging on its parachute, heading for the ocean. Splashdown occurred at mission elapsed time of 10 hrs 41 min 26 sec. As they waited for the *Leonard F. Mason* to recover them, pararescue divers attached a flotation collar to the spacecraft. Three hours later the astronauts scrambled up a net aboard the recovery ship, heading for berthing at Okinawa. Two days after the end of their truncated mission they arrived back at the Cape, where preparations for Gemini 9 were underway. Armstrong told waiting newsmen, 'We had a magnificent flight – for the first seven hours.'

When NASA released films and tapes of the incident, the courage and skills of both the astronauts and the ground control team was evident in the non-panic commentary and matter-of-fact approach. President Johnson – who, like most Americans, followed the coverage of the plight of Armstrong and Scott – also praised the professional attitude as media services across America interrupted regular broadcasts to bring live coverage of the recovery of the astronauts. 'All of us are greatly relieved', the President stated, and added that both astronauts displayed remarkable courage and poise under stress. However, it was also reported that several thousand TV viewers of different stations had complained that their favourite TV shows, such as *Batman* and *The Virginian*, were cancelled in favour of the live pictures from the Pacific Ocean and Mission Control.

To help quash wild rumours about the incident, NASA released a statement that preliminary evaluation of telemetry, film and initial interviews during Armstrong and Scott's debriefing indicated that the problem was caused by a short circuit in a manoeuvring thruster aboard the Gemini spacecraft. Further analysis was being conducted at the McDonnell Douglas plant in St Louis, where the capsule – minus its aft sections, discarded as part of the re-entry sequence – had been taken directly from its landing site in the Pacific Ocean.

During a press conference at the Manned Spacecraft Center, Houston, on 19 March, Flight Director John Hodge reaffirmed his decision to terminate the flight. There was absolutely no doubt in his mind that they had to come down. He first sensed trouble when the tracking ship *Coastal Sentry Quebec* informed them that the spacecraft were undocked when they should not have been. Obviously, they had some kind of problem in space. Once they had stopped the tumbling by using the RCS thrusters, he knew mission rules demanded immediate return, and this is what he ordered, just 15 minutes after Armstrong first reported their situation. Hodge added, 'A thought flashed past me ... gee whiz, what a shame, but thank goodness they got the docking in!' They had come home early, but they had achieved their primary objective.

As the real story of Gemini 8 became apparent, so did the efforts of the crew in coping with their potentially fatal situation. MSC Director Robert Gilruth stated that preliminary data ruled out any possibilities of pilot error. In fact, the astronauts had demonstrated a remarkable level of piloting skill in overcoming the problem and bringing the spacecraft in to a safe landing. Furthermore, NASA Associate Director for Manned Space Flight, Dr George Mueller, joined with Gilruth in praise for the astronauts. He also praised the flight control team, the recovery forces and the anomaly evaluation team, who came up with the answer to the incident less than 72 hours after it had actually occurred in space.

Armstrong and Scott explained their actions onboard the spacecraft at a press conference on 26 March 1966. This followed the presentation of awards to the DoD recovery force – the NASA Group Achievement Award – for prompt action in recovering the two astronauts. President Johnson announced that under the civil service programme, Armstrong would receive a quality increase, and Scott would be promoted from Major to Lieutenant Colonel. The two astronauts were presented with the NASA Exceptional Service Medal for their remarkable work.

Armstrong: 'After seven hours elapsed time, the incident occurred. Dave reported the spacecraft was diverging. We had not heard any thruster activity, nor had we seen any reflection of thruster activity on the spacecraft, and looking at the indicator, although rates were low, attitudes were changing in yaw and roll. This was not considered a problem at the time. It was assumed some anomaly in Agena control systems confused the Agena. We immediately activated our own control systems and shut down Agena's control systems. For about the next three minutes we attempted to control the combination and reduce the rates by different Agena commands, by isolating those that might cause a problem. It then became obvious this was not completely effective. We began to consider there might be some spacecraft control system involved. For the next four minutes the rates were reduced almost completely and we began to think we had the situation under control. Then the problem reoccurred, and rates increased to a point where the structural integrity of the combination of the spacecraft was threatened. We alternated activating control systems and deactivating them. We tried all of them, none successfully. Eventually, rate was reduced to a point where undocking would be safe enough to avoid contact between the spacecraft, as there was rotation in all directions. We wanted to ensure we could get far enough away from the Agena before any recontact was encountered. This took three minutes. Upon undocking the two spacecraft, it became quite evident that there was a problem in the spacecraft control system, and for approximately three minutes we attempted to isolate this problem. The hand-controller was inoperative at this time. We were accelerating to quite high values and activated our re-entry control system and turned the [OAMS] manoeuvring system off. We were successful in returning the control and rates to acceptable values after 30 seconds. One of the entry control systems was deactivated in order to save fuel. Over the next six minutes the rates were reduced to a very low value and, once the spacecraft was stabilised, we slowly reactivated the orbital system and detected No. 8 thruster was failed in the open position. Post-flight analysis indicated it failed to 'on' for three seconds, 'off' for three seconds, 'on' for three minutes, 'off' for four minutes and back on, tying in with our data on the timing. In reactivating our thruster we found that all yaw thrusters were inoperative except No. 8, which had failed to open. Pitch [thrusters] were opening correctly. After long conversations with flight control, we were advised of the recovery plans and were extremely reluctant to give up the flight at that time as there were a lot of unanswered questions we had to work on. But the decision to re-enter was reasonable and we agreed to prepare for recovery.'

In St Louis, McDonnell Douglas engineers conducted detailed examination of the thrusters that had not been jettisoned with the retro sections prior to re-entry. In an isolated control laboratory, the work continued for more than a month, and eventually – well in time for Gemini 9 so as not to interrupt flight schedules – McDonnell Douglas released the results of the evaluation team: 'The valves on thruster No. 8 opening unintentionally was probably caused by an electrical short circuit. There were several locations in the spacecraft at which the fault could have occurred, possibly in a solenoid which opened the fuel and oxidiser valves in the thruster.'

To prevent a repeat of such a hair-raising incident, McDonnell Douglas changed the attitude control circuit switch so that when it was in the 'off' position no power would go to the thrusters. On Gemini 8 and all the spacecraft before it, merely turning off the electronic package did not prevent power going to the thrusters, so they could still fire, as happened on Gemini 8. A short circuit probably occurred as the crew switched off to on and off again, sticking the thruster in the open position.

So ended the first in-flight abort of a manned spaceflight. Despite this, the remaining four missions proved highly successful, and completed a very rewarding programme. Agena 8 was visited in space by the Gemini 10 crew the following July, and was seen to be a very stable vehicle.

Gemini 8 achieved the world's first docking in space – a notable achievement, and an important milestone in America's planned exploration of the Moon. For the Apollo series, rendezvous and docking techniques would prove to be a major factor for success.

Gemini 8 also demonstrated, in stark reality, the dangers of space and how adequate training and correct crew selection were vital for securing safe recovery of a crew from a stranded flight. The mission demonstrated to the world that NASA could cope with unforeseen situations and in-flight emergencies – an image they built upon over the next twenty years.

# The Apollo 13 explosion, 1970

The space programme has seen many notable achievements during its short history. From these, several space explorers have made memorable comments or statements in flight – none more famous, perhaps, than Armstrong's 'giant leap for mankind'. On the tapes of a certain Apollo lunar mission, five words exist that are a classic piece of understatement. They changed the course of the third manned lunar landing mission and the lives of the three astronauts onboard and many of the controllers, engineers and recovery forces who participated in the six-day flight of Apollo 13 in April 1970: 'Houston, we've had a problem.'

This was the first indication that a major incident had occurred onboard the lunar-bound spacecraft, 205,000 miles from Earth, during the evening of the astronauts' third day in space. Over the next four days, the whole world watched and prayed as the astronauts and flight controllers battled against the odds to return Apollo 13. The story of this is one of incredible effort, ingenuity and plain good luck.

The mission emblem for Apollo 13 depicted the Sun God Apollo being carried across the heavens in a horse-drawn chariot. The motto *Ex Luna, Scientia – From the Moon, Knowledge* – epitomised Apollo 13 and the next phase of the US manned lunar programme. The flights of Apollo 11 and 12 had proved that landing on the Moon was possible, that it could be done accurately and that useful work could be performed, and that a crew could be sustained there, if only for a few hours or days. Unfortunately, by the time Apollo 12 flew, both the American political machine and the public at large – whilst openly proud and supportive of the Apollo programme right up to Apollo 11 – had turned their backs on the programme in favour of more pressing needs at home and abroad. The TV camera that failed during the first Apollo 12 lunar surface EVA did nothing to help promote the increased scientific objectives of the Apollo flights. Apollo 20 had been cut from the NASA budget to make room for Skylab, and the scientific community was trying to save Apollo 18 and 19 from the lost mission list as Apollo 13 flew.

The crew for this demanding mission was led by America's most experienced astronaut on his fourth and last flight into space. James A. Lovell Jr (nicknamed 'Shaky' by fellow astronaut Pete Conrad) had narrowly missed selection for the first astronaut group in 1959, but was selected for the second group in September 1962.

The original Apollo 13 crew (*left–right*) Commander Lovell, Command Module Pilot Mattingly and Lunar Module Pilot Haise. Mattingly was removed from the flight 72 hours before launch, due to exposure to German measles – which he never contracted.

He flew twice in Gemini. The first flight – Gemini 7, in December 1965 – set a world endurance record of 14 days, which still stood and would not be broken until the 18-day flight of Soyuz 9 a couple of months after Apollo 13. (Gemini 7 held the record for US astronauts until Skylab 2 in 1973.) Lovell had flown a rendezvous and docking flight on Gemini 12 in November 1966, followed by the historic Apollo 8 lunar orbiting mission at Christmas 1968. He was now making his last flight – the pinnacle of his career – on Apollo 13.

By this time – beginning with '13' – NASA was initiating a policy of flying a highly experienced Commander with a pair of rookies. (Before this mission each crew consisted of three experienced astronauts, including a Command Module Pilot with rendezvous experience.) The two rookies for '13' had both been selected in April 1966. Thomas K. (Ken) Mattingly had performed extensive support work in the Apollo programme, including long hours developing the Apollo spacesuit. He was the Command Module Pilot, who would remain in orbit to conduct landmark tracking and photography while his two colleagues explored the Fra Mauro region of the Moon. The Lunar Module Pilot, who would accompany Lovell to the surface, was Fred W. Haise Jr. He had also performed extensive support work in the Apollo programme (on the back-up crew for Apollo 8 and 11), and was one of the Astronaut Office experts on the Lunar Module and its systems.

The crew was named for the Apollo 13 mission in August 1969, and immediately began specific mission training. As Lovell and Haise had served Apollo 11's back-up crew, and Mattingly had been promoted from the support crew, their normal rotation would have seen them bypass Apollo 12 and 13 and fly Apollo 14. The original Apollo 13 crew of Alan Shepard, Stu Roosa and Ed Mitchell were finding it tough going to meet the Spring 1970 launch time, and were judged by NASA HQ to require more training time due to their overall inexperience. (Shepard was the only flight-qualified astronaut in the crew, with only 15 minutes as America's first spaceman back in 1961. He had been out of crew training from 1963 to 1968 and had two new spacecraft to master – the CSM and the LM.) In June 1969 – following the return of the Apollo 10 mission and before the launch of the Apollo 11 mission – the Lovell crew was approached to fly Apollo 13. Lovell said he thought they could do it, so it was Apollo 13 for which they trained. Thus began an ironic sequence of events associated with that flight. Asked once before the flight if being assigned to a mission with the '13' designation bothered him, Lovell replied, 'The Italians think it's a lucky number, so I vote for them.'

The mission of Apollo 13 was intended to last ten days, and would include two Moon-walks. The first would deploy a geophysical experiment station called ALSEP (Apollo Lunar Surface Experiment Package), and the second was to be a geological traverse to the 370-m wide Cone Crater. Lovell and Haise were to attempt to dig trenches, collect the usual random samples, take photographs of their landing site, and drill into the lunar surface for important core samples. The core sample from the LSD was part of the experiment to plant sensors to measure the heat flow from the Moon's core. Upon the mission's return, scientists would have a third selection of rocks to examine and analyse – which would keep them busy for many years.

Just one week before launch it became known that Mattingly had been exposed to German measles, from the children of a friend of Charles Duke, a member of the back-up crew. Mattingly, it was discovered, had no immunity to the disease, and so there was the predicament of whether to cancel the flight, replace the whole crew with the back-up crew (one of whom had also been exposed to the disease) or replace Mattingly with his back-up John L. (Jack) Swigert. Lovell argued to keep Mattingly on the crew. According to Lovell, Mattingly was one of the most conscientious and hard-working team members of the astronaut group. He tried to present the case to Dr Tom Paine, the NASA Administrator, that German measles was not that bad, and that if Mattingly was unfortunate enough to contract it during the flight, it would be on the way home on the quiet part of the mission. Lovell stated that from his previous experience as CMP on Apollo he knew both he and Haise could bring Apollo home if they had to. In any case, Mattingly did not have the measles at that moment, and might very well not be infected at all. Despite the fact that Lovell was proved right (Mattingly never contract the disease), Dr Paine (who was not an MD and relied on the Flight Surgeons for advice) said that the risk was too great. Swigert, however, had first to prove to Lovell and Haise that he could take Mattingly's place with only a couple of days' proficiency training. On Thursday and Friday, 9 and 10 April, Swigert was put through his paces with Lovell in the CM simulator. On the afternoon of the day before launch, Lovell accepted that Swigert could carry out the job and could fly on '13.'

Lovell later stated that he should have recognised several omens leading up to the flight, perhaps indicating that things were stacking up against them – the flight number, the exchange of Mattingly with Swigert, the planned launch time of 13.13 (Houston time) on 11 April (the major incident that had yet to befall them would occur on 13 April!), and the launch from Pad 39A (3 × 13). In addition, a couple of pre-launch incidents also failed to trigger alarms in the minds of those concerned.

As ground tests on the LM were being conducted at the Cape, poor insulation of the supercritical helium tank in the descent stage was discovered. The flight plan was revised to enable the crew to enter the LM three hours earlier than originally called for, to obtain onboard read-out of the helium tank pressure. This procedure, according to Lovell, provided two benefits during the incident. It allowed the crew to 'shake down' the spider-like spacecraft earlier in their mission. This was one turn of luck that helped the quick power-up of the LM after the explosion and, in retrospect, probably helped to save their lives. It also meant that LM controllers would be at Mission Control just when they were needed most for subsequent incidents and not just prior to the lunar landing phase.

Finally, there was the No. 2 oxygen tank, which had originally been destined for use on Apollo 10. It had been removed prior to that mission for modification, but sustained some damage and had been replaced. The refurbished tank was recycled for Apollo 13 instead. Lovell has always congratulated the Apollo 10 crew for 'getting rid of that tank'. Almost twelve months after it was fixed, factory tested, and reinstalled in '13's Service Module, it was tested again during the countdown demonstration test of 16 March 1970. In normal operation, the liquid oxygen in the tanks was half consumed. Tank No. 1 achieved this consumption, but No. 2 dropped to only 92% of capacity. Engineers tried to clear the problem by piping gaseous oxygen through vent lines at 80 psi to expel the liquid oxygen, but this did not work. Despite the publication of an interim report indicating the problem, tests were repeated on 27 March, only two weeks before launch. According to Lovell, 'The No. 1 again emptied normally, but its idiot twin did not'. Following talks between the contractor and NASA personnel, it was decided to boil off the remaining oxygen in tank No. 2 by using the electrical heater in the tank. This worked, but it took eight hours of 65-volt dc power from ground support equipment to remove the oxygen.

With the benefit of hindsight, Lovell admitted that he should have demanded the replacement of the tank before it flew – as was within his authority as CDR – but he went along with the decision. As a result, he now agrees, he too must take a share of the responsibility for the expensive Apollo 13 failure. This reflects on the direct astronaut involvement that the whole flight-crew had during the early one-shot missions, in which a crew followed the development of 'their' spacecraft almost from construction to flight. This not only gave the crew confidence in their vehicle, it also made them intimately familiar with the operation of its systems from an engineer's viewpoint as well as from that of a pilot.

The Apollo 13 mission finally began at 14.13 EST (13.13 Houston time) on Pad 39A at the Kennedy Space Center. The sixth manned launch of the monstrous Saturn V vehicle carried Lovell, Haise, Swigert and their spacecraft to an appointment on the Moon – or so they thought.

Launch-day breakfast for the Apollo 13 crew (*left–right*) Haise, Lovell, and former back-up Command Module Pilot Jack Swigert, who replaced Mattingly.

The launch of Apollo 13 from Pad 39A (3 × 13) at 13.13 hrs CST on 11 April 1970. During launch, one of the second-stage engines malfunctioned, whereupon the crew thought they had had the glitch for the mission – but they were wrong.

The Saturn family had enjoyed 100% launch success rate as the rocket thundered into the skies above the Cape. With Apollo 13 in progress, it was hoped that this record would continue. However, just $5\frac{1}{2}$ minutes into the mission, the three astronauts inside the CM felt a small vibration. Shortly afterwards, the centre engine of the S-II second stage shut down two minutes earlier than planned. To compensate for this, the other four engines burned for an extra 34 seconds, and the third stage added an extra nine seconds in order to achieve orbital velocity. Lovell thought that this was perhaps the failure that all those omens were leading to and that, from this point on, the flight would be routine and uneventful.

Once in orbit, the crew successfully checked the spacecraft and then ignited the Saturn S-IVB single third-stage engine once more to set off for the Moon. The Command Module (Odyssey) then extracted the LM (Aquarius) from the top of the spent stage of the Saturn, and they moved clear of the spent stage, which was to impact with the Moon. For the next two days the flight was as smooth as Lovell had hoped. Haise succumbed to the effects of space adaptation syndrome, but he soon recovered.

The crew cancelled the mid-course correction manoeuvre, deeming it unnecessary, but late on the second day in space they initiated a 'hybrid' transfer to the Moon. This would allow the combined spacecraft to sweep as low as 70 nautical miles above the surface instead of the 115 miles altitude as on previous missions. This was done to allow the spacecraft to arrive at the desired point on the Moon at the correct time for the most favourable light conditions for landing. The manoeuvre also meant that the spacecraft moved out of the so-called 'free-return trajectory' and would no longer be able to swing around the Moon and return to Earth without further course

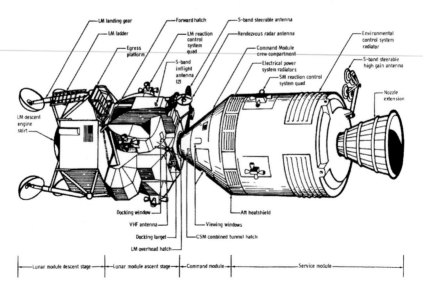

An artist's impression of the Apollo 13 LM and CSM, showing the spacecraft in docked configuration at the time of the explosion.

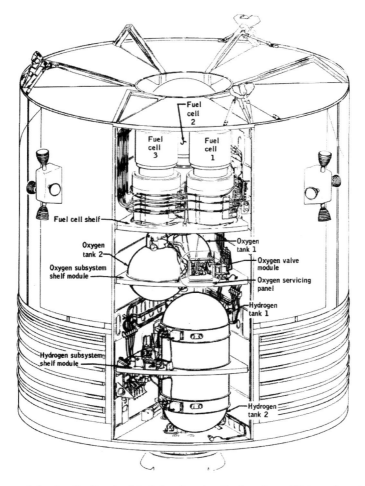

Cutaway of the Apollo Service Module, showing the location of internal equipment.

corrections. This free-return trajectory – used on Apollo 8, 10 and 11 – was a safety precaution, in the event of an engine failure, that provided the crew with a relatively safe and guaranteed ride home. The 'free return' restricted the options for a launch date and lunar landing site. It was forsaken on Apollo 12 to facilitate the required landing approach, and a 'hybrid free-return' trajectory was selected. The S-IVB third stage would burn to aim for a different point near the Moon, still allowing a loop around the Moon but requiring a further burn of the service propulsion system on the SM to put the spacecraft on course for a successful return to Earth. This decision was to have serious consequences for Apollo 13.

The hybrid burn was conducted at 20.54 EST, and was so accurate that no further small corrections were needed. The plan worked satisfactorily for Apollo 12, allowing Conrad and Bean to land near the Surveyor 3 automated probe in the Moon's western hemisphere. Apollo 13 was to land a few hundred miles away. As a

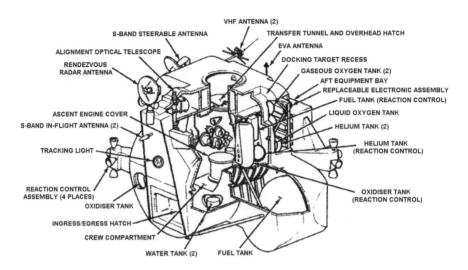

Cutaway of the ascent stage of the Apollo Lunar Module, showing the inside of the crew compartment.

consequence of this burn, if Apollo 13 proved unable to enter lunar orbit, it would loop around the Moon and head for the Earth, but would miss it by 4,000 miles. Apollo 13 was therefore taking a substantial risk, but NASA was, by now, very confident of CSM systems.

During the evening of 13 April, Lovell and Haise entered the Lunar Module for a check on the helium pressure in the suspect tank. They found it at an acceptable level, and spent the next hour showing TV viewers around the inside of the spacecraft that would be their home on the Moon. It was a relaxed tour. Their launch and successful trajectory setting for the Moon had relieved some of the tension of the mission. The previous day, Swigert, usually a calm person, suddenly sent a panic message on the downlink, explaining that in the rush to fly on Apollo 13 he had forgotten to submit his Federal income tax return. He asked for an extension, and after some ribbing from the CapCom he was told that American citizens out of the country were eligible for a 60-day extension – and he was definitely out of the country. 'I assume this applies', Swigert thankfully replied.

The spacecraft had by then reached the point where it no longer had the engine power to simply turn around and return to Earth. They were committed to at least a loop around the Moon if anything went wrong – but what could go wrong? Swigert had reported that he was having difficulty in reading the quantity gauge on one of the oxygen tanks, which had gone high off the scale. One of his regular tasks was to turn on the fans in the tanks to stir the contents to provide accurate readings. This was to be done shortly after the TV show had ended. As the TV show continued, some of the Flight Control positions at Mission Control were vacated. The night shift was soon to come on duty, and the mission was proceeding so close to flight plan that there was little for anybody to do until the astronauts reached the vicinity

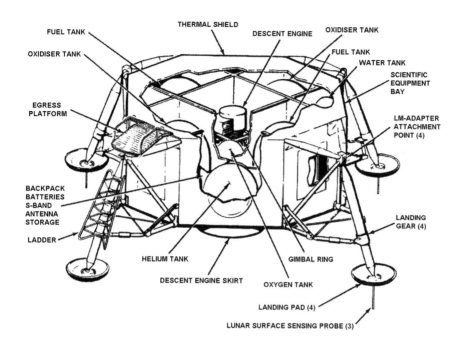

Cutaway of the descent stage of the Apollo Lunar Module, showing the inside of the unpressurised compartment.

of the Moon. Several of the larger TV networks did not even broadcast live, and beamed edited highlights on the late evening news once the crew had bid farewell to the ground. Public interest in the flights to the Moon had dwindled so much that the only really interested viewers were the handful of flight controllers and close family members who had gathered in the viewing room at Mission Control especially for the event.

At GET 55 hrs 46 min, as the crew finished the 49-minute broadcast, Lovell announced, 'This is the crew of Apollo 13 wishing everybody there a nice evening, and we're just about ready to close-out our inspection of Aquarius and get back for a pleasant evening in Odyssey. Good Night.'

Haise was still in the Lunar Module nine minutes later, as Lovell swam through the tunnel to the Command Module to join Swigert as they prepared to close-out the LM and settle down for the night. Seconds later, Lovell afterwards stated, 'the whole roof fell in ... Fred was still in the Lunar Module. Jack was back in the Command Module in the left-hand seat, and I was halfway in between the lower equipment bay wrestling with TV wires and a camera – watching Fred come down – when all three of us heard a rather large bang. Just one bang. Now before that, Fred had actuated a valve that normally gives us that same sound. Since he didn't tell us about it, we all rather jumped up and were sort of worried about it. But it was his joke and we all

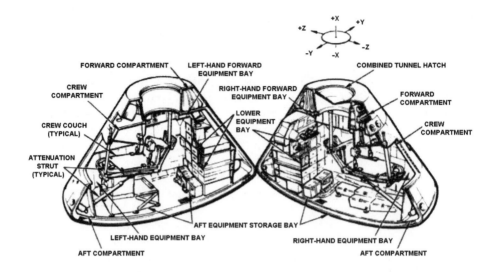

Cutaway of the Apollo Command Module (the centre couch is removed for clarity).

thought it was a lot of fun at the time. So when this bang came, we really didn't get concerned right away. But then I looked up at Fred and he had that expression like it wasn't his fault. We suddenly realised that something else had occurred, but exactly what, we didn't know.'

Haise, from within the confines of the tunnel, mentioned that he felt a vibration. Swigert reported a master alarm and a Main Bus B under-volt (a loss of power) two seconds later, and promptly informed Houston of the situation.

Swigert: 'Hey, we've got a problem here.'
CapCom (Lousma): 'This is Houston, say again please.'
Lovell: 'Houston, we've had a problem. We've had a Main Bus B under-volt and we had a pretty large bang associated with the caution and warning there. We got a Main Bus A under-volt now too. Main A is reading zip right now.'
CapCom (Lousma): 'We'd like you to attempt to reconnect fuel cell 1 to Main A and fuel cell 3 to Main B. We're still trying to come up with some good ideas here for you.'
Lovell: 'Let me give you some readings. Our $O_2$ cryo number 2 tank is reading zero, and it looks to me, looking out of the hatch, that we are venting something. We are venting something out into space. It's gas of some sort.'

Lovell later continued his recollections of those moments in the spacecraft: 'I guess its kind of interesting to know what the feelings of the crew are when something like this happens. When you first hear the bang, you don't know what it is. We've had similar sounds in the spacecraft before that were for nothing, and then I looked out the window and saw this venting. My concern was increasing all the time. It went from 'I wonder what this is going to do to the landing' to 'I wonder if we can get back home again.' When I looked up and saw both oxygen pressures –

one actually at zero and the other one going down – it dawned on me that we were in serious trouble ... A pleasant evening indeed! My first thoughts of disappointment about not landing on the Moon vanished. Now it was strictly a case of survival.'

The bang that they had heard was an explosion in the second liquid oxygen tank located in the Service Module. This tank provided life-supporting oxygen, supplying fuel cells 1 and 2, and enabling them to generate electrical power to operate the CSM systems. This was Apollo's primary power supply. True, there were batteries available in the Command Module, but they were needed for entry during CM independent flight. Besides, they had only a ten-hour maximum lifetime. Even allowing for all the most favourable situations, Apollo 13 was, at the very least, 87 hours from Earth. One of the tanks had dropped to zero in eight seconds and the other was visibly doing the same – and the astronauts knew that this was their source of breathable oxygen.

On top of a building next to Mission Control, a group of amateur astronomers were tracking the course of Apollo 13, as they did during all the lunar flights. As they peered through their telescopes, images on a rigged-up TV screen showed what looked like a variable star and a growing cloud. They had no communications with the spacecraft or with Mission Control, and had no reason to connect their observations with Apollo, as they had been experiencing reception difficulties all night. They went home, unaware of what they had just witnessed until they heard the news.

On the spacecraft, the dropping gauges and venting gas showed the results of the explosion. Fortunately, the astronauts could not see the gaping hole in the side of the Service Module. If they could, then perhaps their morale might have been reduced, and the return flight would have been much more difficult.

The crew's first thought on hearing the bang was that a meteorite had struck the craft. Their reaction to this was likened to submariners closing the hatches to limit the potential flooding of their submarine. Before Lovell spotted the venting, his first action was to order the closing of the hatch in the tunnel that connected the two spacecraft. They tried, but failed to lock the hatch due, it was thought at the time, to its slight misalignment as a result of the effects of the explosion. Once they realised they did not have a cabin leak, they left the hatch open and began to think in terms of using the still attached LM and its independent systems to help them home. Later, when they had to lock the hatch before jettisoning the LM, it closed and locked easily. 'That's the kind of flight it was', said Lovell.

PAO: 'Here in Mission Control we are now looking towards an alternative mission, swinging around the Moon and using the LM power systems, because of the situation that has developed here this evening.'
CapCom: 'We're starting to think about the LM as a lifeboat. We figure we have about 15 minutes-worth of power left in the Command Module, so we want you to start getting over in the LM and getting some power on.'

By then, about an hour had elapsed after the explosion first rocked the spacecraft, and the crew had also been considering the move into the LM. The idea of the LM 'lifeboat' was not new. Indeed, preliminary ideas and suggestions had circulated during evaluation of the lunar orbital rendezvous mode selected in 1962 and continued into 1963, and some of the original specifications had been modified, with this in mind. However, in a 1964 report the contingency plan for using the LM as a lifeboat was finally dropped because, according to the report, 'no single reasonable CSM failure could be identified that would prohibit the use of the service propulsion system'. But that theory had just been proved incorrect.

Fred Haise, one of the astronaut office LM experts, had spent 14 months at the Grumman plant in Bethpage during its development, and had never been shown the LM plans for the mode that they were to use in space. Simulations were performed on the possibility of using the LM as a back-up propulsion system in the event of a failed SPS system. They trained for use on the Descent Propulsion System (DPS) and in some cases the Ascent Propulsion System (APS) engine, but always with a fully operable CM and a partially working SM. The use of an LM with an essentially dead mother-craft had been inconceivable. Apollo 9 had evaluated some of this theory in space (by manoeuvring a docked combination using the DPS, but not the APS), but Apollo 13 was now rewriting the book as it flew.

Almost as soon as the explosion occurred, Mission Control had come alive with flight engineers, astronauts and contractor personnel, who flooded in to lend whatever support they could offer. And so it was for the rest of the mission. Flight Director Gene Kranz linked into the Flight Directors' Loop in MCC and informed his team to save all data and carefully consider the situation. There was no need to 'blow the mission', as he expressed it. At GET 56:32, Grumman's Apollo 13 Flight Director Willard Bischoff put in a call to his head office in Bethpage, New York, to rally support in Mission Control and a group of engineers in New York. It was approaching midnight in New York, and it was to be a long night for people associated with Apollo 13. For most, it was a night that lasted three days, until the astronauts were safely floating in the ocean!

With Apollo 13 approaching 58 hours MET, the spacecraft was well over 210,000 miles away from Earth and moving away at only 2,100 mph. Lovell and Haise were in the LM, activating its systems as quickly as possible. Swigert was in the CM, shutting down its systems – which were apparently undamaged – in order to conserve battery, oxygen and cooling water for the re-entry and landing phase. As the two astronauts began bringing the LM systems to life, Haise also conducted preliminary studies on how they could sustain their LM consumables for the long trip home. Working on the data being displayed, and with his own notes on earlier LMs which he had worked with – notes which he had fortunately taken with him – he began to assess how they could stretch the LM's planned 45-hour life to 90 hours, and how it could support three men instead of two. He also calculated that they had more than enough oxygen, as in addition to the large tank in the descent engine there were also two tanks in the ascent stage. These, plus the two portable life support systems and their PLSS tanks, which were to have been used on the Moon, all added to more than enough air for the trip home for all three men. (At the time of LM jettison it was calculated that they still had 28.5 pounds of oxygen left – a little more than half of the amount that they started with.)

During the flight home, the LM itself was partially powered down to conserve all available power. All that remained active were the controls and supplies for life

Lovell in the Commander position of Lunar Module Aquarius during the mission, from where he controlled the firing of the LM descent engine to refine the flight path.

support, communications, and environmental control, with those for propulsion being used only during important manoeuvres. The power usage was thus reduced to less than 11 amperes per hour. Battery power in the LM totalled 2,181 ampere hours, and with the CM being turned off, apart from essential instruments that could not be reactivated, the LM would recharge the CM batteries.

The first major task performed within the LM, once power had been transferred, was to align the spacecraft to the correct attitude. Swigert transferred the platform inertial reference verbally (by yelling the alignment through the access tunnel) to Lovell and Haise in the LM, before cutting power to the CM. Swigert used battery power to keep the CM alive, while the other two astronauts powered up and aligned the LM gyroscopes to bring them in line with those on the dying CM. The astronauts had to work fast, but they succeeded in aligning the spacecraft at the correct attitude. Lovell later thought of this as the first major turning point after the explosion, as the systems on the LM were less sophisticated than those on the CM and were never used for deep space tracking. Had the crew lost their inertial reference, they would have had to manoeuvre the combined spacecraft to sight on the Sun, Earth and Moon. This would have been hindered by the gaseous cloud which had formed around the spacecraft as a result of the explosion, because it prevented sextant sightings of known stars.

Lovell had experienced this phenomenon during the Apollo 8 mission, when urine dumps masked the stellar background for star sightings as the liquid froze and evaporated in the vacuum. He had also inadvertently dumped the computer memory

Astronauts at Mission Control monitor the progress of their colleagues: (*standing, left–right*) former Apollo 13 CMP Mattingly and support crewman Vance Brand; (*seated, left–right*) astronaut Deke Slayton (Director of Flight Crew Operations), CapCom Jack Lousma and back-up Apollo 13 commander John Young.

at one stage, so he was well accustomed to the navigational difficulties on an Apollo mission when flying a 'programmed' trajectory.

The next obstacle was to return the spacecraft to the free-return trajectory, enabling it to swing around the Moon and head for home. With the CSM inert, however, they could not use the large SPS engine, and had to rely on the smaller descent engine on the LM.

A 38-fps prograde burn would have to be made using the LM DPS engine. The LM would be pushing a larger spacecraft, with a different mass and a different centre of mass. Grumman's engineers had designed the LM with this in mind, and during the Apollo 9 flight, just over a year earlier, Jim McDivitt's crew had operated their LM Spider in a 'tugboat' mode for more than six minutes, pushing their CSM Gumdrop. Apollo 9 proved that it could be accomplished in space, although nobody thought that the method would actually be used on a lunar flight.

The free-return manoeuvre would also ensure that the mission duration was reduced, providing a larger margin for the consumables. Several possibilities were examined, including a very tricky immediate about-turn of 180°, eliminating a lunar flyby and leading to a return and landing in the Indian Ocean. This required a second burn that would reduce the return leg by ten hours and provide a velocity reduction of 860 fps. Questions remained over the capability of the LM to sustain this, and so NASA turned to the specialists and outside contractors who were

Dr Charles Berry, Director of the Medical Research and Operations Directorate, briefs Marilyn Lovell, wife of Apollo 13 Commander Jim Lovell, in the Mission Control viewing room.

beginning to flood into the MCC and related sites to determine if this manoeuvre would work. TRW Systems propulsion engineers stated that the descent engine was designed for a burn duration of some 17 minutes. It could be throttled and could be restarted up to twenty times. In addition, their own engineers had test-fired LM Aquarius' engine before launch, so they had actual performance data on the engine that they were to fire in space, as well as data from hundreds of test firings and previous LM flights. They expressed confidence that the engine would perform as required and as a back-up they completed computer and simulator tests. These tests also suggested that there would be no foreseeable problem in burning the LM engine for the required duration.

Some thought was given to dumping the now useless SM in order to lighten the mass that the LM would be called upon to manoeuvre in space. Willard Bischoff put a call in to LM expert Tom Kelly at Bethpage at GET 66:59, informing him of the status prior to the burns and of NASA's request that all contractors be at the ready during the burn period. Kelly asked why it was necessary to keep the SM attached, since everybody was questioning its reliability, in view of the need to return the astronauts as soon as possible. The reply was three-fold, and included the simple reason that the LM had never fired its engine in space with just the CM attached and there were no data from simulations. No one knew what damage the SM/CM separation system had been subjected to by the explosion, and it would be advisable to wait until they *had* to separate it prior to re-entry. In addition, keeping the SM attached would afford some protection for the all-important CM heat-shield from the cold of space and the additional threat of debris impact from particles generated by the explosion. The conclusion was, therefore, that the SM would remain part of the combination for as long as possible.

The crew asked for more time to prepare for the first burn, which was successfully accomplished at 61:28:43. It imparted an increase of 16 fps to the vehicle, moving it off the hybrid trajectory and back into free return. The astronauts now tried to get some rest, and while Swigert and Lovell tried (unsuccessfully) to relax in the chilly, dark CM, Haise stood watch in the LM.

As they rested, offers of help came in from all around the world to assist in the recovery of the crew. The Soviet Union announced that several vessels were heading for the recovery areas, and that 'the Soviet Government has given orders to all its citizens and members of the armed forces to use all necessary means to render assistance in the rescue of the American astronauts.' Millions around the world began to follow each minute of the mission with intensity, in stark contrast to the apathy of a few hours earlier. Apollo 13 hit world headlines and front-page news, and Lovell's broadcast, which the TV networks had ignored, was now played repeatedly. There was no more TV from '13', so the now-eager networks were starved of live footage of the crisis.

After being ignored by the news media for two days, the Apollo 13 mission became the centre of media activity for the remainder of the mission to return the astronauts to Earth.

Pope Paul, at an audience in front of 10,000 people outside the Vatican, extended prayers and thoughts for the three astronauts. Prayers were also held around the world, in schools, offices and factories. People stopped in the streets to listen to the latest reports and bulletins on the flight's progress. In New York, Times Square's neon display carried a running commentary on the mission's status. The whole world was following Apollo 13. There are only a few missions in human spaceflight that have captured the world's imagination and brought everyone together: Gagarin's flight, Apollo 8, Apollo 11 – and now Apollo 13.

Around America, the space community was coming alive to support the huge effort that was being mounted to bring the three men home. All hopes of a lunar landing had been dashed. All that mattered now was to bring home the stranded astronauts. At North American Rockwell in Downey, California – home of the Apollo CSM – engineers used available data to run computer simulations on emergency situations. At the Massachusetts Institute of Technology – where the Apollo guidance and navigation system was developed – a group of thirty specialists worked through the night with mathematical problems and solutions. In New York, a ten-telephone link was kept open between Mission Control and a special support room set up at Grumman at Bethpage. President Nixon was briefed by former NASA astronauts Mike Collins (of Apollo 11) and Bill Anders (of Apollo 8), and personally telephoned the wives of Lovell and Haise and the parents of bachelor Swigert. On hearing the news, many NASA employees and contractors, on their way home from a long shift, turned around and drove straight back. All knew that they had skills that would be needed to bring the astronauts home. Engineers stopped for speeding explained where they were going, and were escorted by police outriders. Suddenly after almost ignoring the third attempt to put men on the Moon, it seemed as though the whole of America felt personally involved with overcoming the problems of Apollo 13.

Astronauts also flooded into the space centre to carry out support duties and simulator work. Al Shepard and Ed Mitchell operated one LM simulator at JSC, while Gene Cernan and Dave Scott worked on the other one. At the Kennedy Space Center, Florida, back-up Apollo 15 Commander Dick Gordon used the LM simulator there to determine emergency procedures. Ken Mattingly joined the back-up crew of John Young and Charles Duke, simulating procedures, manoeuvres and techniques to prove them before sending them up to the crew in space. The astronauts used identical equipment to simulate as close as possible the real environment that their colleagues were experiencing in space.

CapCom astronauts Joe Kerwin, Vance Brand and Jack Lousma all took turns in reading checklist after checklist up to the crew, and other astronauts wandered in and out offering support, advice and ideas. Astronauts Joe Engle, Ron Evans and Tony England all worked long hours during the flight as the space community became, for a few long days, a larger and much closer family.

It was later estimated that each of the simulators was used for about forty hours during the flight. Teams of simulation and training groups worked around the clock without a break – some of them sleeping on the floors of the offices – to ensure that all commands given to the crew would be provided in minute detail.

Early on Wednesday, 14 April, the crew received the data for the second burn of five minutes to speed up the trajectory and return them more quickly. This was to take place just two hours after they had looped behind the Moon. As Lovell was busy evaluating the procedures for the burn that would bring them home, he observed that his two rookie colleagues were at the windows, setting up cameras to take photographs of the Moon as they made their fleeting pass around the far side. Having flown Apollo 8, Lovell had seen it all before. He said to them, 'Gentlemen, if we don't make this next manoeuvre correctly, you won't get your pictures developed', to which they replied that this was their first and probably only trip out here, and that Lovell had been there before. It was the only chance they had of approaching that close to the Moon and they were not going to waste the opportunity. As it happened, the photographs they took proved to be some of the best images of the far side obtained during any Apollo mission.

At 18:21 CST, 14 April 1970, they slipped behind the Moon and out of radio communication with the Earth. This was a lonely time on any Apollo mission, but for Apollo 13 it was even more so. With their spacecraft in such a state, the astronauts' only link with home was, for the time being, severed. It was a time for the flight controllers to catch their breath, as, with blank screens, they had nothing to do but wait and hope. It was the only chance of a rest that the whole team had from the time of the explosion to splashdown.

On board, there was excitement at flying so close to the Moon, and considerable disappointment in being denied the landing. Given the mission's problems, it was possible that the Apollo programme would be curtailed, and that this would be the last Apollo crew to visit the Moon. (Earlier, Lovell had told his two colleagues, 'Boys, take a good look at the Moon. It's going to be a long time before anybody gets up here again.' However, he was not aware, at the time, that he had been on a 'hot mike' for about 45 minutes, and that his comments were being beamed to Earth. Unfortunately, Lovell was later wrongly accused of sabotaging the Apollo programme with this comment. Dr Tom Paine had to explain that he really did not mean it, and after the flight Lovell was asked to explain his comments before the Senate Space Committee. He said that he never intended to make public those comments, and that they were intended only for his two colleagues. Unfortunately, his comments turned out to be accurate. The setback of Apollo 13 helped seal the fate of Apollo 18 and 19, which were cancelled in September 1970. Apollo 14 did not fly around the Moon until February 1971. The real fear of possibly losing a crew around (or on) the Moon was too much of a risk. The remaining Apollo missions would be stretched out until 1972, and the programme would then yield to the Skylab space station in 1973.)

At 18:49 the spacecraft emerged from the other side of the Moon and re-established continuous communication with Earth.

Apollo 13: 'Houston, Aquarius. The view out there is fantastic. You can see where we're zooming off.'

As they sped away from the Moon, the spent third stage of the Saturn V headed straight for the planned crash-landing to stimulate the seismic instruments left by

Apollo 12, with the shock from impact serving as a probe of the lunar crust. At 19:09 EST on 14 April, the S-IVB struck the Moon at a force equivalent to $11\frac{1}{2}$ tonnes of TNT, and set off a seismic impact signal 20–30 times greater than had the unmanned LM ascent stage from Apollo 12 some months before. The stage hit the Moon at 8,465 fps (just less than 6,000 mph), 74 nautical miles west-northwest of the Apollo 12 site, and provided the only useful scientific data returned from the Apollo 13 experiment programme.

CapCom: 'By the way Aquarius, we see the results from Apollo 12's seismometer. Looks like your booster just hit the Moon and it's rocking a bit!'
Apollo 13: 'Well at least something worked on this flight. I'm sure glad we didn't have an LM impact too!'

The crew had endured many crises onboard during the flight, and there was no time to relax. One of the more serious incidents occurred when the time came to align the platform with known stars in order to burn the engine to speed them homewards. As the astronauts looked out, all they could see was the floating debris that they had been venting since the explosion. It was impossible for them to sight on the required stars, because they could not distinguish them from the sunlight glinting on the debris. Without alignment, the correct burn in the desired direction would not be possible, nor would the eventual alignment for CM entry. Mission Control produced the idea of using the Sun to check the alignment, as it was so distinctive that it could not be mistaken. This they did successfully, and correctly aligned the spacecraft. Yet another procedure had been added to the contingency flight data book!

This time the burn would start with five seconds at 10% throttle, then 21 seconds at 40%, and almost four minutes at full throttle. It added 585 mph to the velocity, and set course for the splashdown point south of Samoa in the Pacific Ocean. The prime recovery carrier, *Iwo Jima*, was already heading for that area.

Life onboard Apollo 13 was difficult to say the least. A flight to the Moon on Apollo was sheer luxury compared with Lovell's two weeks cooped up in Gemini 7. However, even the Apollo Command Module was designed for only a limited duration and a specific goal. Better facilities would be developed with later and larger craft. The Lunar Module was even more basic, as it was designed to support two men for only about three days.

It was also extremely cold in the CM. With the systems turned off there was no internal heat source to keep temperatures at a comfortable level. The inert CM's temperature settled at 38° F, and was so uncomfortable that the astronauts had extreme difficulty in sleeping. They tried to sleep in the LM, which was warmer, but it was still uncomfortable. Lack of sleep contributed to the situation that they were experiencing, and led to concerns over fatigue and the inability of the astronauts to perform even routine chores. The sunlight streaming in through the windows, even with the shades up, did not help. Lovell recollected, 'We were as cold as frogs in a frozen pool, especially Jack who got his feet wet and didn't have the lunar overshoes that Fred and I wore.' It was so cold that water vapour condensed on the cold metal. 'The sight of perspiring walls and wet windows made it seem even colder. We considered putting on our space suits, but in the spacecraft they were bulky and sweaty. Our Teflon-coated in-flight coveralls were cold to touch, and we longed for some good old thermal underwear.' Swigert eventually put on an extra set of longjohns. Haise developed a kidney infection in the latter stages of the mission, and became quite ill. Understandably, all three of them became irritated and very tired. At one point, they pulled off the biomedical monitoring devices, much to the dismay of the Flight Surgeon on the ground who, upon losing all telemetry readings from their vital organs, thought all three astronauts had just died!

The LM did not have hot water to mix the dehydrated foodstuffs, so all of their food was cold. After hours of power-down, part of the spacecraft began freezing over as the intense cold of space penetrated through. The astronauts dozed irregularly, with at least one of them always awake, and they never really had a good sleep period. Late in the flight, Deke Slayton, Director of Flight Crew Operations – one of the few men (himself an astronaut) who could interrupt the communications and talk directly to the crew – came on the line: 'I know that none of you are sleeping worth a damn, because it's so cold. You might care to dig out the medicine kit and take a couple of dexedrine tablets apiece.' Dexedrine is a powerful stimulant, and the medics had been considering its use for some time; but it has the unpleasant side-effect of a severe let-down when it wears off. The crew held off taking the tablets until a couple of hours prior to re-entry when their effect was negligible and they were exhausted.

CapCom: 'If we could figure a way to get a hot cup of coffee to you, it would taste pretty good about now, wouldn't it?'
Apollo 13: 'Yes, it sure would. You don't realise how cold this thing [the spacecraft] gets.'
CapCom: 'Hang in there, it won't be long.'

Water was a problem. Haise had calculated that they would run out of water five hours before splashdown. Again, using his experience on previous LMs, he knew that the Apollo 11 LM ascent stage was not crashed on the Moon as were later ones. Engineering analysis on this vehicle had shown that the LM could sustain 7–8 hours in space without active water cooling, until its guidance system objected to this unplanned roasting. The crew did conserve water, but with the increased fear of dehydration they were closely monitored by the ground medics. Liquid was available from the PLSS backpacks intended for use on the Moon, but the cooling suit liquid was not exactly fit to drink. As it was the crew drank sparingly. Each took six ounces of water a day – one fifth of normal intake – plus fruit juices, cold hot-dog sausages and other wet-pack foods (when they ate at all). They *did* become dehydrated; and by dehydrating himself, Haise aggravated his kidney infection. The crew set some unenviable records which were unbeaten on any other Apollo flight. Lovell lost 14 lbs in weight, and the crew as a whole lost a record 31.5 lbs – almost 50% more than any other Apollo crew. The flight finished with 28.2 lbs of water remaining – 9% in total.

Initial computations by the crew suggested that they had enough lithium hydroxide canisters for the important removal of carbon dioxide from the LM cabin atmosphere, keeping the crew alive. There were four in the LM and four from the

The 'mail box' jury-rigged by the crew upon instruction from controllers at Mission Control to scrub the atmosphere of carbon dioxide inside the spacecraft.

Portable Life Support System (PLSS), including back-ups. But instead of supporting two men for two days, they needed to support three men for four days. 36 hours after the explosion, the build-up of carbon dioxide to a dangerous level threatened the lives of the crew. They would have died from poisoning their own lungs had it not been for an ingenious system emanating from Crew Systems Division at Mission Control, devised by astronaut Tony England and relayed to the crew by radio. They used the CM square canister, which would not fit the LM system's round opening, attaching it to the LM system by using plastic bags, tape, cardboard cue-cards from the unused lunar EVA maps, and linking them to two EVA suit hoses to clean the air in the LM. It was crude and it looked awful, but it worked well.

Two course corrections made by the crew refined the alignment for atmospheric entry as they came sweeping in. As the spacecraft entered Earth's influence of gravity at 08.38 on Wednesday morning, still 216,277 miles from home, tracking indicated the craft would have missed the Earth by 99 miles and sailed on into deep space. The crew used a manual burn alignment technique first devised on Lovell's Apollo 8 mission, but like most of the events and procedures used on Apollo 13 it was never expected that it would be used on a lunar mission. Swigert took care of the time for the burn, telling his fellow astronauts when to ignite the engine and when to stop it. Haise covered pitch, and Lovell took care of the roll and the start and stop buttons. Together, as an integrated but exhausted team, they achieved the necessary burn.

$6\frac{1}{2}$ hours prior to entry they started to prepare the CM for the end of the mission, but found that they could power up the batteries only $2\frac{1}{2}$ hours prior to entry – still well within capacity. They also found that Mission Control was now expert at rewriting the documents and sending them up. It took three days, instead of the usual three months, to develop CM power-up procedure. Lovell commented: 'We found the CM a cold, clammy tin can when we started to power-up. The walls, ceiling, floor, wire harnesses and panels were all covered with droplets of water.' They suspected that the condition was the same behind the panels, and the chance of a short-circuit caused some apprehension in the cabin. But thanks to procedures developed following the Apollo 1 pad fire, no arcing took place. The amount of condensation only became clear when it 'rained' in the CM as they decelerated through the atmosphere!

'Things were getting better all the time', Lovell later commented. 'We were in a different situation now, because normally when you come home you have only the CSM. Now though, we had a dead Service Module, a Command Module that had no power and a Lunar Module. A wonderful vehicle, but it didn't have a heat shield and shortly we'd have to abandon it.'

LM systems had to be powered down as checklist after checklist was read up to Swigert in the CM. 'I may not sound too clear, because I'm holding a flashlight between my teeth.' At least the CM was warming up to a comfortable level.

An artist's impression of the separation of the damaged SM, showing the unique CM/LM configuration never before flown in space. (Courtesy NASA.)

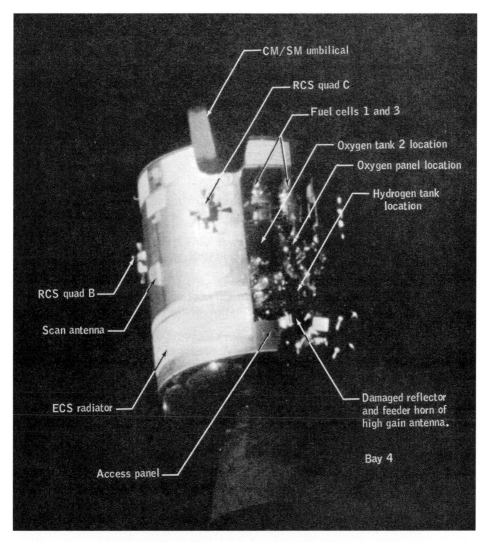

CM/SM umbilical

RCS quad C

Fuel cells 1 and 3

Oxygen tank 2 location

Oxygen panel location

Hydrogen tank location

RCS quad B

Scan antenna

ECS radiator

Access panel

Damaged reflector and feeder horn of high gain antenna.

Bay 4

The damaged Service Module, showing the Sector 4 panel completely blown away, exposing the internal systems of the module.

CapCom: 'You can jettison the Service Module when you're ready. No big rush but anytime.'

Swigert had taped over a switch that was marked 'LM Jettison' so that he would not inadvertently trigger the wrong control when Lovell and Haise were still inside it. Lovell considered this to be an especially good idea!

Swigert later recalled: 'The ground had read up a very nice time-line. We had one nervous moment. Normal procedures required arming the logic bus [the pyrotechnic devices used to sever connection between the SM and the CM] and letting Houston look at the relays. At this time, we didn't have any telemetry with Houston. Fred came up and I said, 'Fred, I'm all ready to jettison the Service Module. I'm just getting ready to arm the pyros.' Fred said, 'I'll get a 'Go' from Houston.' I told him, 'We don't have any telemetry with Houston, so you're just going to have to put your fingers in your ears and stand by.' So I armed the 'A' system and I could hear the relays. Nothing happened, an encouraging sign, so I armed the 'B' systems and nothing happened. So I kind of felt we were home free, the procedure worked well. We used a push–pull method. Jim and Fred were in the LM using the translation controller to give us some velocity. When Jim yelled 'Fire!' I jettisoned the Service Module and it went off in the midst of a lot of debris, which is usual.'

As the Service Module drifted past the windows, the crew snatched hurried photographs and saw for the first time the extent of the damage.

Apollo 13: 'OK, I've got her and there's one whole side of that spacecraft missing! Right by the high gain antenna the whole panel is blown out, almost from the base to the engine. It's really a mess. Man, that's unbelievable! Looks like a lot of debris just hanging out from the side near the S-band antenna.'

As Haise powered down the LM, Lovell floated back into the CM, stating, 'Well I can't say that this week hasn't been filled with excitement.' CapCom replied, 'Well Jim, if you can't take any better care of the spacecraft than that, we might not give

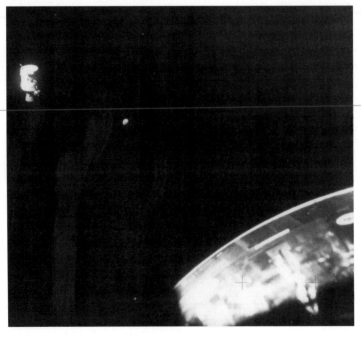

Lost Moon. From the window of the LM and beyond the docked CM the astronauts hurriedly took photographs of the damaged Service Module with the unreached Moon in the background.

An artist's impression of the separation of the SM and subsequent undocking of the empty LM as the CM heads for re-entry and splashdown. (Courtesy NASA.)

you another one.' The CM now contained the three astronauts once more, and was powered by its own batteries for the descent. After ensuring that all their film was out of the LM, and having grabbed a couple of mementoes from the cabin, they closed the hatches and sealed themselves in the CM.

Since the CM engines could not push or pull the spacecraft away, the docking tunnel was used to allow internal air pressure to force the two craft apart. At 11:23 Swigert pushed the button for LM jettison.

CapCom: 'Farewell Aquarius, and we thank you.'
Haise: 'She sure was a great ship.'

Aquarius ended its life, taking a Radioisotope Thermoelectric Generator (RTG) power rod with it, in an unceremonious burn-up over the Pacific Ocean, to be dumped into one of the deepest oceanic trenches. 600 miles south-east of Samoa, the recovery fleet awaited the CM as the vehicle headed through the blackout and on to its parachute deployment.

Grumman's Apollo 13 Lunar Module Flight Director at Mission Control, Willard Bischoff, completed the final entries in his personal flight log. He had monitored the flight from Mission Control, and had relied on the support team at the Grumman LM manufacturing plant in Bethpage to come up with answers to the constant stream of questions that NASA engineers were asking about the capabilities of the Lunar Module. Now the time was approaching to discard the vehicle that had undoubtedly saved the astronauts' lives, and complete the mission:

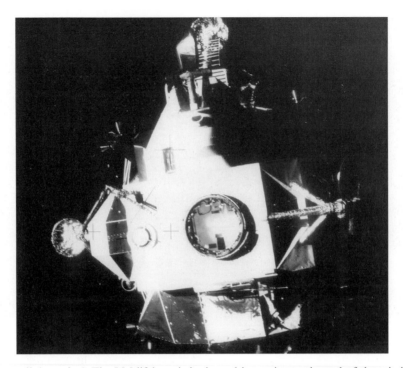

'Farewell Aquarius'. The LM lifeboat is jettisoned just prior to the end of the mission, after the successful completion of its job in returning the astronauts back to near Earth.

GET 141:30: 'LM Jettison.'
142:17: 'EMS checkout OK.'
142:28: 'Lost LM data.'
142:31: 'LM data back – still looks good.'
142:33: 'CM G&N [Guidance and Navigation] is Go!'
142:35: 'Lost LM data again.'
142:36: 'Data back but spotty [intermittent].'

The CM – now on its own – began its fast descent towards the Pacific Ocean, flying over the Indian Ocean and across southern Australia at 24,000 mph, and entered the atmosphere at 400,000 feet. Lovell, Haise and Swigert caught sight of a tiny and distant Moon that sank below the horizon as they began their fiery re-entry.

142:39: 'LOS'. Loss of signal; blackout begins.

For several minutes all that anyone – including the astronauts onboard the CM – could do was wait and hope. Ionised particles – formed due to the speed of the spacecraft entering the thickening layers of the atmosphere – had created a sheath that blocked all radio communications. For those on the ground the minutes seemed like hours as they waited for news that the CM had survived through the re-entry blackout.

The view from Mission Control, showing the TV coverage of 'Apollo on the mains' descending by parachute to the ocean. The central TV monitor on the flight controller's console also shows a cartoon of a Lunar Module lifeboat 'towing' a crippled CSM.

142:45: 'ARIA Aircraft has AOS.' Acquisition of signal; telemetry data from CM.
142:46: 'Voice contact with CM [Swigert].'
142:49.30: 'Drogues out.'
142:50: 'Mains [parachutes] out on TV.'

Onboard the CM, Lovell warned his crew to prepare for a hard landing, having experienced a 'ton-of-bricks' splashdown on the Apollo 8 mission.

142:54:46: 'Splashdown! Stable I [right side up] 413.5 miles from ship.'

Odyssey had one more surprise as it caught the crest of a wave and settled gently into the ocean – finally back on Earth. Inside the capsule the three men lay quietly in their couches and waited for the pararescue divers. The whole world cheered. They were home, exactly 142 hrs 54 min 41 sec after lift-off.

Swimmers attached flotation equipment to the spacecraft, and one by one the three astronauts scrambled out into the rubber raft and were raised up to the waiting helicopter. On the recovery ship, they looked tired and relieved. Too tired for public speeches (those would come later), they were whisked away for medicals and a well-earned rest. Still to come were the trip home, debriefings, Presidential welcomes and family reunions – and the realisation of what they had gone through. Lovell also expressed a personal feeling: 'You do not realise what you have on Earth until you leave it.'

Apollo 13 flight log, 11–17 April 1970

April, Central Standard Time

| | | |
|---|---|---|
| 11 | 13.13 | Launch from Kennedy Space Center, Florida; premature cut-off of one engine of second stage of Saturn V. |
| | 15.48 | S-IVB third stage reignited in translunar insertion burn to place on course for the Moon. |
| | 17.14 | CSM separates, turns around, and extracts LM from S-IVB stage. |
| 12 | 19.54 | Apollo combination changes to hybrid trajectory in preparation for landing approach at Fra Mauro |
| 13 | 21.08 | Oxygen tank No. 2 explodes in SM, 205,000 miles from Earth: 'Houston, we've had a problem … looks like we're venting something.' |
| 14 | 02.43 | LM engine burns to put spacecraft back on free return trajectory. |
| | 18.21 | Apollo 13 swings behind the Moon; Haise and Swigert take photographs; Lovell says 'If we don't get home, you'll never get them developed.' |
| | 18.49 | Spacecraft emerges on the other side of the Moon: 'The view out there is fantastic.' |
| | 19.09 | S-IVB stage strikes the Moon with the equivalent force of 11.5 tons of TNT, and the Apollo 12 seismometer records the impact: 'Well, at least something worked on this flight.' |
| | 20.41 | Aquarius descent engine completes a 4 min 30 sec burn to increase the return speed of the spacecraft. |
| 15 | 22.31 | Mid-course correction to refine the aim at Earth: 'The Earth is whistling in like a freight train.' |
| 17 | 07.15 | The damaged SM is jettisoned and photographed by the crew: 'There's one whole side of that spacecraft missing.' MCC: If you don't take better care, we won't give you another one.' |
| | 10.43 | The crew transfer to the CM and jettison LM Aquarius: 'Farewell Aquarius, and we thank you.' |
| | 12.08 | CM splashes down in the Pacific Ocean. |
| | 12.53 | Apollo 13 astronauts safe onboard the prime recovery ship *Iwo Jima*. |

On 17 April a Review Board was set up to investigate what had happened. It reported its findings on 16 June 1970:

- An electrically initiated fire located in the No. 2 oxygen tank of the Service Module caused the accident, and the cause of the combustion was the ignition of Teflon wire insulation on the fan motor wires inside the tank. The wiring had created an electrical arc.
- A review of telemetry data recorded immediately prior to the accident showed electrical disturbances caused by electrical arcs within the tank. The review of the construction of the tank also revealed potential fire ignition sources such as electrical wiring, unsealed electric motors, rotating aluminium fans and Teflon wire insulation, all of which would burn in supercritical oxygen.
- Tracing the construction history of the tank also pointed to a loosely fitting fill tube assembly, probably caused by jarring during assembly at the prime contractor and which later caused difficulties during normal pre-launch

operations at KSC. Unusual procedures not previously employed during pre-launch testing and checkout for an Apollo mission actually subjected the tank to an unprecedented eight-hour period of heater operations and pressure cycling.

- Two protective thermostatic switches on the heater assembly were also found to be design-rated for 28 volts of dc power instead of the 65 volts dc power actually used at KSC and constructor facilities for a series of tests.
- Finally, this discrepancy was not detected at NASA, North American Rockwell (the prime contractor for CSM) or Beech Aircraft Corp. (the supplier of the tank), or in any review of documentation. This was poor co-ordination by the programme management. Nor was there any requirement to perform qualification or acceptance testing requiring switch cycling under high g loads. The Board stated that 'this was a serious oversight in which all parties shared.'

The Board's recommendations for ensuring that an accident of this type could not happen again on an Apollo SM were:

- Modification of the cryogenic storage system in the SM to include the removal of all wiring and unsealed motors from direct contact with liquid oxygen, where a potential short-circuit could ignite adjacent materials. This was to ensure the minimum use of Teflon, aluminium and other relatively combustible materials in the presence of oxygen.
- The modified cryogenic oxygen storage system should in future be subject to a rigorous requalification programme, including awareness of potential operational problems.
- A review and modification of caution and warning systems aboard the Apollo spacecraft and in Mission Control Center. This was to include different levels between expected normal operating ranges and master alarm trip levels, prevention of one alarm blocking a second in the same sub-system, and the provision of a second level of visual and audible alarms, which could not be easily overlooked.
- Consumable and emergency equipment in the LM and the CM should be reviewed to enhance their potential for use in a 'lifeboat' mode. System compatibility was also to be reviewed between the CM and LM sub-systems and components.
- NASA should also complete additional reviews of LM power system anomalies including further investigation and explanation of any significant anomalies that occur during launch preparations in critical sub-systems. Critical decisions involving the flightworthiness of such sub-systems should be supported by the participation of an expert in that sub-system.
- NASA should also conduct a complete re-examination of all its spacecraft, launch vehicles and ground support equipment that contain high-density oxygen. Further research on materials capability, ignition and combustion of strong oxidisers at various g levels and on the characteristics of supercritical fluids should lead to new design standards where necessary.

- NASA should also review all Apollo sub-systems as well as the engineering organisations responsible for them at both Houston and its prime contractors to ensure adequate understanding and control in areas of engineering and manufacture.

On 16 June 1970 the Review Board Chairman Dr Edgar Cortright stated to the House Committee on Science and Astronautics that 'the Apollo 13 accident is a harsh reminder of the immense difficulty of this undertaking [exploration of the Moon]. The total Apollo system of ground complexes, launch vehicle, and spacecraft constitutes the most ambitious and demanding engineering development ever undertaken by man. For these missions to succeed, both men and equipment must perform near perfection. That this system has already resulted in two successful lunar explorations is a tribute to those men and women who conceived, designed, built and flew it. Perfection is not only difficult to achieve, but difficult to maintain. The imperfection in Apollo 13 constituted a near-disaster, averted only by outstanding performances on the part of the crew and the ground control team who supported them. The Board feels that the Apollo 13 accident holds important lessons which, when applied to future missions, will contribute to the safety and effectiveness of manned spaceflight.'

It was also noted that in 1965 the CM had undergone many improvements, which included raising the permissible voltage in the heaters in the oxygen tanks from 28 to 65 volts dc. Unfortunately, due to an oversight, the thermostatic switches on these

Apollo 13 controllers at Mission Control celebrate the successful return of the Apollo 13 astronauts.

Jim Lovell reads the newspaper headlines of his own rescue at the end of his fourth and final mission into space.

heaters were not modified to suit the change. During one final test on the launch pad, the heaters were on for eight hours. This subjected the wiring in the vicinity of the heaters to very high temperatures (1,000° F), which subsequently severely degraded the Teflon insulation on the fan motor assembly wires. The thermostatic switches started to open while powered by 65 volts dc, and were probably welded shut. Later tests were undertaken but warning signs were not acted upon, allowing the tank to be filled with oxygen and becoming in effect a potential bomb, which exploded on 13 April 1970 as the spacecraft approached the Moon. From Apollo 14 on, a third oxygen tank was added in a different location, and an isolation valve and auxiliary battery were installed in the SM.

In terms of lunar exploration, Apollo 13 was a failure, but in other ways it clearly demonstrated the resourcefulness, teamwork and spirit that NASA could command at that time. Apollo 13 was very much a 'successful failure'.

After splashdown, Grumman – the builders of the Lunar Module – sent Rockwell – the constructors of the CSM – a tongue-in-cheek breakdown and recovery bill, which Rockwell countered with a towing charge bill! It was smiles all round.

The Moon landing was lost, but at the same time the recovery of the crew had showed NASA at its very best. The next mission – Apollo 14 – would not fly until 1971, and only four more teams of astronauts would walk on the Moon during the Apollo programme. NASA was already planning a new generation of spacecraft – the Space Shuttle, which, it was hoped, would be a reusable and reliable vehicle for frequent and safe access to space. Apollo 13 had demonstrated that the exploration of space was still a high-risk venture, and although it was a failure in that the lunar landing was not achieved, it was termed a 'successful failure' due to NASA's recovery of the astronauts from the brink of a major disaster. Sixteen years later a different disaster and a different NASA would not be so fortunate.

# Mir: fire and a collision, 1997

After greeting the resident crew at the entrance hatch of the Russian space station Mir, the first things that a new crew-member noticed when boarding the station for the first time were the noise, the humidity and the smell. Mir had been in space a long time. Images of a pristine, organised, spacious environment were a thing of the past.

Unlike the Shuttle – which remained in orbit for only a week to 17 days – Mir was a long-term research platform. The first element of the Mir complex, the Base Block, had been orbited in the spring of 1986, and the structure gradually expanded over the years, with extra modules, tons of equipment and succeeding teams of cosmonauts.

The first section to be added to the Mir complex was the astrophysical module, Kvant 1, which was launched in 1987. This was accessed via a docking hatch at one end of the cylindrical Base Block. On the other side of Kvant, one of two Soyuz ferries was docked. These brought the cosmonauts to Mir and would take them back to Earth at the end of their mission.

At the opposite end of the central module was the hatch leading into the multiple-docking module, or Node, where four scientific research modules had been attached over the years. Kvant 2 (the augmentation/airlock module) was added in 1989. Kristall (the technology module, to which the Shuttle Docking Module was attached in 1995) was added in 1990. Spektr (the remote sensing module) was added in 1995, and Priroda (another remote sensing module) in 1996. They were arranged around the four radial docking ports on the Node. Each module was linked through open hatches to the main module and connected by cables and ducting to the power and supply sub-systems of the Mir complex. A sixth hatch led to the second Soyuz ferry spacecraft, used for the remaining three cosmonauts, at the opposite end of the complex to the first one.

Mir was designed for an operational life of five years, but by 1997 it was entering its eleventh year in orbit. It was beginning to show signs of age and, in addition, nothing ever seemed to be thrown away. Although fresh supplies had been brought up and tons of rubbish had been taken away by automated Progress cargo supply ships, Mir was still full of unwanted equipment. It bore little resemblance to the pristine training modules on Earth. This was a real, living, space station.

Far from the idea of a platform from which to conduct scientific experiments, aimed at furthering our knowledge of the space environment, much of the cosmonauts' time was taken up with hours of housekeeping and maintenance. They were constantly forced to compromise between the power supply levels generated by the solar arrays on the hull of the complex and the power requirements of the range of scientific experiments and onboard systems.

From 1986 until 1995, Mir provided the Russians with a valuable training aid for a sustained, routine life in space. The longest stay – by cosmonaut Dr Valery Polyakov – was 14 months. This was ideal preparation for building larger, more complex space stations and for providing baseline data for extended flights to Mars in the distant future. The Americans, busily revising the plans for the International Space Station, had no such resource. Indeed, the only space station they had had was Skylab, built from surplus Apollo lunar hardware and launched in 1973. Three crews had lived aboard Skylab for 28, 59 and 84 days, finally surpassing the 1965 Gemini 7 duration record of 14 days. Since 1981 and the start of Shuttle orbital operations, America's only option for duration flight had been up to 17-day missions (operating with two twelve-hour shifts) with the Shuttle/Spacelab module. What NASA needed was space station experience. The opportunity came with the creation of the joint US/USSR co-operation programme in 1992.

American astronaut Jerry Linenger arrived onboard the Russian Mir space station on 15 January 1997. He was the fourth of seven NASA astronauts who would all become part of Russian resident crews, rather than members of visiting crews. Norman Thagard had begun the series, spending three months on Mir in 1995. He was followed, in 1996, by Shannon Lucid, who began a two-year continuous occupancy of the space station by American astronauts.

Lucid had been replaced by John Blaha, who was on board Mir when Linenger floated through the docking hatch from *Atlantis* and into his new home. Mike Foale succeeded Linenger, and astronauts David Wolf and, finally, Andy Thomas, would complete the programme. The seven astronauts would provide much-needed baseline data for the Americans to build upon towards the ISS. They were to experience first-hand what it was like to live week after week in sustained orbital flight. Unlike Shuttle flights, they would be cut off from familiar surroundings in more than just the physical sense, as they were to fly as part of a Russian national mission. They were part of a Russian crew, under the command of a cosmonaut, and under the control of Mission Control, Moscow, not Houston. This was to be a very different experience, and not even the months of mission training in Russia prepared the Americans for the cultural shock they were all to encounter on Mir.

The character of both Linenger and Foale was dramatically challenged – much more so than the others – during their combined eight-month residency on Mir in 1997. During their stay, two of the most serious events in human spaceflight history occurred onboard Mir. These put at risk both the success of the mission and the lives of the crew.

Linenger's first month onboard the space station passed without incident. He had arrived as a member of the STS-81 crew, and joined the twenty-second (EO-22) Russian Mir Expedition (resident, or long-duration) crew. He exchanged his custom-

Mir space complex in orbit, January 1997. The central core 'module (Base Block) supports a Soyuz spacecraft (centre foreground) and a Progress unmanned freighter docked to Kvant 1 (background). Clockwise from top are the modules Spektr, Kristall with the Shuttle docking module (at right), Kvant 2 with the YMK manned manoeuvring unit stored outside (bottom), and Priroda.

built seat liner in the Soyuz descent module for that of John Blaha, who was going home on the Shuttle at the end of his tour.

During the five days of joint operations, Blaha spent around two hours each night briefing Linenger on the layout and facilities of the space complex. He also wrapped up his own experiments, as part of the NASA Board Engineer 3 programme, and assisted Linenger in setting up his programme, NASA Board Engineer 4. After the handover period was completed, Blaha and the STS-81 crew departed Mir, and landed in Florida on 22 January.

During the rest of January and the first week of February, Linenger devised a routine for his own experiments. He also tried to become part of the station team, with his Russian colleagues, Commander Valeri Korzun and Flight Engineer Alexandr Kaleri. On 7 February the Mir crew moved the docked Soyuz TM-24 spacecraft from Mir's front docking to the rear docking port. Linenger described this short flight in the Soyuz as 'an afternoon spin in a spaceship'. This manoeuvre had occurred many times during the long life of the space station, and was carried out to free the main port for the next Soyuz crew. The Soyuz was the transport to and from Mir for Russian-launched crews, and was also a 'space lifeboat,' available in case the crew needed to make a quick exit and emergency return to Earth. In order to redock TM-24, the automated Progress M-33 cargo vessel was undocked from Mir's rear port and manoeuvred to a new parking orbit, where it would remain until it could redock after the departure of the Russian EO-22 crew.

Soyuz TM-25 – launched on 10 February – carried the replacement Russian crew for the twenty-third main expedition (EO-23), Commander Vasili Tsibliyev and Flight Engineer Alexandr Lazutkin. Also on board was German astronaut Reinhold Ewald, who was to complete a 20-day stay on Mir before returning home with Korzun and Kaleri, leaving Linenger to continue his resident stay with Tsibliyev and Lazutkin.

On 12 February Soyuz TM-25's automatic rendezvous and docking system (Kurs) failed during the final approach to Mir. The automatic system took the Soyuz to within 8 feet, and then shut down. The vehicle then moved in to about 20 inches before it began to draw back in response to the failed system. Tsibliyev was ordered to take manual control, which the crew had not been trained to do so close to the docking port. With less than 60 seconds to go to orbital sunset, when any docking is difficult, Tsibliyev nudged the docking probe into Mir's drogue to dock successfully. (As a cost-cutting measure, the Russians would remove the expensive Ukrainian Kurs system from Soyuz and Progress spacecraft in 1999.) The cosmonauts would have to complete manual docking with the Russian TORU docking system. Progress cargo spacecraft would also use the TORU system, but would be controlled by the cosmonauts from inside the Mir complex, via visual observations and TV cameras mounted on the approaching Progress vehicle. The decision to test this system during the EO-23 mission would have a dramatic effect on future Mir operations.

For three weeks the six cosmonauts were to perform joint experiments throughout the space station. These would be in addition to the scientific experiments, routine maintenance and housekeeping that accompanies every Mir mission. One of these mission chores occurred during the night of 19 February when the crew encountered problems with one of the six gyrodynes of the station's attitude control system. The faulty unit was located in the Kvant 2 module, and was repaired. However, events of a far more serious nature would overtake this minor problem a little more than four days later.

On 24 February a joint NASA/Russian press release stated: 'A problem with an oxygen-generating device on the Mir space station last night set off fire alarms and caused minor damage to some hardware on the station.' The report went on to say that none of the six cosmonauts were injured, and performed well in handling the fire. The fire reportedly burned for 90 seconds, exposing the crew to heavy smoke for 5–7 minutes and forcing them to don face-masks. After the fire was extinguished, Linenger, a physician, completed a physical examination of all crew-members. Ground-based medical personnel advised them all to wear face-masks and goggles until Mir's atmosphere could be analysed.

Phase One Shuttle–Mir Programme Director, astronaut Frank Culbertson, added that it was unfortunate that the incident had occurred, and that 'Russian management and operations specialists have been very informative as to what happened, and we are working closely with them on evaluating the health of the crew and how best to respond to the damage.' He also said that the ground support on both sides had been 'excellent'.

Mir systems continued to operate normally after the fire, and the crew spent the next day cleaning up the area of the Kvant 1 module in which it had occurred. It was several weeks before the full story of the fire became known.

The accident originated in the lithium perchlorate canisters (candles) which are burned to produce additional oxygen for a larger station crew. A core crew of up to three cosmonauts onboard Mir draws upon oxygen provided by water hydrolysis, which also purifies the air in the station. The candles supplement this hydrolysis system when more than three cosmonauts are onboard. These candles last for around twenty minutes, provide sufficient oxygen for one person for one day, and are changed periodically. Initial reports from the Russians indicated that they thought there was a crack in the shell casing of the oxygen generator, allowing a leak into the actual hardware holding the cartridge. This exposed the metal holding device to extremely high temperatures.

On the evening of the fire the six cosmonauts were all together around the communal table in the Base Block of Mir, enjoying a meal and listening to a music cassette over the deafening noise of the workings of the space station. Although a nuisance, the 'humming' reminded them that the station was 'alive', providing heat, light, air to breathe, and power throughout the complex.

During casual conversation between the six cosmonauts, resident Commander Korzun asked new Flight Engineer Lazutkin to change one of the candles for a fresh one. On his first flight in space, Lazutkin was happy to oblige. He floated over the table and into the Kvant 1 module, where the Elektron unit was located, selected a new candle, lit it and put it in its holder. This was a normal, routine procedure that had been completed countless times before. Suddenly he screamed, in a combination of shock and pain, as a jet of flame spurted from the canister and across the width of the module. Grabbing the nearest bungee cord, he quickly pulled himself back into the main module to inform the others of the dramatic events. By then the other five cosmonauts had been alerted by the fire alarms, and could see the flames, the melting metal around the surrounding walls, and the black smoke filling the area.

A fire on any vehicle is dangerous, but in the controlled environment of a spacecraft it is even more life-threatening. The carefully controlled environment is essential to sustain life on board, and 'jumping overboard' is not an option because there is only the airless vacuum of space. This was another reason why Soyuz spacecraft were docked to the station, so that in the event of an emergency the crew could rapidly escape. The problem in this case was that the flames blocked the path to one of the Soyuz vehicles.

Reacting instantly, Korzun grabbed a wall-mounted fire extinguisher and dived into the Kvant module, where the fire was still burning. After a couple of moments, and with smoke billowing around him, Linenger reached in and pulled at Korzun's

Linenger wears a full Russian face mask following the fire and billowing smoke.

leg to determine whether he was still conscious. Turning to the American, his face black, Korzun gasped for more extinguishers, an oxygen mask and goggles. He told his colleagues to also put on protective gear, as the fumes might be toxic. The heat increased onboard Mir, and while the cosmonauts were wearing the protective face-masks and goggles, they were also wearing little actual clothing – only shorts!

During a lecture in 1998 Lazutkin relived the experience, describing the event as both tragic and comic. Korzun's firing of a jet of water onto the fire from an extinguisher reminded Lazutkin of a mill worker standing in front of a furnace. They almost abandoned the station until they realised they had only one copy of the latest Soyuz retro-burn schedule. This was updated by the ground, and was sent to the onboard computer on Mir. The cosmonauts then printed it and put the most recent version in the Soyuz descent module. The schedules were updated daily and enabled the most accurate calculations based on the current orbital parameters to be available to the cosmonauts when manually entering re-entry programs into the Soyuz computer in preparation for descent.

With two Soyuz spacecraft docked they needed a second copy, but they had not printed one. As other crew-members frantically floated around Mir, trying to find extra fire extinguishers, Kaleri activated the program to print a second schedule. It amazed Linenger that at the height of the fire, and with billowing smoke, the cosmonaut was trying to work on his computer, not realising what Kaleri was trying to do. The smoke even floated into the Soyuz craft docked at each end of the Mir complex, which would have made breathing in the closed confines of the spacecraft very uncomfortable had they chosen to evacuate the station.

They also realised later that, even with a second copy of the retro-burn schedule, if both Soyuz spacecraft had separated from Mir, the emergency return would have been equally hazardous. No one had thought of the need to provide separate data for two Soyuz craft. The two vehicles would have been using the same retro-burn data and flying the same return trajectory to re-enter the atmosphere at the same time, heading for the same landing site. This had never been done before, and the risk of collision this would have caused could easily have turned an emergency into a disaster.

They later confirmed that the fire had burned for about 90 seconds, but it seemed much longer. (In later press conferences, Linenger actually described a 3-foot flame that lasted for 15 minutes!) The three extinguishers that were used seemed to be of little use, as the candle burned itself out.

Linenger recounted the experience in an air-to-ground interview the following month: 'You had to react to the situation, you had to keep your head about you, so I guess it was just a matter of survival. Without getting that fire out, there was no way to get to one of the Soyuz capsules. The smoke was the most surprising thing to me. I did not expect smoke to spread so quickly – about ten times faster than I would expect. The smoke was immediate, and it was dense. I could see the five fingers of my hand and I could see a shadowy figure of the person in front of me. But I could not make him out. Where he was, he could not see his hands in front of his face. It was very surprising how rapidly the smoke spread throughout the complex.'

Linenger went on to explain that the crew put on the Russian oxygen respirators, and that without them they would have been unable to breathe. The respirators were activated by chemical reaction, whereupon the wearer strapped a mask against his face, flipped a switch on the attached oxygen bottle, and took a couple of quick breaths. For full flow of oxygen the humidity from the breath was supposed to activate full oxygen supply. For Ewald, nothing happened with his first mask, and for Linenger the first one he wore took too long, so he grabbed a second. Linenger did not want to risk taking a breath in the fume-filled compartment, and all six cosmonauts were relieved that the respirators worked well to protect them.

After the fire was out and when the thick smoke stopped billowing from the unit, Linenger became very concerned about the health of the crew. He set up emergency equipment around the station in case a serious respiratory ailment arose due to inhalation of toxic fumes. He immediately examined each crew-member (including himself), and did so again after one, 24 and 48 hours. He also determined oxygen saturation in their blood, and checked their lungs. He concluded that their quick action in donning the protective gear had prevented any problems.

The crew narrated a video of the fire damage showing the damaged canister and the crew wearing face-masks and goggles, and downlinked this to the ground. Damage had been confined to the immediate area and had affected insulation panels and support brackets, but the structural integrity of the station had not been compromised. The crew spent several days cleaning up the mess, wiping bulkheads and removing the damaged equipment, which was later sent back to Earth for analysis.

After blaming Lazutkin for errors in changing the canister, the Russians later determined that the most probable cause of the fire was frayed fabric from a ground worker's overalls becoming caught in the unit during pre-flight handling, which ignited when the canister was used.

Following the fire, attention shifted to the landing preparations for Korzun, Kaleri and Ewald. They would bring samples taken after the fire back to Earth to be analysed by life science teams in Russia, the US and Germany. The three men landed in Russia on 2 March.

If the remaining crew-members onboard Mir were hoping that life would settle back into a routine after the fire, they were to be disappointed. Other problems seemed to plague them. The first was the attempted redocking of the Progress M-33 cargo craft on 4 March. Now that the Soyuz TM-24 crew had departed, the Progress could be docked to the rear port. This was to be the test of the Russian built TORU docking system, but it was unsuccessful. According to Tsibliyev, the cargo craft would not respond to his commands, and actually passed within 750 feet of Mir. All future attempts to redock the Progress were abandoned to conserve propellants for its deorbit burn.

On 7 March it was reported that the primary Elektron unit, housed in the larger Kvant 2 module, had failed. This was the system designed to separate oxygen from the onboard waste water system and return it to the cabin atmosphere using electrolysis. The crew was instructed to activate the second unit in the Kvant 1 astrophysics module (where the fire had occurred), which they did. Almost immediately they reported that the second unit was showing higher levels of hydrogen than it should have been. It would have been shut off automatically by a gas analyser measurement after six hours.

Based on this information, the crew was instructed to shut down the unit and activate the solid fuel oxygen candles. There were enough candles onboard Mir to support a three-person crew for two months, and a new Progress, M-34, was scheduled to bring up extra candles and replacement parts for the primary Elektron unit in early April. The crew was also instructed to position fans around the station, to evenly distribute the oxygen supply throughout the modules, before activating additional candles. After the incident of the fire on the previous occasion, the crew expressed concern in the further use of candles, but there was no other choice.

With the primary Elektron 2 system in Kvant 2 suffering a filter bypass problem, and the second unit – Elektron 1 in Kvant 1 – continuing to have pumping problems, NASA Phase 1 Director Culbertson issued an update on the situation. He expressed confidence that the crew had the ability to make the repairs once the spares were delivered by the next Progress. The reliability of these spares was a problem, however, with some having to be manufactured before loading onto Progress. NASA was also asked to add hardware to STS-84, the next Shuttle–Mir docking mission. This was not easy so late into the preparations for a Shuttle launch, as the cargo manifests for every Shuttle mission are determined weeks or months before lift-off. Late additions are not normally allowed, as the mass and centre of gravity for launch, plus the emergency and scheduled landing, are all programmed into the computers well before lift-off.

To ensure that no repeat of the fire occurred, the crew was to use modified procedures for lighting the candles and for dealing with any future fire.

Lazutkin and Tsibliyev inside the cluttered main Base Block.

Modifications to fire-fighting procedures were also evaluated. A second cosmonaut was to stand with an extinguisher, while the first changed and ignited the new candles. The Russians considered that a second fire was not probable, and after tests of candles on the ground, those on Mir were thought to be reasonably safe. The crew was not so sure!

On 19 March the Spektr primary attitude control sensor (Omega) failed, prompting the motion control system computer to automatically switch to its back-up system – a process that took three minutes. During this transfer the station gyrodynes controlled the orientation and movements of the station around the three axes. However, when the back-up system came on line, the rotation was beyond the range of the gyrodynes. The crew therefore turned off the attitude control system to place Mir in 'free drift', and then used the onboard thrusters to stabilise its attitude. Because the station was in gravity gradient, in which the centre line of the station always pointed to Earth, the solar arrays were unable to lock onto the Sun and charge the onboard batteries. To conserve power, the gyrodynes and other equipment (including some scientific experiments) were turned off. Ground control then uplinked new data to the crew to restart the gyrodynes and manoeuvre the station to regain solar lock-on. There were also plans to reroute cabling to a different

Omega unit in one of the other modules. During the first week of April a far more serious problem occurred. A leak in one of Kvant 2's thermal cooling loops had been detected by ground control. The system was leaking ethylene glycol into the station's atmosphere. This troubled Linenger. 'Inhaling ethylene glycol caused the concern that I have as a crew physician', he reported on 11 April. 'We wear respirator face-masks when we are doing repairs in Kvant, so that lessens some of the effects. But we're suffering from congestion, which I'm sure is due to some of the fumes.'

Lazutkin had suffered an allergic reaction after some of the glycol had floated into his eye. Linenger later revealed that had allergic reactions in space when exposed in a closed environment, although they had been advised by both the Americans and the Russians that the vapours from the glycol (which was of the same type as that used as a coolant in car engines) would not harm humans, and that no ground test revealed allergy symptoms. To help compensate for the loss of cooling in the loop, the orientation of the complex was altered to place the Kvant 2 module in the shadow of Kvant 1. The crew began working on the leak, using a special sealant and waterproof cloth. Once it was repaired they redirected some of the cooling ducts into the Base Block to lower internal temperatures. A pressure drop had been found in one of the Kvant 1 cooling loops, which had triggered the shut-down of the Vozdukh carbon dioxide scrubbing system that purified the air. This increased the internal temperature to 104° F in some parts of the Base Block. As a consequence, the crew reduced their use of the exercise treadmill and the Chibis pneumatic vacuum suit, used to simulate the pull of gravity on lower limbs, as part of the in-flight countermeasures to prepare for returning to Earth.

Linenger noted that the increase in temperature had affected both the crew and some of the onboard equipment, reducing the ability to remove moisture from the air in particular. The situation could become very complex at times, and he suggested a need to 'fix our problems'. He also added: 'Out here at the frontier, I expected the unexpected. We've been getting some of that.'

The launch of Progress M-34 to Mir, on 6 April, was a vital step in allowing the mission to continue. It carried much needed replacement equipment, repair tools, spare fire extinguishers, face-masks and lithium hydroxide canisters. In case the Progress automatic docking system were to fail, Tsibliyev had practised operating the TORU manual docking system from inside Mir. This time, however, the automatic system operated successfully, and Progress M-34 docked at the rear port on 8 April. Commander Tsibliyev had not been called upon to test his skills.

Within three hours of the docking, the crew began unloading the cargo craft. The 110-lb oxygen tank was one of the first items to be put into use, to replenish the cabin atmosphere. With the crew starting to repair Mir, Russian Deputy Lead Flight Director, Viktor Blagov, told reporters that the Americans had exaggerated the situation onboard the station. He added that it was not life-threatening!

On 13 April it was reported that the toilet system on Mir had broken down, and that the crew had had to revert to using the back-up system. This was the system used on early spaceflight, consisting of a condom-style leak-proof sack for urine, and adhesive-lipped bags for solid waste, with a bacterial tablet inside to prevent explosion from the build-up of gases. The crew managed to fix the toilet's manual

The view from the Node through one of the hatches, showing the darkened interior of the connecting module and the air-conditioning hoses snaking between the modules.

mode, but this did not allow the 'separator' section (that converted urine to water) or the 'conserving' section (which added chemicals to the urine, preventing microbial growth) to function.

Linenger highlighted the crew's concerns in an e-mail from Mir: 'Yes, oxygen is important. Scrubbing the air of carbon dioxide is vital. Supplying cooling to the equipment – necessary. But problems with the toilet jump right to the top of the list. Our priority is to fix the toilet.'

Through April and May, the problems onboard the station were either repaired or at least held in check. By 9 May the toilet had been fixed, although the crew was not allowed to drink the water until samples had been analysed on Earth. The next Shuttle launched would carry Linenger's replacement, Mike Foale, and its other important cargo element was 110 gallons of water. The crew on the station had already stored 66 gallons of water contaminated with ethylene glycol from the coolant loops. Tsibliyev tried to obtain permission to dump the unwanted liquid through the *Atlantis* water dump system when it docked, but NASA refused – not surprisingly – as they wanted more time to evaluate the effects of handling and dumping such a quantity of contaminated liquid. The contaminated water was to be stored until sample analysis determined that it could either be recycled or put in a Progress and incinerated during re-entry. One of the problems in storing this and other unused or unwanted equipment on Mir was revealed during Linenger's TV tours around the station. Equipment cluttered passageways and access to other areas, hampering operations during a normal working day, let alone in an emergency situation.

On 15 May *Atlantis* – with Mike Foale on board – launched on the STS-84 mission and headed for a rendezvous with Mir. Docking was achieved on 17 May, and after the traditional handover and welcoming ceremonies, Linenger showed Foale around his new home. This included the Spektr module, where the American researchers conducted most of their science experiments, and where they slept.

With a combined crew of ten astronauts and cosmonauts onboard Mir, the repaired systems underwent an additional qualification period for the five days that *Atlantis* was docked. On 22 May, with Linenger onboard *Atlantis* and Foale on Mir, the two craft undocked. *Atlantis* landed back at the Kennedy Space Center on 24 May.

Three weeks after his return, Linenger gave a press conference about his experiences on Mir. He stated that the main lesson he had learned was to have flexibility in what needed to be accomplished. A crew also had to be responsive to

Change of crew on Mir. STS-84 core crew delivers Mike Foale and picks up Jerry Linenger. The combined crew of 10 are (*front, left–right*) Jerry Linenger, Vasily Tsibliyev, Charlie Precourt, Alexandr Lazutkin, Mike Foale, (*back, left–right*) Ed Lu, Eileen Collins, Jean-Françoise Clervoy, Elena Kondakova and Carlos Noriega.

events happening in real time, using less than ideal hardware to effect jury-rigged repairs, until the next supply craft came up with the necessary items. With regard to the fire, his first response was to 'open a window', but he realised that 'this was ridiculous, you can't open a window.' He recalled his Emergency Room training, where he was taught to take his own pulse first and then attend to the patient. Once he was sure about his own condition, he helped the rest of the crew fight the fire.

Back on Mir, Foale was settling into his routine and continuing with his scientific programme. He noticed that his Russian colleagues were engaged in almost constant daily repair work around the station. The cosmonauts finally found the source of the leak in the VGK loop. They repaired it, and confirmed that it was at last holding pressure. It had taken almost six months to find the leak, and it was in such an awkward place that it was suggested that there must be fewer inaccessible places on future spacecraft.

In his training, Foale had taken care to understand Russian rather than merely have a passing acquaintance with the language. He had noted his earlier colleagues' difficulties in both settling into prolonged orbital life and in living amicably with the Russian crew-members, and he was determined to overcome these barriers. He wanted to seek out opportunities to help where he could, without imposing himself on his new Commander. For the first month he observed both cosmonauts, and although occupied by his scientific experiments and observations, he felt underused. He could quite easily assist in some of the more mundane, routine chores around the station, but he realised that this would not be easy to accomplish.

Glycol levels were still being recorded in the modules. Foale took regular samples, but the levels were so tiny that it was determined that it was not a threat to the crew. By 13 June the repairs were completed and the residual glycol had been mopped up. Tsibliyev and Lazutkin – called 'real heroes' by Foale – continued to reconnect the back-up Elektron unit and the Vozdukh carbon dioxide removal system to the coolant loops in Kvant. However, they encountered problems with connectors that they could not find or that would not fit, which further frustrated their efforts.

By his fourth week in space Foale had settled in well, and was obtaining good scientific results from his experiments. He was sleeping well, and eating – including some treats of chocolate sent up on the docked Progress – was pleasant. He commented on life onboard Mir in an interview on 12 June: 'The view is fantastic and the work is interesting, so I am having a good time. Mir is comfortable; it's better than most camping trips. Everything's going rather well. It's a lot easier than I expected.' A NASA Mir systems specialist added the observation, on 20 June, that the previous week had been 'the quietest one for months'.

Foale had expected difficulties in maintaining a regular communications link with his personal mail through the Russian Mission Control Centre (TSUP – 'Soup'), near Moscow, and he soon established radio links with amateur radio enthusiasts, known internationally as 'radio hams'. In this way he was able to receive e-mail or voice messages, with contacts around the world. As Mir uses little power to maintain communications, this network proved to be an unexpected bonus over the next few months. The additional line of communication also allowed Foale's family to learn of his activities, and his impressions of his prolonged orbital life-style.

At first Foale was a little surprised at Tsibliyev's remoteness. He was accustomed to the American style of command, in which the Commander encouraged a team rapport and a development of joint knowledge and trust. This was not the case with his new Russian Commander, who was very much a reserved character. This may have been due, in part, to his experiences with Foale's predecessor, Linenger, who had kept himself very much to himself and had not found it easy to blend in with the Russian crew.

Over the first few weeks some small approaches were made, and the three men gradually became more relaxed in each other's company, but usually only at meal and relaxation times. As he observed the two cosmonauts trying to keep ahead of the station repairs and maintenance, Foale could see that they were falling behind and neglecting other important items. Being careful not to invade the authority of his Commander, he said that he was willing and able to take over some of the regular cleaning and maintenance tasks. He surprised the cosmonauts with his knowledge of some of the station systems, and his long hours of study in Russian and in Mir hardware were beginning to pay off.

Support for his help also came from the civilian engineer, Lazutkin, who tried to convince Tsibliyev that Foale's help would benefit all of them. The Commander, unconvinced, was worried about the reaction from Ground Control. Foale was amazed that this discussion took place openly in front of him. The subject was raised again a couple of days later when Tsibliyev was complaining to ground control that the two cosmonauts had not had a free weekend in three months. They had not had time to clean up the water and ethylene glycol, and all the other maintenance tasks were accumulating. Foale, who was nearby, took the opportunity to say, in Russian, that he could do it. After a long pause, ground control (with much laughter in the control room) said that if he wished to do so then he could. A few hours later the NASA liaison controller, Keith Zimmerman, wanted confirmation that Foale had offered to do 'Russian work', to which he simply confirmed.

When Foale had offered to help out earlier, his colleagues had commented that they were Russian and would do the heavy work, while Foale was 'a soft American poodle and not of the same kind'. As a 'foreigner' he was there for research, not Russian work. According to his father, former RAF Air Commodore Colin Foale, who later wrote an account of his sons experiences on Mir (*Waystation to the Stars*, Headline, 1999), the comments were like a red rag to a bull. Michael Foale had always had a style of logical thinking and an organised approach to anything he attempted. With an Anglo-American parentage, even his fellow American astronauts thought of him as English, or European. But this multicultural background and upbringing certainly helped win support for his increased workload on Mir, and the Russians eventually considered his attitude to be more English than American.

Almost immediately, Tsibliyev began trusting Foale to take over general station work tasks, and eventually even ground control began to involve him in the Russians' tasks. Foale felt that he had made a significant step forward in the development of international relations with this achievement. Tsibliyev was especially relieved when Foale began to prepare a software program to receive and organise the incoming data uplink and messages. The more important communications went directly to Tsibliyev, so he no longer had to separate them from the mundane messages.

The crew gradually accepted Foale, and events were soon to prove that this was a wise decision.

On the morning of 24 June the Progress M-34 was undocked from Mir's rear port for a short automated flight. After one day, the plan was to have Tsibliyev test the TORU docking system, to redock Progress to Mir. This was to have been followed by the launch of Progress M-35 on 27 June, carrying fresh supplies to Mir. The following day, Progress M-34 was to be undocked for the last time to head for re-entry and destruction in the atmosphere; and the day after that, Progress M-25 would automatically dock with the rear port and resupply the crew on the station.

When the plans to use the TORU system for docking Progress freighters had been developed, they included options for using old, departing freighters for the purpose of systems tests and docking practise. There was no indication that this was planned for M-34, but an increase in radio traffic over several days prior to 25 June indicated that Tsibliyev had been instructed to check the TORU system and was being instructed by ground control. The entire programme of manoeuvres fell within the range of ground tracking stations, which would support the re-docking.

It was clear to Foale and Lazutkin that Tsibliyev was not comfortable with the forthcoming manual docking. During the earlier, abortive attempt, the Commander was unable to receive clear TV images of Mir from the camera onboard Progress M-33. He was flying blind, and commanded the thrusters to change the approach of Progress to pass the station, as he felt it was approaching much too quickly.

For several days prior to this new attempt, Tsibliyev worked on the TORU system, but despite words of confidence and support from Lazutkin he was already showing signs of tension. Foale assumed the procedure had been well devised by the experts on the ground. He had not received any briefing on the docking, as it was a Russian operation, but he would help his Commander when he could. Many months of working with the Russians had already shown him that he was told only what he needed to know and nothing more – unlike the American system, in which an astronaut becomes familiar with almost every nut and bolt. It was different, but it was just the Russian way.

On the morning of the docking, Tsibliyev set up the equipment one more time. He had dual joystick controls to command changes in the approach and attitude of Progress, and a TV screen on which he would see Mir as viewed from the approaching vehicle. Behind him, his actions, and the view from the TV screen, were to be recorded by video camera for later post-flight analysis. The crew later stated that watching these tapes of the event was like watching a horror movie.

Lazutkin was at one of the Base Block windows, and had a laser rangefinder to point at the Progress as it approached. The readings would provide Tsibliyev with precise closing ranges of the spacecraft, as the onboard TV camera could not provide this vital information. Foale was sent to the Kvant 1 window when Progress came close to the docking port on the astrophysical module's far end. Unfortunately for both men, the observation windows were rather poorly designed and had restricted fields of view. Neither Foale nor Lazutkin could see Progress begin its approach run.

Tsibliyev had not been advised when to start the docking approach, nor in what position both spacecraft should have been in relationship to the Earth. As the Progress thrusters were fired, the spacecraft temporarily passed outside the range of radio reception. Tsibliyev's TV screens showed Mir, with docking lights shining and Earth moving in the background. For a while the size of Mir's image seemed to remain constant, but it gradually grew larger as Progress closed in on the station.

As Tsibliyev watched the screen he saw Kvant's docking port slide towards the bottom of the screen and the Spektr module come into view – and into the path of Progress. He executed preplanned braking manoeuvres, but Progress still headed for Mir, which was by then more towards the docking Node than to the planned approach to Kvant. Tsibliyev commanded continuous braking on Progress at 820 feet. He had only seconds to react to events and induce commands to deal with them, and yet everything seemed to move very slowly at first.

At the 164-feet call from his Commander, Lazutkin could still not see Progress. Then suddenly, he saw it. Realising it was very close and moving at high velocity, he also noticed that it was on a direct collision course with the Node. A collision was unavoidable, and he shouted to Foale, 'Michael. Go to the Soyuz, *Now!* Vasily, *Stop!*' Tsibliyev was frantically operating the hand control to try to prevent the imminent collision. Foale was floating back from the Kvant, unable to see Progress, when he received Lazutkin's call. He reacted immediately, diving into the Node and into the Soyuz. As he did so a very loud thump rocked the station, vibrating through the structure. Normal Progress docking speed is 1 fps. Progress M-34 hit Spektr at 10–15 fps!

The crew immediately felt a change in air pressure in their ears. Pressure warning sirens sounded around the station, accompanied by loud hissing noises. Foale had only seconds to separate the vent hoses and power cables that snaked through the hatches to Soyuz, so that they could close the hatch and use the spacecraft as a lifeboat. Lazutkin had seen Progress hit Spektr, so he raced to the Node to attempt to close the hatch to the leaking module, to be joined by Foale.

As with all modules on Mir, the cables and hoses passed through the open hatchways and therefore hindered closure. As Foale disconnected the power cables, Lazutkin floated into the Spektr to find the main junction box from which the cables were routed. The hissing inside Spektr was louder. There was no time to disconnect all the cables, so Lazutkin used a knife to cut the last few, with showers of sparks flying from the live cables.

The two men then tried to close the Spektr hatch, but the lower pressure in the module pulled it inwards and away from the hatchway. They quickly grabbed an external hatch cover which, because of the different air pressures, was sucked against the opening, sealing the hatchway. They had done it – but only just.

Tsibliyev had calculated that from the time of the impact they had twenty minutes to evaluate the station. By the time the hatch cover was in place, they had just eight minutes left. When the gauges recorded a stabilisation in pressure they realised they had only one leak, which for the time being was sealed off.

Luckily, Progress did not hit the Node; but it struck the solar arrays, which deflected it onto one of the large radiator panels on the hull of the Spektr, which

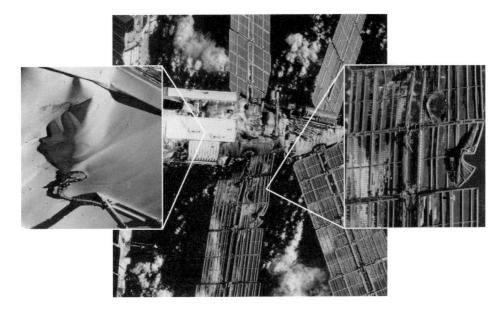

Damage caused by the impact of the unmanned Progress resupply vehicle on 25 June 1997: (*centre*) a general view of the damage to the Spektr module; (*left*) close-up of the Spektr radiator panel; (*right*) close-up of the solar array.

absorbed most of the momentum. Progress M-34 continued past Mir and was by them some distance away from the station. Had the cargo vessel hit the Node, then the rupture in the station's hull would have been more severe and the crew would probably have died before they could do anything to prevent the atmosphere escaping. They would also have had to pass through the Node to reach the Soyuz on the other side.

The vital decision to sever the power cables from Spektr had taken out 40% of the power supply drawn from the solar arrays on the outside of the module. The collision had also applied a turning force (torque) to the station, which the onboard stabilising devices could not control. This also had the effect of moving all the remaining connected solar arrays away from the Sun. Mir was now quickly losing power.

On Earth, Mission Control in Moscow analysed the facts before reporting to the media. NASA reported that the crew was safe and well and that the agency was working with their Russian colleagues, discussing plans to restore power and control to the station. A later e-mail from Mike Foale to his wife, Rhonda, indicated that that the ruptured hull had been isolated. Mike was fine, but had lost all his personal belongings from his quarters in the Spektr module. She was not expecting a further e-mail from him as he was 'probably very busy'.

The Russians wanted to analyse all the data from the robotic craft before commanding Progress through automatic entry to its destruction. Several working groups, and a full inquiry under the chairmanship of Yuri Koptev, Head of the

Russian Space Agency (RKA), were formed to examine different aspects of the incident. Koptev also emphasised that Mir operations would not be interrupted, that the crew would not be evacuated and brought home, and that Russia would continue to fulfil its obligations within the framework of international co-operation.

Tsibliyev was ordered to remain at his communications post throughout the crisis. Mir's batteries were now draining towards redline levels, and the crew had to shut down all non-essential equipment to conserve as much power as possible until they could realign the solar arrays. Ground control could programme the station thrusters to fire counter to the tumbling motion around two axes, but before they could do this they needed to know the rate and duration of the rotation before all electrical power was lost. There were no onboard instruments that could provide this information, so Foale improvised by holding his thumb out at arm's length and timing the passage of stars that passed it – which he calculated was about 1° per second as the station rotated around its central axis. As soon as Mission Control in Moscow received this information, they commanded the thrusters to counter the roll and stop the station. They succeeded, but the active solar panels were not aligned with the Sun. According to Lazutkin, the station was now orbiting in a silence that was deafening. Foale, always upbeat, found the new silence of the station exhilarating, and enjoyed the very peaceful view of orbital sunrise. Noticing that Tsibliyev seemed very depressed, Foale commented that without the problem they had just lived through, they would not be able to enjoy this unique experience of beauty and peace. Tsibliyev was not convinced. To him, it was still a terrible day.

The crew gradually managed to twist the Base Block arrays on their internal mountings to obtain a better lock on the Sun in later passes. But they were still only able to maintain power during daylight passes, without generating any surplus to replenish the batteries for the night-time passes. The electricity was needed not only for light and heat, but also for air, and the air ventilation system was needed to prevent a build-up of carbon dioxide. Without the movement of air around the station, pockets of carbon dioxide would gather around the crew's faces, possibly leading to loss of consciousness and eventual suffocation. For the next few hours the crew dared not sleep, and constantly watched each other, fanning their faces with maps and sheets of paper from the logbooks, and moving between modules. With only torches available for the night-time passes, it was also becoming colder.

Ground control instructed Tsibliyev to try to move the station to a better Sun angle by using the thrusters on the Soyuz. Small thrusts from the engines in the spacecraft, located at one of the extremities of the complex, would be enough to nudge the arrays back into solar alignment. It sounded simple, but it would require a great deal of lengthy practice to succeed, and the crew was already showing signs of fatigue.

They had to determine the direction that the station needed to be moved and also when to stop it to lock on to the Sun. Once that was achieved they could attempt to induce a small spin to catch the sunlight for as long as possible. Finding the Sun was the first trick. Foale deduced that the most efficient way to use the solar arrays that they still had available was to gently nudge the station until Spektr pointed at the

Sun. They could then try to spin the station around the centre line of the Spektr module and the Kvant 2, docked opposite – the station's 'Y' axis.

Aligning the station with the Sun would take time – longer than a daytime pass. The alignment manoeuvre had to start during the night-time pass, but the problem with this was that there was no Sun with which to align. Foale used his astronomical knowledge to align several stars, which could be used to direct the station, thus allowing Spektr to point at the Sun during the next day's pass. It sounded easy, but it was not.

Tsibliyev entered Soyuz to command the thrusters. Foale's view from the window in the station was at a different orientation from the directions of thrust that Tsibliyev would make on the Soyuz thrusters, so they had to be very careful to relay precise and accurate directional instructions to the Commander. The duration of the burn was guesswork, and Tsibliyev continually reminded Foale of the extra use of precious fuel in the Soyuz – fuel that they might need to use to return home. In fact, Soyuz carried around 900 lbs of fuel for manoeuvres, with a separate tank for the de-orbit burn, and at the end of the sequence only 22 lbs had been used.

It was a slow process, but eventually they arrived at the correct position and heard the arrays' automatic response in locking on to the Sun. Foale wanted to speed up the rotation to secure stability by firing ten pulses of the Soyuz engines, but Tsibliyev, with fuel economy in mind, allowed him three. Such differences in thinking prolonged the whole manoeuvre, but there was no friction between the crewmen. Two hours later the arrays were again misaligned, but by that time they had recharged the batteries for up to 30 hours of power. Although the station swung in and out of alignment, this could be corrected using the stored battery power.

After 24 hours of high-tension work, the crew could at last get some sleep, albeit in rotation. On the ground, plans were being formulated about how to seal the leak in the Spektr module.

Initial reports indicated that the early blame was placed totally on the two Russian cosmonauts – particularly on Tsibliyev, who had controlled the docking. It soon became clear that this was not the full story. Other contributing factors were the planning of the event, the training of the cosmonauts and the hardware and software used. On 26 June, Russian Public TV reported that it was a fault in the Ukrainian electronic guidance system that had caused the collision. The following day, Ukrainian TV replied that it was the Russian-developed docking equipment, to which their electronics were attached, that had failed. They also pointed out that such reports were 'detrimental to co-operation between the two countries'.

Having survived the initial collision and secured the relative safety of the station, the crew remained onboard the station to try to repair and salvage what they could and to prolong the orbital life of the complex. On Gemini 8 in 1966 the crew survived an in-flight emergency and was back on Earth ten hours after leaving. In 1970, Apollo 13 was two days out from Earth when an onboard explosion ripped out the side of the spacecraft, but the crew was back on Earth in four very long days. But the crew on Mir were not planning to return home for at least eight weeks. If anything else threatened their survival, however, they could use the still-attached Soyuz to effect a recover – if they had time to reach it and detach it.

Over the coming weeks the skill and character of all three men on Mir would be constantly challenged as new problems arose. During the two weeks following the collision, power was gradually restored. The Vozdukh carbon dioxide atmosphere scrubbing system was also working again after several days' reliance on the American-supplied lithium hydroxide canisters, along with the Elektron oxygen-generating unit in Kvant 2.

On the ground, plans were being evaluated for sending repair equipment to Mir on the next Progress launch. This equipment included a special plate through which 22 cables would be routed. The idea was to adapt the spare Konus docking drogues which were kept aboard the station for module docking and repositioning. At the end of the drogue was a detachable end-cap, through which the cables could be fed and then attached to the plate. The drogue and plate would then be fitted over the hatchway into Spektr during an internal space-walk (IVA). Connections with the solar arrays outside the hull of the module would then be restored, to feed further power to the battery system on the main complex. Tsibliyev and Lazutkin would work in the depressurised Node in their Orlan EVA suits, sealed off from the rest of the station but not venturing outside. Foale would wear his own Sokol pressure suit and stay in the Soyuz Descent Module. The IVA was being rehearsed in ground simulations by several Russian cosmonauts, American astronauts, and EVA specialists from both countries.

The ground and station teams still had much to do to bring Mir back to full working order, but at least they could look forward to the launch of the Progress and plan for the IVA. This was to be followed by a trip outside to inspect the hull damage and to evaluate whether a repair would be possible. A few days after the incident a Russian news agency estimated that the collision had already cost $3 million – a figure that was likely to rise when the cost of repairs was added.

On 26 June, Foale told Zimmerman, on the radio, that most of his personal equipment – including medical kits, exercise shoes, treadmill harness, shaver and toothpaste – were lost on Spektr. He requested replacements to be sent up on the next Progress, along with the NASA Life Sciences equipment, the computer hard drives and a printer. Although he had two laptops with him, Foale also requested that some replacement equipment be loaded onto Progress. By the end of the month his equipment was on the way to the Baikonur launch site in Kazakhstan, to be placed onboard the Progress.

The following day Foale had an opportunity to send a long letter (written on 26 June), over the radio ham network, to his wife, Rhonda. Knowing it would be available to a wider audience, Foale chose his words carefully, and tried not to provide too much insight into the drama that the crew had played out before the full facts were released to the media. He knew that NASA would keep Rhonda informed, and that she in turn would tell his family in England. He wrote (in part): 'I hope you've not been fretting. We are very thankful to God that the impact was not further away from our centre of mass and the Node, causing a much bigger torque on the Node with very different consequences [total rupture]. I have much more to say about the impossible situation Vassy [Tsibliyev] was in to do the docking. There was no way, in my opinion, that the docking could be achieved from the original set-up of the Progress in relation to Mir. But that is all hindsight now ... I feel guilty that I did not pay attention to the rendezvous planning, but that was not my job. I was to measure the range with a laser ... I didn't get to do that before the Progress hit us ... I have lost all my personal stuff, pictures of you and the kids, I'm afraid. But I am still happy with the outcome, nevertheless ... Love, Mike.'

On 29 June the Moscow TV programme *Segodnyaov* broadcast an interview with the Mir crew. During the interview, Lazutkin described how previous simulations during training had triggered his automatic reaction to the depressurisation. Foale was beginning to be recognised as a key member of the team in their efforts to save Mir. It was noted how well he fitted in with the crew. Tsibliyev stated that Foale 'can rightly be described as a true number-two onboard engineer, helping us with everything that needs to be done. In my opinion he is more knowledgeable than many other US astronauts.'

A letter dated 29 June – from Rhonda Foale to Mike Foale's family in Cambridge – revealed some of the closeness that by then existed between the three men on Mir as they became a more cohesive team after the accident: 'Talked to Mike this morning and he sounded very well. He joked about Sasha [Lazutkin] and Vasily [Tsibliyev] having to scrounge supplies for him. They tease him, calling him a vagrant, a street person.'

Foale, now without an official sleeping berth, was having to use one of the other modules – which was damp and cold – with his head protruding into the Node for fresh air. In an e-mail – also dated 29 June – to his wife, he recounted his memories of when the space station lost all power after they had sealed off Spektr: 'Vassy found a container of mysterious red liquid. It was in the back of the fridge, and showed up while he was trying to figure out what had spoiled [due to loss of power]. A very nice surprise when we tasted it, and we told each other stories ... I will remember being in total darkness, no power, no fans, and all in front of the big window, looking at incredibly complex, swirling auroras, with the galaxy showering down on them, with nothing else for us to do. I know, you are thinking the red liquid has got to us.'

As the crew prepared for the arrival of the Progress cargo craft, they continued to manage the onboard systems and prepare equipment for the IVA. In the first days of July, the emotional strain and physical effort of keeping Mir in alignment with the Sun, while preparing for the repairs, was showing in further messages from Foale:

July 2: 'We lost control again, and once again we were rushing from one dark module to another, trying to look out windows and point the solar arrays. The station dies very quickly if we don't do this. [Foale considered this to be much like being on a yacht in a rough sea, with changing winds, and having to rush around putting the sails up or down before a storm comes.] Life on Mir is characterised by long periods of monotonous, serene calm and short interludes of extreme frenzy. I am learning.'

July 3: 'We are all tired now [after early rises and IVA preparations]. Another power supply burned up. I have spent the last hours cobbling tape and wire together.

The smell of burning insulation was a big shock to me when I came to the Base Block. Sasha and Vassy had fallen asleep [as] they were so tired, until I shouted '*Fire!*' [We] found smouldering wires ... I got to do real, onboard engineer stuff today [supporting IVA preparations]. The flight feels totally different for me now after the collision, since I am now just being an astronaut – not much science. It's like a different flight.'

All three men were becoming very tired, but they received a welcome boost when the Progress M-35 docked automatically, without incident, on 7 July. After several days of tests, the M-34 Progress that had collided with Mir was sent, by ground control, to its destruction over the Pacific Ocean.

July 8: 'We are now unpacking the Progress – the smell was heavenly, of fresh apples. Have opened two work packages for me and found my hygiene stuff ... and chocolate.'

Foale had always enjoyed hampers from home when he was at boarding school, and this was just the same. The morale boost to the crew was the major benefit, with letters and photographs from home, the repair equipment, fresh food, and a new video player. Foale could continue translating American films for his colleagues (*Apollo 13*, starring Tom Hanks as Jim Lovell, being one of their favourites!). Foale also enjoyed his first cup of tea in two months, and the e-mails from home – telling him that his family were well – helped him to concentrate on the task ahead.

On 13 July, during a regular exercise period and later sleep periods, Tsibliyev showed the first sign of an irregular heartbeat (cardiac arrhythmia). The next day, under instructions from ground physicians, he conducted more exercise. He was told to rest and await instructions from the ground, which raised doubts about his ability to perform the IVA. The stress of the mission workload, the feeling of responsibility for the collision and the lack of adequate rest, was building up. By 15 July the IVA had been postponed, and although his general health was good, the medical support staff prescribed a course of strong cardiac and sedative drugs over the next ten days. In response to a string of medical questions, Tsibliyev replied that he had been conscious of the problem 'for a few days'.

At this time the Russians approached NASA to allow Foale to conduct the IVA with Lazutkin. Foale had trained in EVA during his role as back-up to Linenger (who performed the first American EVA from Mir during the previous April) and had performed his own EVA during his third mission on STS-63 in February 1995. The following day permission was given, and Foale began training with Lazutkin for the IVA operation. At the same time, Foale and Lazutkin were concerned about the state of their Commander's health and his mood after yet another blow. They allowed him extra rest as they prepared for the IVA.

Two days before the event, Lazutkin was in the Node to begin the final preparations. Foale was taking a sleep period, and Tsibliyev was in the Base Block. Lazutkin was to disconnect 100 of the 140 cables that ran through the Node, to save time during the IVA itself. The last 40 would be disconnected just prior to starting the IVA. It was a complicated operation, needing constant referral to the procedures book and to the actual hardware in front of him. Lazutkin knew he had to leave some cables connected, as they were important for keeping the station alive. He thought at one stage that he had pulled the wrong plug, but then realised that he had not. He relaxed, and continued working – and then froze again, as he again feared that he had disconnected the wrong cable. He remained silent, waiting for the worst to happen – but nothing did happen. Certain that he had pulled the wrong lead, Lazutkin floated back into the Base Block to tell Tsibliyev that 'I have pulled the wrong plug. Has anything happened?' Before any reply they received the answer in the form of an alarm.

Lazutkin had accidentally disconnected the communication connection between the gyrodynes that stabilised the station. He tried to re-establish the connection, but it would take time to recover, and the station was already beginning to drift out of alignment and tumble in space. He told Tsibliyev that he would have rather shot himself than disconnect the wrong plug. His Commander decided not to take him up on his suggestion, and took the news with resignation. By this time, Foale was wide awake, and together they analysed their latest challenge.

For the next three hours the gyrodynes powered down as the batteries died and sent differing signals to the stations' computers – trying to turn it one way and then another. What resulted was a multi-axis spin, as a dark space station began to become colder. They were suddenly back to the conditions in which they had found themselves just after the collision, but this time without stored electricity. Communication was only possible via the Soyuz.

Ground control did not like what they heard in Tsibliyev's report. Information supplied to the media was that the station had lost attitude control and was spinning out of control in orbit. All thoughts of the IVA were postponed, and it was hinted that there had been human error. Ground control was powerless to help them because of the communications failure. They suggested trying what had become known as the 'Michael Solution', for which they were to use the Soyuz to move the station back into position to lock on the arrays with the Sun.

With Tsibliyev in Soyuz and Foale at a window, they tested the Soyuz thrusters, which gave a loud, machine-gun sound, frightening all three men. Tsibliyev turned everything off in the Soyuz and shot through the hatch to the Node. Ground control was also concerned with the unexpected, rapid firings of the thrusters. If Soyuz was also damaged, their escape route was in danger.

It was thought that the automatic thrusters on the Soyuz were attempting to compensate for the erratic Mir attitude control system in trying to stabilise the station. The crew was told to turn everything off and switch the control rockets from the automatic to the manual system. This time it worked, and the crew used their previous experience to gradually place Mir in the correct attitude for the batteries to once more recharge.

On Earth, Foale's father Colin had received a telephone call from a TV newsdesk, asking for his comments on the latest events on Mir. Not aware of any new developments, he asked for more details. He was told that all they knew was that somebody had disconnected a plug that had caused the station to 'spin chaotically. It has lost altitude.' Colin Foale immediately picked up on this last word: 'Altitude! Are you sure Moscow said 'altitude'? Could it not have been 'attitude'?' They rechecked, and confirmed that Mir was not tumbling out of orbit, but rather, it was just spinning. This was serious enough, but was much more preferable to an uncontrolled re-entry!

The consequence of these events was the postponement of the IVA. It was decided to allow a fresh Mir crew to conduct the IVA and later EVA operations. The new crew (EO-24) of Commander Anatoly Solovyov and Flight Engineer Pavel Vinogradov were to be launched in early August, and the EO-23 crew (Tsibliyev and Lazutkin) would return on 20 August. Foale would remain onboard Mir until the next scheduled Shuttle docking in September. The French cosmonaut, Leopold Eyharts (who was scheduled to be a member of the EO-24 crew for a three-week mission before returning with the EO-23 cosmonauts), was to be reassigned to the next long-duration crew, EO-25, in 1998. Foale considered that delaying the repairs until the new crew arrived was the correct decision. The three men onboard Mir were disappointed, but they were all becoming tired after several weeks of the constant strain of simply keeping Mir working.

One of the problems they had encountered inside Mir – both in preparing for the IVA and in keeping Mir operational – was in organising the interior. This frequently involved the relocation of unused equipment and the movement of good batteries from one module to another. They found that, although the Sun had charged batteries in one module, it had not done so in a second. The cosmonauts therefore hauled the very bulky charged batteries, swapped them with the uncharged batteries and hauled those back for charging. Foale had briefly mentioned how tired they all were. Hauling 'huge' pieces of equipment out of the Progress and Kvant, through the cramped confines of the Base Block and into the modules for storage, had taken three hours. It had not helped when ground control asked them to find, at the last minute, items from different storage bags: 'These things will have to be stored in modules that have no power, are $+5°$ C, and are covered in water condensation. A hard day, but it is never dull here!'

The large items were only just squeezed through the hatches of Mir. These were designed with no real thought as to whether such a large item would damage walls, ducting, or even the cosmonauts' fingers. Several times, the cosmonauts caught their hands and ripped the skin as they tried to manoeuvre objects around the station. Trying to halt the objects' momentum was also difficult. On many occasions they almost felt like crying in frustration and pain, and from tiredness. They also had to take care when loading items back into Progress for disposal. The 3-foot, wobbly bags of contaminated water, containing the glycol, could easily affect the flight stability of the Progress as it backed away from the Mir, which could have resulted in a second collision.

On 31 July, NASA Shuttle–Mir Program Manager Frank Culbertson and Russian Phase 1 Director Valeri Ryumin announced at a press briefing that they would fly David Wolf – rather than Wendy Lawrence, as scheduled – as the next US crew-member on Mir. The two space agencies had mutually decided to in future assign Americans who were trained for EVA, in case of the need to perform repair EVAs. The removal of Lawrence was not due to a lack of EVA experience (Wolf had not performed an EVA either); it was simply her size. She was just too small to fit into the Orlan EVA suit. In order to allow Wolf more time to train in a Russian suit, it was decided to delay the launch of the Shuttle by ten days until the end of September. This meant that Foale would not be returning home for another eight weeks.

As Tsibliyev and Lazutkin began their preparations for return to Earth, the Commander took time to reflect upon their months on Mir: 'We have had a fire, Elektron oxygen failure, gyrodyne spin-down, a very dangerous collision, an almost catastrophic plug-pull, and I may have had a heart attack. What else is there left for us to have?' Foale – not wanting to tempt fate, since he still had two months left on the station with the new crew – replied: 'Well, we still haven't been invaded by aliens!' Tsibliyev merely rolled his eyes in amazement. What else could one expect from a Westerner?

On 30 July, Foale was interviewed on CBS TV News, and was questioned about the fears in America of sending further astronauts to Mir. Foale replied (in part): 'I think it would be a great shame [not to do so] ... because the United States is getting a lot of experience out of this. We're seeing a station that's been through eleven years of life and has worked pretty well for a long period of time, and now is older and is up against money constraints and having to come up with new solutions to some problems, with some difficulty I think. We will come across those lessons ourselves on the International Space Station in a few years from now. So this experience is really very valuable for us now, and I don't think it's a great price to pay ... to have a permanent presence of an astronaut onboard Mir ... Though the conditions are hard, and they're not as optimum as they could be for science experiments and other things – what we're learning in terms of operations and how to work together is just absolutely priceless.' Foale was also asked questions about the state of equipment, experiment results (such as blood samples) and debris inside the Spektr module. He replied that all of the experiment samples had been contained before the accident, so there should not be a problem. His drink bags and shampoo might have burst and frozen in slivers of ice, but these were also not thought to pose a problem.

On 7 August, Soyuz TM-26, with the EO-24 crew, docked with the aft port of Mir, recently vacated by Progress M-35, now full of rubbish. This would be redocked after the departure of the EO-23 cosmonauts. Throughout the following the two crews worked together to prepare Tsibliyev and Lazutkin for the return to Earth and for the IVA which was to be completed by Solovyov and Vinogradov.

On 14 August, after an emotional farewell, the two departing cosmonauts undocked from Mir for the last time and prepared to return home. It had been an eventful mission for them, but it was not yet over. The in-flight tape recorder did not work, despite receiving a thump from one of the crew. But they were more concerned that the engines would fire, that the Soyuz would separate its modules and that the parachutes would work. They were too accustomed to failure on this mission, but this time, all went well – at least until the landing. The chasing helicopters advised them that they were expected to land on the edge of a salt desert area. They braced themselves for landing, expecting to be cushioned by the landing retro-rockets. Then there was a heavy bump as the capsule hit the ground. The retro had not worked! It was determined that, had the third crew-member position been occupied, he would have received a severe injury during the hard landing.

Mike Foale in a pre-launch crew photograph of the Mir 24 crew of Anatoly Solovyov and Pavel Vinogradov, who in August 1997 took over the residence of Mir from the Mir 23 crew of Tsibliyev and Lazutkin.

Back on Mir, Foale was settling in with his new crew. Their first task was to redock the Soyuz to the other end of Mir, to allow the Progress to redock with the station. Foale enjoyed the 45-minute trip, and photographed the outside of Mir as they passed by. However, after he re-entered Mir the automatic docking of the Progress was cancelled due to an error by Mission Control. The crew was ordered to go straight to Soyuz, as it was not known how close the Progress would pass to the station! The crew had decided to take no chances, and were already in Soyuz by the end of the call!

When Solovyov offered Foale the chance to watch the cargo craft pass by, he declined. He thought it too frightening, and would feel much better if they all closed their eyes so that they could at least *feel* safe! The Russian – having temporarily forgotten his earlier experience with a rogue Progress – found this very amusing. Foale was pleased to see that Solovyov had taken his advice in positioning a wrench to disconnect an important hose ducting to the Node, should the need arise to vacate the station quickly. Lessons were being learned. The Progress was redocked successfully – almost – a few days later. Just prior to docking, the Mir computer crashed. It was fortunate that Solovyov had taken manual control at a much later stage than during the previous attempt in June, as this time the Progress was moving much more slowly.

The computer failure led to another loss of attitude control, and it was a feeling of *dèja vu* for Foale. He shrugged his shoulders when Solovyov wanted to consult the ground on what to do next to recover the station's attitude control. Having experienced this three times before, Foale knew what needed to be done, but his new Commander was still not sure. When Solovyov communicated with the ground, he was told to follow Foale's instructions, as he had gained much experience of this situation. Solovyov accepted this advice, and from then on Foale knew that he would be comfortable with his new crew for the rest of the mission.

On 22 August the crew performed their IVA. Foale assisted Solovyov and Vinogradov into the EVA suits, and then went into the Soyuz in case the crew needed a quick escape from the station. It was Foale who heard Vinogradov mention that he could feel air flowing across his gloved hand as the decompression proceeded. He quickly told him stop moving his hand, as it appeared that his suit had a leak. For a few anxious moments he tried to attract the attention of the Flight Engineer, and then with the help of ground control and his Commander, who had realised the danger. Vinogradov stopped wiggling his hand as Solovyov raised the pressure in the Node. After a new glove had been found, the operation continued. Had it not been noticed by Foale, Vinogradov could easily have died due to the slow leak. The two cosmonauts worked well together, having practised the procedure in the water tank simulators on Earth. They successfully reconnected the power cables through the new hatch plate, and also managed to retrieve some of Foale's personal items.

Mike Foale inside the Soyuz TM descent module during training. He assumed this position during the internal EVA during the inspection of the inside of the Spektr module by two Russian cosmonauts.

Mike Foale prepares to climb into the Russian Orlan EVA suit during training at NASA JSC, Houston. His training on Russian EVA hardware was instrumental in his selection to perform an EVA to inspect the outside of the Spektr module with Solovyov.

For some weeks there had been talk of Foale conducting an EVA to inspect the outside of the Spektr module. On 4 September the final approval was received for the trip outside. The EVA, on 6 September, was to inspect the surface of the module in an attempt to pinpoint the puncture in the hull. Foale and Solovyov were to prepare the work area, and the repairs were to be carried out during a later EVA. They did not discover the source of the puncture, but the EVA provided information on the damage to the hull of the station, and allowed future planning for later EVAs to be refined.

With the success of his EVA behind him, Foale looked forward to returning home on the Shuttle. However, the mission was not yet over. On 14 September the USAF Space Command indicated that the orbit of an obsolete 370-lb US military satellite would intersect the path of Mir, to within 3,300 feet. At a combined orbital speed of more than 30,000 mph, should it hit Mir there was nothing the crew could do. Leaving nothing to chance, the three men once more positioned themselves in the Soyuz, as the expected orbital paths intersected. It was a long wait before ground control informed them that the satellite had passed by safely.

As the date for the Shuttle launch approached, the computer on Mir crashed again – but again the crew were able to restore it. A computer crash of this kind, as the Shuttle made its final approach, would send Mir into an uncontrolled spin, and it would be impossible to link the vehicles. For Foale, yet another mishap was too much to think about, and – as he observed in an e-mail on 15 September – the forthcoming docking of the Shuttle made his desire to return home as soon as possible even more important: 'I am feeling a bit fed-up today. Currently we have no hot water, hot food and no toilet running for at least twenty-four hours. This power loss is the direct result of a computer failure. We changed computers with an older one, stored away from many years ago. We are on the edge of being without flight control at all. I now realise I really do want to come home soon.' Foale was becoming very tired and worn out after battling problem after problem. However, he still found time to provide a formal record of why he felt his replacement, David Wolf, should continue with the work on Mir, at a time when a meeting of the US House Science Committee was discussing that very question: 'This flight ... has been one of the hardest things I have ever attempted in my life. I have to remember what John F. Kennedy said when I was about four years old, forgive me if I get it wrong. He said, 'We do not attempt things because they are easy, but because they are hard,

Near the end of his four months onboard Mir, after wearing Russian crew attire Mike Foale wears clothes representing the STS-86 crew with whom he would return to Earth. The prospect of soon returning home is clearly shown in his expression.

and in that way we achieve greatness'.' He continued, stressing that although this co-operation was not always easy, it was reaping great rewards. He stated that he really valued his time on Mir, and would 'always remember the last three or four months with great clarity and nostalgia.'

*Atlantis* was launched on 25 September, and docked with Mir, without incident, two days later. For the next five days, Foale guided Wolf around the complex and prepared himself for his return home. He was not eager to leave – it had become home – but he longed to see his family again, and repeatedly looked out of the windows to ensure that the Shuttle had not left without him. On 5 October 1997 he was back on Earth, at the end of one of the most eventful episodes in space since Apollo 13, and in which the drama had continued every day for almost nine months. Foale was correct in stating that it had been a difficult mission, but with the construction of the ISS due to begin a year later, it was an experience that would be a valuable asset to the operation of the new station, if the lessons of Mir had truly been learned.

On 27 September 1997 – the day that STS-86 left the pad to collect Mike Foale, – comments from former astronauts and cosmonauts indicated that there was now a general agreement on the cause of the Progress collision in the previous June. Prior to this, most of the blame had rested on the Russian crew – in particular, upon Commander Tsibliyev. Both cosmonauts had their full pay for the flight withheld, and had expected repercussions when they landed.

Mike Foale is reunited with his family at the end of a long $4\frac{1}{2}$-month mission. He is joined by his wife Rhonda and by their 5-year old daughter Jenna and 3-year old son Ian – and an appropriate 'Welcome Home' banner hangs in the background.

Former Apollo astronaut General Tom Stafford was assigned by NASA administrator Dan Goldin to investigate the problems on Mir. His report pointed to a series of incidents that all contributed to the final accident. It was not the fault of one person or element, but a combination of several actions of a variety of people and by different hardware and software.

Stafford noted that the only visual reference available to Tsibliyev was the Progress video camera, which pointed 'down' towards Mir against an Earth background. He had experienced the difficulties of such a docking during his Gemini 9 mission more than thirty years earlier, and had advised flight planners that such a docking was extremely difficult due to the lighting conditions, and should be attempted only in the case of an emergency. He also noted that the performance of the thrusters on board the Progress was low, and that Tsibliyev had not received adequate training for such a docking. He had not practised such a manoeuvre for 130 days.

During the STS-86 pre-launch Press Conference, former Russian cosmonaut and Director of the Shuttle–Mir Programme, Valeri Ryumin, clearly pointed out that Tsibliyev had decided that the relative velocities of Progress and Mir were acceptable for docking and had still made the decision to dock. He had not updated the rates, which was part of the TORU docking system software package, and had also performed the braking manoeuvre burns much later than he should have done. However, Ryumin said that he would absorb Stafford's recommendations.

Deputy Director of the Cosmonaut Training Centre, former cosmonaut General Yuri Glazkov, supported Ryumin's comments, but then added that there was a series of problems that would necessitate future improvements. These included crew training, the increased range of information available to the crew and the speed that it was delivered, and the state of the crew's health while performing difficult and complex tasks. He too appreciated Stafford's input into the investigation.

It was clear that Tsibliyev was not totally to blame. The docking was a late addition to the flight plan, partly to speed up the change from the costly Ukrainian KURS system to the cheaper Russian TORU system. Tsibliyev had not had an opportunity to undergo sustained training on the docking profile, and the set-up at the beginning of the approach was incorrect. The information that was presented to him during the manoeuvre was limited and restricting, and neither Foale nor Lazutkin could obtain a clear view of Progress from the start of the approach. Lazutkin managed to catch only a glimpse of the vehicle as it approached the station, heading for a collision, but by then it was too late. Neither did ground control ensure that full communication was available during the approach, when they could have monitored the display readings from both vehicles. In addition, pressure to achieve a success was placed on both the Commander and on Mission Control. Tsibliyev had never practised manual docking in a Soyuz close to Mir, but he had had to carry out this sequence of events to dock the TM-25 spacecraft. And yet he was still ordered to perform a manual docking of Progress four months later, with no real opportunity to train for it.

After much consideration, Russian officials eventually conceded that the collision was not the fault of the crew, who eventually received the post-flight awards that were due to them.

Foale became a respected 'cosmonaut', and almost a Russian hero. His cheerful and optimistic character and his clear thinking under stress helped him to survive the 18 weeks of prolonged, relentless problems. His English background, American training, and understanding of Russian ideology, helped when he needed them most – a true international space explorer. In April 1998 the Russians awarded him a citation for his skill and his contribution in saving Mir. For Foale, being accepted as a full member of the Russian crew, rather than just as a scientist or guest researcher, provided more than enough satisfaction.

# Summary

Working and living safely in space depends not only on the skills of the flight crew and ground support teams but also on the reliability of the hardware designed to sustain life. The addition of the human crew to a spacecraft increases the complexity of the design and need to sustain the life support system in whatever emergency situation the crew encounters.

All missions are provided with a medical kit and drugs to help with minor ailments and injuries. Skylab astronauts were provided with extra training in dental care, whilst Shuttle crews are trained in the recovery of incumbent International Space Station crew-members should any of them be injured or fall ill and have to return to Earth. There are also procedures for the return of a deceased crew-member, should the need arise.

The incident of the STS-96 crew in encountering a 'bad air' situation on the ISS in 1999, highlighted the need to maintain an adequate monitoring system to provide early warning of the change of the spacecraft's atmosphere. This was also demonstrated during the Apollo 13 situation when the crew had to construct a new lithium hydroxide scrubber system in the LM when it was found that the CM canisters were incompatible with the LM system.

Medical assistance can also be derived from the support and encouragement of other crew-members. During Vladimir Vasyutin's illness aboard Salyut 7 in 1985, his colleagues supported and encouraged him and shared his workload, until it became clear that because of his temperature of 104° F there was no choice but to cut short the mission and return him to Earth.

The strain of a mission is demanding enough, but when accidents *do* occur the additional stress can cause further health risks to the crew-members concerned. During the Apollo 13 mission, the lack of sleep due to the vehicle being so cold was having an effect on the responses of the three astronauts. The disappointment of losing the Moon-landing added to the stress of surviving the trip home, having to perform engine burns, and preparing for re-entry, all of which contributed to the crew's understandably terse comments towards the end of the flight.

This was again demonstrated during the sequence of events on Mir in 1997, when not just one immediate incident threatened the crew, but repeated failures challenged the cosmonauts over a period of several months. In these events the training and preparation of the ground crews in supporting and understanding the predicament in which the flight crew find themselves, is as important as the determination of the astronauts or cosmonauts in space to continue with the will to survive.

Apart from medical incidents, equipment and systems failure has been the most frequent cause of mission anomalies, accidents and disasters. The several docking failures by Soyuz spacecraft during the 1970s and early 1980s led not only to the cosmonauts' frustration in not being able to fulfil the mission for which they had trained for many months, but also placed them in sometimes life-threatening situations when trying to return to Earth throughout often hazardous re-entry and landing sequences – the splashdown of Soyuz 23 and the system malfunctions of Soyuz 33 and T-5 being the most notable.

In both the American and Russian programmes the desire to complete the assigned mission can sometimes also have an effect on the health and safety of the crew in space. After Apollo 14, Alan Shepard often gave the impression that nothing would have stopped him from landing on the Moon – whether it be a faulty docking system or a malfunctioning landing radar. However, during the climb to Cone Crater at Fra Mauro, the physical effort required to reach the target, combined with the astronauts' insufficient knowledge concerning their whereabouts, due to the lack of references on the surface, contributed to the tiredness of both Shepard and Mitchell. Disappointment in not reaching the rim of the crater was compounded when they later discovered how close they had approached without knowing that they had done so.

An additional safety concern on the Moon was that an overzealous astronaut might fall and rip his suit, and die in the Moon dust. There was no way for the remaining astronaut to recover the body and hoist it up into the LM to bring it home. Of course, the other concern was the ignition of both the LM ascent engine and the Service Module engine to bring them home. Had either of those failed on any of lunar missions, then the astronauts would have become permanent monuments to the conquest of space.

A further hazardous procedure – soon to be a frequent aspect of the International Space Station – is rendezvous and docking. Any attempt to link together two objects while both are moving is a difficult procedure, and attempting to accomplish this in space is even more hazardous. Although travelling at 17,500 mph in Earth orbit, all objects move relative to each other. It is therefore not necessarily the high speed that causes the problems but the small differences in approach speed of two vehicles – as witnessed during the Kvant incident in 1987 and the Mir incident in 1997 – and the mass of the vehicles.

Chemical batteries, fuel cells and solar arrays have all powered crewed spacecraft, and if these fail, all vehicle power can be lost. The most notable fuel cell failures have been on Gemini 5, Apollo 13, and three Shuttle missions (STS-2, STS-44 and STS-83); battery limitations were the primary reason for the quick return of Soyuz 15, 23 and 25; and solar array difficulties have plagued Soyuz 1, Skylab orbital workshop, and Mir.

Travelling into space has never been easy, and never *will* be easy. Providing suitable equipment and hardware to sustain the vehicle for the duration of its mission, and the health and well-being of the human crew, is a continuing challenge for the ISS and beyond.

# Return from space

# Overview

The recovery of a crew from space is probably the most demanding part of a mission. It can be difficult to control, and everyone must be ready for unexpected occurrences. For to the crew it is perhaps the most important element of their flight – and rightly so. After years of training, and having survived the launch and completed all mission objectives, they expect to return home to reap the benefits and, perhaps, begin the cycle again.

The Soviets (and latterly, the Russians), having such a large land mass and limited ocean resources, opted for land recovery of all space crews from the very beginning of their programme. For the Vostok series, ejector seats were provided for the cosmonaut to descend by individual parachute, separate from his descending capsule. When the vehicle evolved into the multi-crew Voskhod, such a system became impractical and was replaced by a retro-pack above the capsule, which was fired at the last second to slow the vehicle's parachute descent to a cushioned landing. The retro-pack system was adapted for Soyuz, with the solid rockets fired from the base of the Descent Module, seconds before 'dustdown' on the steppes of Kazakhstan. Cosmonauts have also trained for water recovery, and in 1976 a crew landed in a lake.

During the 1980s, cosmonauts trained for runway landings of returning Buran shuttle vehicles. Runway landings of Shuttle-type craft returning from space evolved four decades earlier. During the 1940s, the X-series of rocket planes first landed at Muroc (now Edwards) Air Force Base in California. Short-winged rocket aircraft landings on runways were always fraught with hazard, and had their share of mishaps. The X-series of aircraft, the Skyrockets, and later the X-15 and the lifting body series, developed the concept and experience of shuttle-type landing through to 1975. Two years later, in 1977, the Shuttle *Enterprise* continued the evolution, and Shuttle flights began runway returns from orbit in 1981. Despite some problems with weather (which hampers landing strip visibility during recovery) and hardware difficulties with tyres, brakes and steering, the Shuttle system has proved that a runway landing from Earth orbit is possible. It is also highly desirable for the reusability of a space vehicle.

Prior to the Shuttle, between 1961 and 1975, American astronauts landed in the ocean, as this was a larger and potentially softer landing area than land. Ocean landings were adopted for Mercury, Gemini and Apollo vehicles and, despite some small mishaps, no astronaut lost his life (although there was for one close call) throughout the whole ocean recovery programme.

During the flight of Aurora 7, excessive use of control fuel forced Scott Carpenter to overshoot his planned landing site, and he spent more than two hours in his life raft in the middle of the ocean, awaiting pick-up (see p. 352). Some of the Gemini and Apollo astronauts (including former naval officers) were seasick during the wait for pick-up, and several got their feet wet during the winch-up to the helicopter. As payloads for rockets, Gemini and Apollo were well adapted. As spacecraft they were excellent. But they were certainly not boats.

The transition from zero g to 1 g in a few minutes caused stowed equipment to move, and some crew-members received bruises – a painful reminder that they were back on Earth. Recovery was a very large, complex and expensive naval operation. It was also highly successful, with only one vacated spacecraft lost at sea (but even that was recovered, more than 30 years later!).

## 1961: LIBERTY BELL 7

The Mercury astronauts all decided to number their spacecraft after the seven members of their group, and gave them personal identification that would reflect national pride or the spirit of the team effort to put Americans into orbit. Virgil 'Gus' Grissom decided to name his spacecraft number 11 'Liberty Bell 7', because of its bell-like shape and its stirring connections with American history and national pride. One of the engineers even suggested painting a crack on the side of the spacecraft, as on the real Liberty Bell. (The actual Liberty Bell was originally cast in London to commemorate 50 years of Pennsylvania as a British colony (1682–1732). The Bell reached Philadelphia, Pennslyvania, in 1752, and just one month after its arrival it suffered a large crack from the stroke of its clapper. After being recast by a local firm, it was later used to announce that the Continental Congress was in session. The Bell was also rung to proclaim the first public reading of the Declaration of Independence. A second crack appeared in 1835, but because of the historic importance of the bell it was not recast again. In 1976 the Bell was mounted in the Liberty Bell Pavilion built for the American bicentennial celebrations in Philadelphia. The bell also appears on the reverse of US 50 cent coin).

Grissom flew the second manned sub-orbital flight in the American programme on 21 July 1961, a flight of 15 min 37 sec from Cape Canaveral, landing in the Atlantic Ocean. The flight progressed smoothly, and it was only after splashdown, 303 nautical miles down range of the Cape, that it almost ended in disaster.

At 21,000 feet, through his small observation window Grissom saw the drogue parachute deploy on schedule. At 12,500 feet the main parachute deployed from the capsule, approximately 1,000 feet higher than planned. Grissom noticed a six-inch L-

Mercury astronaut Virgil 'Gus' Grissom, fully suited, stands in front of his Liberty Bell 7 spacecraft and waits for technicians to make final adjustments to the spacecraft prior to entering for launch on 21 July 1961.

shaped tear in the main parachute, but it did not spread, and so he relaxed and continued his assignments for the descent phase.

Grissom: 'There's the drogue 'chute.'
Grissom: 'There goes the main 'chute. It's reefed. Main 'chute is good. Rate of descent is coming down.'
Grissom: 'Main 'chute is good. Landing bag [light] is on green.'

The descent rate of the capsule soon slowed to 28 fps. Grissom continued to dump his unwanted control fuel and call his instrument panel readings to the ground. After hearing the clunk confirming landing bag deployment underneath him, he removed the suit oxygen hose and opened the visor, but intentionally left his suit ventilation hose attached to the spacecraft connections. He also began to establish communications with his prime recovery ship *Randolph* (call-sign Atlantic Ship) and the recovery helicopters *Hunt Club 1* (Prime), *Card File 23* and *Card File 9*.

Grissom: 'Atlantic Ship CapCom, this is Liberty Bell 7. Confirm auto fuel and manual fuel has been dumped and I'm in the process of putting the arming pins back in the door at this time. I'm passing through 6,000 feet, everything is good. I'm going to open my faceplate. Hello, I can't get one door pin back in. I've tried and tried and I can't get it back in. Do you have any word from the recovery troops?'
Card File 23: 'Liberty Bell 7, we are heading directly toward you.'

The spacecraft hit the Atlantic Ocean at GET 15 min 37 sec, some 262.5 nautical miles from the Cape Canaveral launch pad.

Grissom: 'Roger, my condition is good. The capsule is floating, slowly coming vertical ... have actuated the rescue aids. The reserve 'chute is jettisoned.'

Grissom later commented that his first impression of the splashdown was of a milder impact than he had expected. Liberty Bell then lurched in the water until Grissom was lying on his left side and face down. Soon the capsule righted itself, and as his window cleared he jettisoned the reserve parachute pack and activated the rescue aids on his capsule, now rolling heavily in a rough swell. Most importantly, the vehicle remained watertight.

By this time, recovery helicopters were heading for the astronaut, and they asked if he was ready for pick-up. Grissom had completed many of the procedures for recovery, had disconnected his helmet and had performed a personal check to prepare for leaving the vehicle. The neck dam – a rubber collar to prevent water from seeping into his suit – was difficult to unroll, and he fiddled around with the suit collar to ensure that it would maintain his buoyancy in the event that he had to exit Liberty Bell in a hurry. Grissom informed the helicopter pilots that he would be ready in five minutes, after recording the rest of his cockpit displays. Despite some uncomfortable moments with his suit and the build-up of heat in the capsule, he completed the task and asked the pilots to pick him up.

Grissom: 'OK, Hunt Club. This is Liberty Bell 7. Are you ready for the pickup?'
Hunt Club 1: 'This is affirmative.'
Grissom: 'OK, latch on, then give me a call and I'll power down and blow the hatch. I've unplugged my suit so I'm kinda warm now. Now if you tell me you're ready for me to blow, I'll have to take my helmet off, power down and then blow the hatch.'
Hunt Club 1: 'One, Roger. And when you blow the hatch, the collar will already be down there waiting for you.'
Grissom: 'Ah. Roger.'

Grissom then removed the security pin from the hatch detonator and lay back in his couch, awaiting rescue. Then disaster struck, as he later recalled. 'I was just laying there, minding my own business, when I heard a dull thud.' To his surprise and horror, the hatch cover of Liberty Bell suddenly blew out, and the sea immediately began pouring into the capsule. He was faced with the reality that his spacecraft was shipping water and would soon sink.

Grissom was certain that he had not accidentally touched the trigger that blew the hatch, but he had no time to consider the matter. He had to get out of the capsule –

and fast. Luckily he had had the forethought to unstrap himself from the seat harness, so he grabbed the instrument panel, dived through the small hatch and swam away from the sinking capsule.

The rescue helicopter pilot above him reported that as he was trying to sever the whip antenna from the capsule prior to attempting to secure the capsule for pick-up, he suddenly saw the hatch blow and the astronaut scramble out into the sea. The pilot left Grissom to his own devices while he tried to rescue the sinking capsule – an action based on his training experiences when he had noticed that the Mercury astronauts seemed to be at home in the water during water recovery training. The pilot then hooked the capsule with the shepherd's hook recovery device. By then the helicopter was almost in the ocean itself: three wheels were actually underwater. The capsule was completely submerged, but secured. Engine strain information was recorded on the control panel of the helicopter, indicating contamination of oil due to the extra lifting weight of Liberty Bell. It was an indicator that would later prove to have shown a false reading. Liberty Bell was now so full of water, including a filled landing bag that was acting as a sea anchor, that it weighted more than over 5,000 lbs – more than 1,000 lbs over the helicopter's lifting limit. Unable to raise the capsule clear of the water, which would have emptied the bag, the helicopter was fast

A US Marine Corps helicopter unsuccessfully attempts to recover Liberty Bell 7 from the Atlantic Ocean at the end of the mission. The hatch opened prematurely, allowing water to flood the capsule, although Grissom managed to escape. 38 years elapsed before the capsule was recovered from the ocean bed.

approaching the point where it too could be lost. Yielding to the inevitable, the capsule was cut loose and allowed to sink into the Atlantic Ocean to a depth of 2,800 fathoms. A marker was to be dropped to attempt a later recovery of the capsule, but this was not immediately attempted.

While the helicopter was trying to rescue the capsule, Grissom was still in the sea trying to keep himself afloat. After leaving the capsule he immediately ensured that he was not tangled in any lines that might drag him under, and then noticed that the pick-up helicopter was having difficulty in lifting his capsule. On swimming back, he saw that the cable was firmly attached and that there was nothing he could do to help. He was alarmed when the pilot moved away, taking with him the personnel safety harness that was to lift him out of the water, because he was by now not as buoyant in the swell as when he first entered the ocean. He had already noticed air leaking from his suit, which reduced its flotation capability, and he had failed to seal his suit inlet valve, and water was beginning to seep into the suit. As the second recovery helicopter moved in, churning up prop-wash, swimming became more difficult. Constantly bobbing under water, the astronaut was frantically looking for pararescue divers to help him tread water until pick-up.

The second helicopter then moved in and threw him a 'horse-collar' lifeline, whereupon the scared and angry astronaut thrust for the sling, donning it backwards – although by this time it did not matter! After two further dunkings the astronaut

A US Marine Corps helicopter lifts Gus Grissom, a waterlogged astronaut, out of the Atlantic Ocean at the end of America's second mission into space.

was lifted safely into the helicopter, where he immediately donned a life preserver. He had been in the water only about five minutes, but as helicopters flew around him – apparently not wishing to pick him up, but to just take photographs – it had seemed like an eternity to him.

On the carrier *Randolph*, Grissom was very tired, but he insisted on carrying out the debriefing. He thanked fellow astronaut Walter Schirra for his work on the development of the suit neck-dam, which Grissom believed had helped save his life. Extra training on water egress was highlighted in the debriefing on the carrier and later at Grand Bahamas, as was the development of specific emergency recovery procedures. Examination of the cause of the hatch blowing was hampered by the fact that the capsule lay on the ocean floor. Before the Mercury flights, the contractor, Mississippi–Honeywell, had completed a series of tests on the hatch and its assemblies. They included high- and low-temperature tests, a long shock test, salt sprays and water immersion tests. Never did a hatch 'just blow'.

NASA assigned a special committee – including astronaut Schirra – to investigate the problem. During this investigation an even more severe test programme was carried out on the hatch design, without premature detonations occurring. The group also examined the cockpit layout to determine if accidental activation by a pilot was possible. Schirra later wrote that only a very remote possibility existed that an astronaut could activate the plunger. The exact cause was never discovered. The capsule was recovered in 1999, and it appeared that Grissom was correct: the hatch just blew. There was a metal seal around the hatch, and it was bent, which indicates ripple-fired pyros and not the simultaneous pyros that were planned.

Speculation was rife: an exterior lanyard could have snarled, triggering the device; a ring seal might have been omitted from the plunger, thus reducing the level of pressure needed to activate the system; static electricity from the helicopter could have fired the cover ... But with no hardware to examine, all NASA could do was to amend procedures to ensure that such an occurrence would not be repeated. An astronaut would not touch the plunger until receiving confirmation that the helicopter had safely snared the capsule.

Liberty Bell was the last capsule in the programme in which retrieval of the craft by helicopter was planned. All future spacecraft would be hoisted directly onboard the recovery ship, and the helicopters would concentrate on the recovery of the crew and safing the empty spacecraft prior to pick-up. Grissom maintained – right up to his tragic death in 1967 – that he did not touch anything. He often pointed to the lack of injury to his hand as evidence. John Glenn, Walter Schirra and Gordon Cooper had all received slight injuries when using the plunger on their vehicles.

The joke was soon spread around the Cape that it was the last time they would launch a spacecraft with a 'crack' in it! For Grissom, it was a very lucky escape, but five and a half years later he would not be so fortunate. Despite some suggestions and media reports blaming him for losing the capsule, the confidence shown in his abilities was reflected in his assignment to the first Gemini crew in 1964 (flown as Gemini 3 in 1965) and to the first Apollo crew in 1966.

Several years after the incident there were discussions outside NASA to recover the capsule and display it in a museum. Grissom's widow was not opposed to the recovery, but did not wish it to be put on public display. However, there was a strong desire to retrieve the capsule, as it was the only manned spacecraft which America had not recovered. It was planned that the Kansas Cosmosphere and Space Center, at Hutchinson, Kansas, would display the recovered capsule, regardless of its condition.

In 1992 and 1993, attempts were made to find Liberty Bell, but without success. The recovery team tried again in April 1999, but concerns were expressed that actually bringing the capsule to the surface might destroy it after so long a period at the bottom of the ocean.

On 1 May 1999 Liberty Bell 7 was found at the first of a possible 88 target sites, 2,850 fathoms down on the seabed. TV cameras on the remote submersible *Magellan* revealed the capsule for the first time, almost 38 years after it had been pictured below *Hunt Club 1*. The plan was to have the submersible attach a recovery cable to the spacecraft and then winch it to the surface. However, the Atlantic Ocean was not about to give up Liberty Bell so readily. On the surface, huge waves were hindering the recovery attempts, and the umbilical that connected the submersible to the prime recovery vessel, *Needham Tide*, snapped. On 4 May the submersible sank to lie next to the spacecraft that it was trying to recover. The recovery team returned to Port Canaveral to stock up with new, more robust recovery equipment for the retrieval of both the spacecraft and the submersible.

The new search began on 10 July 1999, and Liberty Bell was located again the following day. It took a further nine days to prepare the spacecraft for lifting to the surface. On 20 July 1999 – the 30th anniversary of the Apollo 11 moon landing – Liberty Bell 7 was finally winched aboard the prime recovery ship *Ocean Project*. On board were Guenter Wendt – the former McDonnell Douglas Pad Leader who had overseen Grissom's entry into the spacecraft prior to the start of the Mercury Redstone 3 mission – and Jim Lewis, the now retired US Marine Corps helicopter pilot who had attempted to lift the flooded spacecraft from the ocean at the end of the mission.

Liberty Bell 7 returned to Cape Canaveral the next day, exactly 38 years after its launch. On hand to witness the arrival was Grissom's brother Lowell, but Betty Grissom did not attend the return of her husband's first spacecraft. Externally it still looked in good shape, with some marine encrustation around its cylindrical parachute compartment. The inside, however, was the worse for wear, with corrosion and silt, presenting the Kansas Cosmosphere and Space Center with a considerable task of restoration. The work would take six months, after which the spacecraft would tour American museums and then be assigned a permanent place in Kansas. The only American spacecraft that had carried a crew, but had not been recovered, was finally back on dry land.

## 1962: AURORA 7

The highly successful flight of John Glenn and Friendship 7 on 20 February 1962 had demonstrated that an astronaut could be more than just a passenger along for the ride. As a result, mission planners adjusted the flight plan of the next mission,

Mercury Atlas 7, to allow for more control by the pilot. In addition, a programme of scientific experiments was added which would not jeopardise either pilot safety or the success of the mission.

On 24 May 1962, astronaut Scott Carpenter was launched aboard Mercury Atlas 7, bearing the name Aurora 7, to become the second American and fourth man to orbit the Earth. After launch and capsule separation, John Glenn had used more than 5 lbs of fuel in turning the spacecraft around to fly its normal attitude using the automatic control system, so Carpenter opted for the 'flyby' wire system and used only 1.6 lbs of fuel.

During 1959 the Mercury automatic stabilisation and control system was under development, and was in effect the spacecraft's autopilot. A small computer would compare inputs of electronic sensory information with deviations from the preset reference points on gyroscopes with the horizon. In response to these the autopilot would then trigger commands to small thrusters to vent hot gas to maintain the balance of the spacecraft. Although termed a 'simple' system, it was still considered very complicated, and was causing some design problems.

An alternative to the automatic system was the direct mechanical linkage to a second completely independent and redundant RCS, allowing the pilot to manually adjust the attitude of his spacecraft in orbit as desired. However, although clearly demonstrating the fundamental Mercury objective of testing man's capability in space, it was also an extremely uneconomical guzzler of fuel supplies.

To help solve the problem of this thirsty manual thruster control, a wired jumper from the astronaut's hand-control to the automatic system allowed the pilot to electronically switch on or off the tiny valves that supplied hydrogen peroxide gas to the automatic thruster chambers. This system completely bypassed the autopilot, and instead relied on the astronaut's senses and response. By allowing the pilot to aid or even interrupt the automatic system around the pitch, yaw and roll axes, it was a semiautomatic system, and was referred to as the 'flyby' wire.

In orbit onboard Aurora 7, Carpenter was working on a very busy experiment schedule, with added manual manoeuvre inputs to the orientation system. He was so busy that he inadvertently activated the highly sensitive control jet six times, which in turn activated the automatic and manual systems at the same time, burning more fuel reserves. By the end of his first orbit he had only 42% in the manual tanks and 45% control fuel supply in the automatic tanks. Throughout his second orbit, ground communicators frequently reminded him to conserve fuel as he continued to fulfil his experiment objectives.

By the beginning of his third and last orbit, Carpenter began a long period of drift through almost a complete orbit to conserve fuel for re-entry and landing. During the free drift, as the spacecraft entered sunrise his hand bumped the capsule hatch, whereupon a cloud of particles flew past the hatch window. He immediately turned the spacecraft around to follow the particles, and by banging on the wall of the spacecraft more particles were produced. They were similar to Glenn's mysterious 'fireflies,' and Carpenter deduced that since the exterior of the capsule was evidently covered with frost, the particles became his 'frostflies'.

Despite the excessive use of fuel on the first two orbits, Carpenter managed to conserve fuel on the third. At the flight control centre in Florida, Mission Director Chris Kraft believed that the astronaut had sufficient fuel to make a normal, safe entry on either the automatic or the manual control system. When the tracking site in Hawaii instructed Carpenter to begin his pre-retrofire countdown and switch from manual to automatic control, the astronaut informed ground control that he was in fact behind in his preparations due to his attempt to verify the origins of the particles outside the spacecraft. If this was not bad enough as Carpenter tried to switch to automatic control he found that the system would not hold in the required orientation. As he tried to solve the problem, he activated to flyby wire system, and fell further behind in his checklist. As he did so he forgot to disengage the manual system, and for the next ten minutes the fuel continued to be used from both systems.

As the spacecraft passed out of range of Hawaii, Carpenter was urged to complete as much of the checklist as possible as he finally managed to align the spacecraft for re-entry. Because the automatic system was presenting difficulties, he checked that it was turned off, and had to manually push the button to ignite the solid-fuel retro-rockets strapped to the heat shield behind him. All three fired successfully, but were three seconds late

Aurora 7 was actually canted at retro-fire $25°$ to the right, and thus the firing of the retros was not exactly against the direction of flight to slow the spacecraft down sufficiently to aim for the planned landing point. Instead, the spacecraft would overshoot by 175 miles. The late firing of the retros added another 15 miles to the trajectory, and the lack of correct thrust added a further 60 miles, resulting in a 250-mile overshoot.

Following retro-fire, Carpenter realised that the manual control system was still turned on, quickly turned off the flyby wire, and checked the manual system, which read no more than 6% fuel (in effect, no fuel) for control input. The astronaut flipped back to flyby wire, which recorded 15% – Carpenter believed that it was less than that. During the drifting glide he kept his hands off the controls to conserve as much fuel as possible for the critical tumble manoeuvre. Attitude control indications seemed useless, but there was in any case no fuel for the system. It was 10 minutes before the 0.05 g point, where Carpenter would again begin to feel his weight. It was a long 10 minutes.

Within seconds of reaching the 0.05 g barrier, Aurora 7 began to vibrate badly. A quick switch to the auxiliary dumping mode steadied the vehicle, and the spacecraft was engulfed in its blazing return to Earth. Inside the fireball Carpenter encountered g forces for longer than expected, and at one point, like Glenn before him, he expected the capsule to burn up at any moment.

After exiting the plasma sheath the spacecraft was 'swinging ' beyond the $10°$ limits of tolerance. Carpenter reached up to arm the drogue parachute, but held off while Aurora 7 rode out severe oscillations. Parachute deployment and descent was normal and, with lesser jolt than expected, Aurora 7 splashed down. Tracking indicated the position to within a few miles, and over the radio Gus Grissom indicated that the rescue team would take about an hour to reach him. It was then that Carpenter realised that he had overshot the landing area. He noticed drops of water on his tape recorder, and thought that he might experience the same fate as

A Navy pararescue diver assists MA7 Pilot Scott Carpenter during recovery operations.

Grissom on Liberty Bell 7, which lost the side hatch. He could not find a leak, but realised the capsule was listing to the left and floating rather deeply. It would be dangerous to try to exit through the main hatch, so in temperatures of 101° inside the cramped capsule he took off his helmet and struggled upward through the neck of the spacecraft to the top hatch. He left his suit coolant hose attached as long as he could and, making sure he had his rescue pack with him, he gently slid into the ocean to inflate the life raft and after a few moments settled in for a long wait. The recovery beacon was operating and the sea dye marker was working.

Those following the flight on TV and radio they knew Aurora 7 was down but not that Carpenter was safe. The spacecraft radio was not transmitting, and there was no radio in the life raft. Waiting for the rescue teams from the approaching aircraft, Carpenter studied seaweed floating by, and 'a black fish that was just as friendly as could be'. After 36 minutes he saw small aircraft approaching and knew he had been found. Twenty minutes later – just over an hour after landing – he saw several aircraft circle the area and two frogmen bail out of an SC-54 transport plane. One of the frogmen missed the landing area by some distance and disengeged his parachute harness to swim under the waves towards the astronaut, who was completely surprised to see him appear next to the raft.

Inflating two other rafts, the pararescue divers waited with Carpenter for pick-up by helicopter. After three hours Carpenter was finally lifted – after a dunking in the ocean – to the recovery helicopter. In Mission Control, Chris Kraft was furious about the delayed retro-fire and Carpenter's apparent lack of attention to the flight plan during the later stages of the mission. Kraft reportedly stated: 'That son-of-a-bitch is never going to fly for me again.' And he did not. After a short time spent working on the development of the Apollo LM, Carpenter took a leave of absence to work on the US Navy Sealab Man-in-the-Sea habitation programme in the mid-1960s, before leaving NASA in 1967 to work full time on the Navy Sealab project.

### 1969: SOYUZ 5

On 18 January 1969, Boris Volynov was nearing the end of this three-day spaceflight. On 15 January, Volynov had been launched with cosmonauts Alexei Yeliseyev and Yevgeni Khrunov to dock with the previous orbiter Soyuz 4, containing Vladimir Shatalov. The following day, Soyuz 5 acted as a passive target as Soyuz 4 docked to it. Later, both Yeliseyev and Khrunov, wearing spacesuits, transferred from Soyuz 5 to Soyuz 4 by EVA. On 17 January Soyuz 4 undocked and returned to Earth with the three cosmonauts, and the next day Volynov, now alone in Soyuz 5, prepared for his own re-entry.

Following his flight plan, Volynov strapped himself in his couch in the Soyuz Descent Module and commanded the jettisoning of the Orbital Module located in front of him. When the Equipment Module located behind him was commanded to jettison, defective explosive bolts failed to fire correctly and did not completely separate the EM, which thus blocked the re-entry heat shield on the base of the DM. Unclean separations had occurred with much smaller units on Vostok and on John Glenn's Mercury spacecraft (see p. 213), but these were much smaller than the mass of the Soyuz EM and disintegrated during re-entry.

As a result of the added mass of the EM, Volynov lost control of Soyuz 5, which began to tumble in space, finally stabilising itself with the thinnest part of the spacecraft, and with the part with the least heat protection facing forward. As the

Cosmonaut Boris Volynov onboard Soyuz 5 in January 1969. The spacecraft was oriented for re-entry very late, and could have burned up, resulting in the fiery death of the cosmonaut, although luckily he survived the ordeal.

Apollo 15 splashdown in the Pacific Ocean, 7 August 1971. One of three ringsail parachutes failed to deploy, resulting in a harder landing than with the normal three parachutes.

combination began to enter the atmosphere, the structure began to fail. The added stress and heat on the support struts between the two modules finally caused them to burn through and snap allowing the EM to finally fall away to burn up during re-entry.

Onboard Soyuz 5, Volynov could only sit and wait as the automatic orientation system struggled to regain control before re-entry began. The DM sought the most aerodynamically stable position as it entered the atmosphere, which was with the heat shield forward – fortunately, before it was too late.

Re-entry was survived, but the parachutes deployed on partially, almost causing a repeat of the fatal Soyuz 1 landing. A failure in the soft-landing rockets in the base of the DM caused a harder than normal landing, almost wrecking the capsule. When the search crews found Soyuz 5 it was empty. Volynov was found, with several broken teeth, in a nearby peasant hut trying to keep warm while awaiting his rescuers.

## 1971: APOLLO 15

On 7 August 1971 the Apollo 15 astronauts were inside the CM under three ringsail parachutes which had deployed correctly at the end of their mission to the Moon. Suddenly, one of the canopies collapsed. The crew was informed that they need not worry, as they could land safely, but more heavily, with two parachutes. Landing velocity with three parachutes was 19 mph, rather than to 22 mph with two parachutes. As they emerged from cloud the CM narrowly missed one of the recovery helicopters. The CM made a hard splashdown in the Pacific Ocean, 353 miles north of Hawaii. In the CM Lunar Module, Pilot Jim Irwin released the parachutes on impact, which he should not have done, as there was no wind to drag the capsule through the ocean. The parachute that failed was the only one not recovered.

During 6–7 October 1971, in a Manned Spaceflight Management Council Meeting at NASA Headquarters in Washington DC, the Apollo 15 parachute anomaly was discussed, along with its impact on follow-on missions. The failure was attributed to RCS fuel (monomethly hydrazine) being expelled during depletion firings (the emptying of fuel tanks before landing). Corrective action included landing with RCS propellent onboard; and for landing abort, leaving the oxidiser onboard but increasing the time delay allowed for rapid propellent dump, avoiding contact with the parachute riser and suspension lines.

The decision to retain the fuel onboard for landing in the ocean would be a contributing factor during the inadvertent firing of RCS jets near an open inlet value in the CM during the 1975 ASTP landing.

## 1975: APOLLO–SOYUZ TEST PROJECT

In July 1975, America performed the last of the series of spaceflights with one-shot spacecraft, begun with Al Shepard and Mercury 3 in May 1961. Apollo 18 was the American half of the joint US/USSR manned docking mission, the Apollo–Soyuz Test Project. Launched on 15 July, the two vehicles docked on the 17 July for a period of joint experiments before the Soyuz landed on 20 July, leaving American astronauts to fly the Apollo vehicle in orbit, for Earth resources studies.

On 24 July the three-man Apollo crew completed America's 31st and last planned ocean splashdown. The final three American astronauts to fly in space until the advent of Shuttle in 1981 were Commander Tom Stafford, CMP Vance Brand and Docking Module Pilot Deke Slayton.

The mission had proceeded so smoothly from the day of launch that, even before the crew had returned to Earth, most observers were already reporting a successful flight. Most American splashdowns since the days of Mercury had been routine. The Apollo recoveries were a major feature on TV and were so 'routine' that when something *did* go wrong it came as a shock to the crew and controllers.

Preparations for the mid-afternoon recovery of Apollo 18 began as soon as the crew awoke on the last day in orbit, at 07.30 CST. They were informed that weather at the primary landing area in the Pacific Ocean was favourable, with visibility of 10 miles, winds at 17 knots, scattered cloud cover at 2,000 feet and a maximum wave height of more than 3 feet. Operationally, all proceeded like clockwork. The CSM engine burned for 7.8 seconds retrograde at 15:37:47 CST to begin the descent to Earth.

Within six and a half minutes, the CM carrying the crew separated from the SM and began its descent through the atmosphere. Both Slayton (who had waited 16 years for this, his first and only spaceflight) and Brand (flying his first mission after a nine-year wait) commented on the build-up of gravity forces as the CM plummeted towards the ocean far below. For Stafford, the mission Commander, this was all very familiar, as this was his fourth flight in ten years. The crew would be the last American astronauts planned to be recovered by parachute and splashdown (though ISS astronauts now train for emergency Soyuz parachute recovery) at 16:18:24 CST on 24 July, approximately 4.5 miles from the prime recovery vessel *New Orleans*, some 270 miles west of Hawaii.

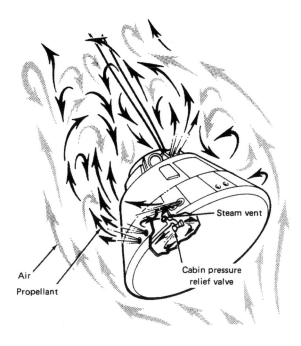

A schematic showing how RCS exhaust gases entered into the CM air intake during the descent phase, causing the three astronauts inside to experience breathing difficulties.

As the Apollo CM hit the water, signalling the end of an historic era in American manned spaceflight, it seemed to the flight controllers to have been a routine re-entry and splashdown, with the only remaining task being the pick-up of the crew. But the astronauts had inhaled nitrogen tetroxide fumes during their descent, and had encountered the most serious mishap so far during the recovery of an American crew since Grissom almost drowned in 1961.

Descent had proceeded as planned until about 50,000 feet. As the CM dropped through the atmosphere at 300 mph, the crew was supposed to activate two switches that armed the Earth landing system, at 30,000 feet. These would automatically jettison the apex cover, programme shut-down of the RCS system and instigate drogue and main parachute deployment. The change in not dumping RCS fuel came as a result of the loss of one main parachute on Apollo 15 (see above). The crew was seated with Stafford in the centre, Brand to his left and Slayton to his right. Due to a higher level of noise in the CM from the firing of the CM thrusters and the passage of the vehicle through the atmosphere, the required steps the crew had to complete were not carried out smoothly. Stafford read from the entry checklist as they approached. Brand, the CM pilot, then activated the necessary switches. Brand later explained: 'At 30k [feet] we normally arm the ELS Auto and ELS Logic. That didn't get done. Probably due to a combination of circumstances, I didn't hear it called out. Maybe it wasn't called out. Any case, 30k to 24k [feet] we passed through the regime very quickly. I looked at the altimeter at 24k [feet] and didn't see the expected apex cover come off or the drogues come out. So, I think at 23k [feet], I hit the two manual switches, one for the apex the other for the drogues. They came out.'

When the failure of the ELS function was noticed, Slayton told Brand to depress the manual deployment button on the control panel. The apex cover was jettisoned, followed three seconds later by the two 16.5-foot drogue parachutes, slowing the vehicle to 170 mph. Due to the manual deployment of the drogues the CM swayed, causing the RCS to work vigorously to counteract the motion. Stafford noticed this continuation of the thruster firing, indicating that the crew had failed to set the RCS command switch to the 'off' position, and he immediately cut propellant flow to the thrusters. Firing stopped 30 seconds later, as fuel and oxidiser that had already passed the isolation valve continued to power the thrusters. Brand continued: 'That same instant, the cabin seemed to flood with a noxious gas, very high in concentration, it seemed to us. Tom [Stafford] said he could see it. I don't remember for sure now if I was seeing it, but I certainly knew it was there. I was feeling it and smelling it. It irritated the skin a little bit and the eyes a little bit and of course you could smell it. We started coughing.'

At 24,000 feet the pressure relief valve opened automatically, drawing in fresh air and the unwanted gases from the roll thrusters located only two feet from the valve. The crew put the oxygen supply into the high-flow mode and activated the automatic switch to deploy the main parachutes – an action that completely deactivated all thrusters. Despite the gas fumes, they continued working through their checklist to the best of their ability. Parachute deployment occurred without incident, but the capsule hit the ocean hard and immediately flipped over to the Stable 2 position,

During a telephone conversation with President Ford, the three ASTP astronauts – Stafford, Slayton and Brand – show signs of sore eyes infected by the toxic gas during descent.

apex-down, with the crew hanging by their seat restraint straps and looking into the depths of the ocean.

Brand was coughing the most, as he was nearest the duct opening. He noticed that Slayton was feeling nauseous, and reminded Stafford to break out the oxygen masks, which were stored in lockers behind the seats – a very awkward position, with the CM face down. To reach them, Stafford had to release his straps and carefully lower himself down to the pointed end of the capsule, then clamber back behind the couches to reach to the masks. The crew had activated the self-righting system of three large floatation bags, but these took three minutes and they knew they could not wait that long to obtain fresh air supplies. As Stafford later recalled: 'For some reason I was more tolerant to the bad atmosphere and I just thought 'get those damn masks – and don't fall down the tunnel.' I came loose and had to crawl and bend over to get the masks. I knew that I had a toxic hypoxia and I started to grunt-breathe to make sure I got pressure in my lungs to keep my head clear. I looked over at Vance [Brand] and he was just hanging in his straps. He was unconscious.'

When Stafford put the mask over Brand's face he began to come round after about a minute of unconsciousness. The spacecraft soon righted itself, and Stafford opened the vent valve, allowing an in-rush of fresh air. The remaining fumes soon dissipated. Their upturned position had prevented the cleansing of the cabin air. The hatch was eventually unlatched, and fresh air soon cleared the atmosphere.

Recovery of the last CM was in Skylab fashion, with the vehicle – with the three astronauts still inside – being hoisted onboard the recovery ship *New Orleans*. As they emerged they showed no visible signs of the encounter with the gas. TV viewers and onlookers saw no signs of discomfort apart from the rubbing of watering eyes, and they appeared comfortable and pleased with their achievements during their telephone call from President Ford. The eye problem was explained away as a result of coming into bright sunlight on the carrier's deck after the relative dark of the CM. No one seemed to notice Deke Slayton's casual comment that they were pleased to see the recovery team through the CM window 'after we picked up a little smoke on the way'.

As the post-flight medical examination was conducted on board the *New Orleans*, the three astronauts complained of discomfort and chest pains when they tried to breathe deeply. Additional examinations revealed that all three were suffering from irritation of the respiratory tract, and they were given cortisone to relieve it. When they arrived in Honolulu the next day, they were admitted to Tripler Army Medical Hospital for observation, instead of immediately flying to Houston. After three days in hospital they were released, and underwent ten days of medical observations while resting in Hawaii, where their wives joined them. Fortunately, none of them developed any complications. They recovered quickly and gained a clean bill of health, although there was concern when on 19 August NASA announced that Slayton was to undergo surgery to investigate a spot on his lung.

After surgery on 26 August – when the 'spot' was found to be benign – medical staff reported that it was not a result of the exposure to gas, but had developed before the flight and might not have been discovered but for the medical examination and X-rays. Had it been discovered before the flight it may have prevented Slayton from flying, repeating the loss of his Mercury flight in 1962 due to a heart murmur that grounded him for ten years.

In the ASTP medical report, published in 1977, it was revealed that Stafford actually had trouble seeing Slayton, Brand and the control panel dials in front of him through the yellowish-brown coloured smoke, which to Stafford smelled like RCS exhaust. The smoke cleared very fast, and probably accounted for the reason that neither Slayton nor Brand reported seeing it. The report stated that the total time of exposure to the oxidiser vapour was 4 min 40 sec from closing the RCS isolation valves to the crew putting on their oxygen masks after landing.

The medical report also stated that Brand was unconscious for about 50 seconds, probably due to the combined affects of exposure to the toxic effects and the effects of landing in Stable 2 with his feet lower than his head, which pooled blood in his lower limbs and led to his fainting.

As the spacecraft hatch was opening after being hoisted onto the recovery vessel approximately 40 minutes after landing, a humid and mouldy smell was detected. This was normal on previous Apollo CM recoveries, but there was no reported odour of the irritant gas.

During tours of the US and USSR, none of the three men showed adverse signs due to gas inhalation. Brand later flew three missions on the Shuttle, while Stafford and Slayton returned to administration positions in the USAF and NASA respectively. The last men to fly Apollo became almost the first to suffer serious injury during parachute descent. The Apollo 18 capsule was the only flight in which the crew had performed manual deployment of parachutes. Had the three astronauts been unconscious, they could have hit the ocean at 300 mph!

Investigations into the cause of the accident identified several contributing factors. Poor communications were reported, but these were due to transmitting conditions and the high load of recovery communications required during that phase of the mission. In addition, the initial comments about the fumes and yellow coloured gas were not considered to be unusual. Previous crews had reported similar conditions, and Stafford had even reported odours during the re-entry of his Gemini flights in 1965 and 1966, and during his return from the Moon on Apollo 10 in 1969. Engineers suspected that these odours originated from seared sections of the CM heat shield or from explosive charges used to open the parachute compartments. No immediate concern arose from the Apollo 18 reports, as there was no reason to suspect that the odours on this flight were any different from those on earlier missions.

Due to severe coughing and intercom noise, the crew had difficulty in communicating with each other, let alone the controllers. Thus the extent of their plight could not be conveyed sufficiently until it was over and until they were undergoing post-flight medicals and mission debriefing. Doctors later attributed Brand's period of unconsciousness to orthostatic hypertension – temporary low blood pressure in the head – experienced by several astronauts during the sudden change from weightlessness through the high g entry to the normal 1 g on landing.

NASA reviewed recordings of the descent, the crew's communications and the telemetry received during the period of parachute recovery, to evaluate the sequence of events. It appeared that the crew's attention had been distracted by interference in communications (identified as a loud unexplained squeal in the headsets), which caused a delay in the activation of the two landing switches of the ELS. This system completed several automatic cycles in sequence, but only after it had been activated by the crew.

As the CM dropped through the Earth's atmosphere, the crew activated the parachute recovery system as normal, and it functioned perfectly. However, they failed to isolate the RCS when required to do so, which would have been sequenced automatically had it been activated. One of the thruster outlets was located very near a 2.5-inch diameter cabin pressure relief valve, which sucked air into the CM and was used to equalise the pressure inside the descending capsule. It was the oxidiser ($N_2O_2$) of the thrusters that entered the cabin and that the astronauts saw, smelled and inhaled. This gas is a very toxic agent that can be lethal if inhaled in large quantities. It can cause severe lung disease, and prolonged exposure has been known to leave victims vulnerable to pneumonia and other respiratory illnesses. The ASTP astronauts were very fortunate that they were not more severely affected.

Normally, an investigation of this kind would result in changes to spacecraft down the production line, to ensure that such an event will not occur again, but Apollo 18 was the last of an era. The Shuttle was the next manned spacecraft that astronauts were to use, and a different landing profile was to be flown, to a runway landing rather than by parachute to an ocean recovery. Nevertheless, NASA was mindful of the findings when positioning the RCS thrusters and inlet valves on the Shuttle orbiter.

## 1976: SOYUZ 23

In the summer of 1976 the Soviet Union was utilising the Salyut 5 military-oriented space station of the first generation of the Salyut series. This station would also prove to be the final one of the series to reach orbit and host a cosmonaut crew. Others were planned, and were manufactured, but none would fly as a manned variant, as the second-generation Salyut 6 would appear in 1977. During the long-duration Soyuz 21 mission in August, an acrid odour in the station's air supply system forced an immediate evacuation and recovery of the two-man crew. This premature evacuation of the station left much of the programme of work incomplete. The flight plan of Salyut 5's next crew would have the added task of completing necessary repairs on the station's systems, as well as picking up the previous crew's tasks. It was a challenging mission. The crew had received EVA training, but there were no indications that any EVAs were planned during this mission. Valeri Rozhdestvensky thought that his previous training as a Navy diver would be useful when executing their tasks on orbit. Using large water tanks is an effective method of simulating EVA operations, and is used by both astronauts and cosmonauts in their training programmes. Ironically, it was not in space that these skills were almost called upon!

Vyacheslav Zudov and Rozhdestvensky were launched on Soyuz 23 on 14 October 1976, for a mission of two weeks, resuming occupation of Salyut 5. Initially all went well, as the crew completed launch and initial approach to Salyut during the first 24 hours of flight. The approach and docking was not so successful, however (see p. 234). At 328 feet, the Commander took manual control. The Soviets then issued a report that during the approach to the station there was 'a malfunction in the automatic approach to Salyut 5, in the rendezvous and approach electronics on board the Soyuz.' When attempts failed to correct the fault, the docking was reluctantly cancelled due to low fuel reserves, and immediate plans for re-entry were initiated. Soyuz 23 was a two-day ferry design relying on chemical batteries, rather than solar panels, for power generation.

Previously in a cheerful and talkative mood, the crew fell quiet, accepted the cancellation, and powered-down their capsule to conserve as much of the limited energy as possible. This included using only one onboard light, and keeping radio silence. Discussions of the failure with ground control had indicated that Zudov might not have even have been able to take manual control from 328 feet. Comparison with Soyuz 15 (1974) – in which control was not possible between 98 and 164 feet – indicated a more serious problem with the equipment.

The announcement of the immediate return of the crew indicated Soviet confidence in their handling of an in-flight emergency, following Soyuz 15 and the 1975 '5 April anomaly'. What was expected was the quick and routine recovery of the two cosmonauts and an investigation into how the docking equipment had once again failed the Soviet programme. What actually happened next was more dramatic than the events in orbit.

At 20.02 Moscow Time, during the 33rd revolution, the command to ignite the descent engines on the Equipment Module was initiated, and the spacecraft slowed to begin the re-entry sequence. Due to the nature of the orbital path, a daylight landing was not available, so the crew was forced to make a night-time descent. Some 23 minutes later, following jettisoning of the Equipment and Orbital Modules, the capsule containing the disappointed cosmonauts began its fiery re-entry through the atmosphere. Eventually, the recovery parachute opened and the capsule dropped to Earth. Normally, as the capsule slowly falls to the ground, this phase of the Soyuz re-entry profile is the least eventful. The emergency return of Soyuz 23 had already dictated a landing in darkness – bad enough for the search and rescue teams, but at the landing site conditions were even worse, with high winds and falling snow!

On board the capsule, swinging beneath its recovery parachute, the two cosmonauts braced themselves for a soft landing on the steppes of Kazakhstan, as had eighteen other Soyuz crews since 1968. Upon entry into the Earth's atmosphere, the crew calibrates a rudimentary altimeter, which is used to gauge the final approach to the ground. Their seats contain compressible hydraulic rams to take the force of landing. These are pumped up before re-entry. Despite the deceleration by parachute and the firing of retro-rockets seconds before touchdown, the landing impact can be quite severe and not as soft as the term implies.

At an altimeter height of 164 feet, Zudov gave the order to his crewmate Rozhdesventsky to be ready for the landing. Both cosmonauts braced themselves and waited for the Descent Module to hit the ground with a thud. Much to their surprise, it splashed down in water.

The high winds encountered by the capsule had caused it to be blown off course, and it had drifted over Lake Tangiz, 130 miles south west of Tselinograd. The first (and so far only) splashdown in the Soviet manned programme occurred at 20.46 MT. The closest any Soviet crew previously come to a splashdown before was in April 1971 during the Soyuz 10 mission, when the capsule landed just 50 yards from a lake.

All Soyuz crews train for water recovery, resembling the method used on the Apollo programme, in which divers attach flotation collars to the spacecraft and the crew are evacuated from the capsule for helicopter recovery. For Soyuz 23, the severe weather seriously hampered these normal recovery operations.

Just minutes after splashdown the cosmonauts were well aware of the conditions outside. Inside the capsule it was dark and bitterly cold –20° C (–4° F). After bouncing around for 15 minutes, the crew felt a 'powerful explosion' and then a list to the side, almost sinking in the partly frozen water. The parachute separation device had malfunctioned, becoming waterlogged and effectively acting as a sea anchor. With the parachute submerged and stretched to its full length, the capsule was stuck in the middle of the lake, with its hatch underwater. In the freezing temperatures the hatch could not be opened. Salt water corroded electrical circuits, preventing the cosmonauts from using the radio to inform the rescue teams that they were alive.

Soyuz 23 crew Zudov and Rozhdestvensky became the first cosmonauts to splash down at the end of their mission. Landing in a lake during a blizzard, they almost froze to death during the prolonged recovery operation.

Despite a blizzard, helicopters battled to keep station over the capsule in the choppy water, shining searchlights to aid the divers, who were trying to attach flotation devices to the inverted capsule. Freezing temperatures of –15 to –17° C (5 to –1° F) further hampered operations. For more than nine hours the cosmonauts waited inside their frozen Descent Module. Most of the reserve battery power had been used during re-entry, and frost was forming on the inside of the capsule. To help keep warm, both men donned all of the survival clothing that they had available. They soon removed their Sokol–K (Falcon) Soyuz pressure suits, which were fine for surviving a pressure failure but useless in the freezing temperatures they now faced. They put on sweaters and thermal garments intended for arctic-like conditions in a survival situation on dry land, but which probably saved the cosmonauts lives in the refrigerator in which they were now floating.

With no communications with the inside of the capsule, the recovery crew thought the cosmonauts must be dead. It was just bad luck that the spacecraft had descended into a lake in the middle of the primary Soyuz recovery zone. Rescue by raft was impossible on the half-frozen lake, and the freezing temperatures meant that divers could not long survive in the water. The high winds buffeted the helicopters, and in the darkness, searchlights were almost ineffective due to swirling snow.

Almost nine hours after the spacecraft returned, the first rays of dawn on the horizon provided some light. The helicopters tried to lift the capsule out of the water, and in doing so unknowingly knocked Zudov out of his recumbent seat liner and smashed his watch against the instrument panel. The lifting attempt failed, and the spacecraft sank back into the water. Earlier, while helping Zudov out of his pressure garment, Rozhdestvensky had also hit his watch on the instrument panel as he slipped. Later, he commented on the irony of the name of the watch – Ocean – following the splashdown, and also thought that his pre-launch comment calling upon his former skills as a naval diver was a bad omen.

The quickest way to reach the shore was to drag the capsule out of the water with a sling around the Descent Module. This was a rough ride for the two men inside, who were bruised and jostled as the experience further drained their strength as they finally reached dry ground. It took all their strength to open the manhole cover upper hatch, much to the surprise of the rescue crew, who had been expecting to find two frozen bodies inside. The announcement of the recovery was not issued until ten hours after the landing, indicating the difficulties encountered.

Cosmonaut Chief Vladimir Shatalov later praised the 'high courage and heroism' of the rescue teams. (It has been reported that a number of the rescuers died while saving the cosmonauts, but this has never been officially confirmed). Shatalov continued: 'All the search and rescue systems were in complete readiness. Everything possible was done to get to the crew quickly. One hardly needs to say how difficult and how hard the work was, in this situation.' A further indication of the difficult recovery came in a statement by Tass, the Soviet news agency commenting on the flight. It reported: 'At all stages of the flight and after the landing, the crew acted in a confident way, efficiently discharging their duties.'

Neither cosmonaut apparently suffered any ill effects from the flight, although neither flew a second mission. Both of them continued cosmonaut training for some years, and were assigned to flight crews that were reassigned before launch. The demise of the military space stations to be operated with a crew also saw the disbanding of the military cosmonaut Almaz training group. The difficulties with the docking equipment were to continue to plague the Soviets for several years, although the Soyuz 24 crew did manage to spend 18 days on Salyut 5 in 1977.

The valuable lesson learned from Soyuz 23 was that under adverse conditions the Soyuz recovery and support programme functioned well, eventually, and safely rescued the stranded cosmonauts from what could have been fatal circumstances.


## 1985: STS 51-D

One of the more dramatic Shuttle landings occurred during mission 51-D in April 1985. In returning orbiter *Discovery* to Florida's purpose-built Shuttle Landing Facility (SLF), Commander Karol Bobko had the added problem of bringing back the first Shuttle through a cross-wind. This was the sixteenth landing in the programme, and although the hardware was in place to provide nose-wheel steering capability to help compensate for cross-wind forces, it had yet to be flight-certified for operational use. For this landing all that Bobko could do was to utilise the differential braking to keep *Discovery* on the centre line of the SLF. Data recorded that the cross-wind he experienced was only 8 knots, but loads on the main landing

A view of the tyre damage caused during the STS 51-D landing in April 1985.

gear in keeping the vehicle straight caused one of the tyres to blow out, a second to sustain bad erosion, and a brake system to seize.

Fortunately the blow-out came at the end of the roll-out, and the vehicle completed wheel-stop a few seconds later. Had the blow-out occurred earlier, when the vehicle was travelling at a higher speed, then it may have resulted in veering off the 100-m wide SLF into the scrubland at the side of the strip.

There had been some concern with the braking system on returning orbiters for some time, and as a result of this incident, NASA decided to use the SLF as an emergency recovery site only until the nose-wheel steering capability was fully introduced. Edwards AFB in California (the site of the first nine Shuttle landings) – with its large expanse of dry lakebeds providing a margin of safety for veering off the landing strip – resumed the role of primary recovery site.

Following the *Challenger* accident, the issues of brakes, steering and arrester systems were addressed in the Return To Flight programme. With the delays to the programme as a result of the accident, and the implementation of recommendations by the review board, no Shuttle landed at the Cape until 1990. In 1992 the introduction of the parachute braking device was initiated on the new replacement orbiter, *Endeavour*. This both slowed the vehicle during the roll-out and, by trailing behind, served to hold it on the centre line. The brakes would not be used until the vehicle had slowed. They were upgraded to cope with greater loads and higher temperatures, and with these improvements KSC became the primary recovery site. No orbiter has ever encountered a serious landing problem, even in cross-winds. Since the approach and landing tests in 1977, the landing of orbiters has been relatively free of incidents.

# The Soyuz 1 landing accident, 1967

During the spring of 1967, NASA and America were trying to come to terms with the tragic loss of three astronauts in the Apollo 1 pad fire at the beginning of the year (see p. 97). The end of the decade was fast approaching, as was President John F. Kennedy's deadline for landing a man on the Moon and returning him safely to Earth. In those dark days of 1967, the Moon seemed further away than it had when the President had set it as America's goal.

1967 was also to be an important year for the Soviet Union. It would be the 50th anniversary of the Bolshevik Revolution of October 1917, and also the 10th anniversary of the launch of the world's first artificial satellite in October 1957. For some time the Soviets had been preparing their own spacecraft intended for human lunar exploration, and with the Americans grounded until at least 1968, the chance to once more demonstrate the leadership that they believed they had in the Space Race, by mounting another space spectacular was attractive. It was simply an opportunity not to be missed.

On 23 April 1967, a brief and typically vague Tass press release stated that at 03.35 Moscow Time a manned spacecraft – Soyuz 1 – had been launched from Baikonur Cosmodrome in Soviet Central Asia, carrying Air Force Colonel Vladimir Komarov, Pilot–Cosmonaut and Hero of the Soviet Union, an honour he received after flying Voskhod 1.

According to the press agency, the objectives of the flight were 'to test the new piloted spaceship and check the ship's systems and elements in conditions of space flight.' In addition, Komarov was to 'conduct extended scientific and physical/technical experiments and studies in conditions of spaceflight [as well as] continue medical and biological studies of various factors of spaceflight on the human organism.'

No other details were released, but Western observers had been expecting a new launch from the Soviets for months. No Russian had orbited the Earth since Voskhod 2 in March 1965, and since then NASA had completed ten highly successful Gemini flights. Rumours of the impending flight of a new multi-crew spacecraft – perhaps a 'Moon-ship' – were confirmed with the launch of Soyuz 1.

The Tass report further stated that by the completion of his fifth orbit, Komarov had reported that the flight programme was being completed successfully and that he felt well. Nothing more was heard apart from the statement that radio communications were being maintained. Between 13.30 MT and 21.30 MT – while the spacecraft was out of contact for an extended period – the cosmonaut rested.

Communication was re-established following the rest period. Komarov was reported to be continuing with the flight programme and was still feeling well. There were also rumours that a second crew was about to launch to join Komarov in orbit on the 24 April, but there was no official confirmation of a planned second launch. Another statement, issued at 04.50 MT on 24 April, again indicated that all was well with both the flight and the cosmonaut. Nothing further was heard for the next nine hours.

Then, at 12.27 MT, Tass and Radio Moscow issued a joint statement that 'Cosmonaut Vladimir Komarov has perished while completing the test flight of the spacecraft Soyuz 1.'

A later report revealed that he had died while trying to land his spacecraft. The report continued: 'When the programme was completed, Colonel Komarov was told to stop the flight and land. After carrying out all the operations connected to initiating the landing procedure, the spacecraft safely passed the most difficult braking phase in the dense layers of the atmosphere and had fully decelerated from orbital speed. However, according to preliminary reports, when the main canopy of the parachute opened at an altitude of seven kilometres, the straps of the parachute became twisted and the spacecraft descended at great speed, which resulted in Komarov's death.'

The report continued to state that during the test flight – which had lasted more than 24 hours – Komarov had fulfilled the systems testing requirements for the new spacecraft, and had been able to carry out all of the planned scientific experiments. He had also been able to manoeuvre the spacecraft, and 'tested its main systems in different regimes and estimated technical characteristics of the new ship.'

It was a tragic and sudden loss to both the Soviets and the space community. For the first time, someone had died during the execution of a space mission. From the initial reports it appeared that all had proceeded satisfactorily up to the time that Komarov had deployed the landing parachutes. He had lost his life as a result of the high-velocity impact with the Earth, caused by the failed parachutes.

The Soviets joined America in mourning one of their space pioneers. An investigation was set up and, as events transpired, no cosmonaut would venture into space again until October 1968 – ironically, the same month that American astronauts finally returned to orbit after the Apollo 1 pad fire.

However, despite official reports on the accident, rumours began to emerge that Soyuz 1 had encountered difficulties soon after launch, and that there was more to the story than a relatively simple and extremely unlucky parachute failure. Over the next 30 years, snippets of information were released or discovered, to reveal the full picture of the maiden flight of the Soyuz spacecraft. During the same time period, Soyuz and its variants would became the workhorse of the Soviet/Russian human spaceflight programme.

Following the end of the Second World War in 1945, the world witnessed a great rivalry between the Soviet Union and America, the two great 'superpowers'. The Cold War would last for almost the next 50 years, and the Space Race evolved from this rivalry. Oneupmanship was the name of the game, even though the technology and resources did not always support the political claims.

In demonstrations of technological superiority, the Soviets had repeatedly launched 'space spectaculars' to beat the Americans in the new Space Age. The first satellite (1957), the first living creature in space (1957), the first probe to hit the Moon (1959), the first man in space (1961), the first woman in space (1963), the first multi-person crew (1964) and the first person to walk in space (1965), were all records that fell to the Soviet Union. In response to these 'technological' achievements, Presidents Dwight D. Eisenhower, John F. Kennedy and Lyndon B. Johnson successively turned to America's efforts in space to demonstrate that the future freedom of the world lay in the conquest of the cosmos.

The Soviet leadership promoted the great advantages of socialism with each succeeding spaceshot, and when America stated that it was going to the Moon by the end of the 1960s, then so would the Soviets, although not so publicly. With the creation of the American Gemini series, Soviet leaders pushed for a rival spacecraft to beat the Americans in rendezvous and docking, multi-crew, long-duration and spacewalking objectives.

The development of Soyuz (Union) began in the early 1960s as a multi-role spacecraft, carrying a crew of up to three cosmonauts and able to support human flights to the Moon and various operations in Earth orbit. Designed by Sergei Korolyov's OKB-1 design bureau, it was a successor to the Vostok one-person spacecraft that had won the historic 'firsts' of putting a man and then a woman into space for the Soviets, ahead of the Americans.

It was initially planned that Soyuz would follow the Vostok series around 1964. Originally conceived as a three-part space complex, Soyuz A was to be the three-person spacecraft, Soyuz B an orbital manoeuvring vehicle, and Soyuz V would be a propellant tanker. In the documents on this proposal, the Soyuz B would be launched first, followed by a series of Soyuz V tankers to supply the manoeuvrable stage. When this was completed, cosmonauts would launch, dock with the Soyuz B, and then probably be sent on a circumlunar mission. A landing was not part of the plans. That would come later.

However, the ambitious plans for Soyuz had been scaled down in 1963 and 1964. OKB-1 finished a complete redesign of the Soyuz system. The spacecraft would now form part of the revised lunar programme that would see manned circumlunar and landing missions supported by automated spacecraft. However, this would not be ready before 1966, after the American Gemini flights had been completed. In response, the designers were ordered to strip out Vostok and launch a series of missions that would not only take the glory from the American Gemini missions and once more demonstrate the superiority of Soviet technology in leading the world to a new future, but in addition allow a test of the EVA pressure suit planned for the lunar mission.

Voskhod, as the 'new' spacecraft became known, actually flew with only two crews, in 1964 and 1965. Although it achieved the first multi-person spaceflight and then the first spacewalk, the spacecraft had to be stripped of its emergency ejection system and most of its internal components to accommodate the extra crew seats and the airlock. Around this time there was a change of political leaders in the Kremlin, and Chief Designer Korolyov – the central driving force behind the early Soviet programme – died on the operating table in January 1966.

Korolyov's successor – his former deputy, Vasily Mishin – was able to convince the new leadership that to continue to fly the old and very limited Voskhod hardware would not benefit the programme, and would in fact see the Soviets lag further behind the Americans. Mishin was instructed to cancel further Voskhod flights in order to devote all of his bureau's resources to the development of the lunar programme and to strive towards the maiden launch of the first manned Soyuz spacecraft.

Even before the launch of Komarov on Soyuz, there were doubts about its integrity. At that time there were no ground-based facilities for testing Soviet manned spacecraft. All that could be done was to send them unmanned into space, to test systems. However, three previous automated test flights had each revealed various problems. The first – on 28 November 1966 – was given the cover name Cosmos 133 upon entering orbit, to disguise its true objective. It was a failure from the start. Its attitude could not be controlled on orbit, with the main engine running without the use of stabilising engines. This burned up precious fuel, and affected the firing of the main engine. When trying to align the spacecraft for entry at the end of the mission, controllers found that they had the same problem, and after considerable effort to begin the entry sequence it was found that the descent path was too flat and that it would overshoot mainland Russia. The trajectory would bring the recovery capsule down in China, and the self-destruct system was activated and the spacecraft destroyed to prevent this happening.

Just over two weeks later, on 14 December, the second test failed on the launch pad when a huge explosion destroyed the launch vehicle. The automatic systems had aborted the launch countdown sequence just seconds before ignition of the main engines. Almost immediately, the support trusses of the launch tower were swung back around the vehicle, and ground crews began to leave their observation bunkers. Suddenly there was a loud bang as the solid rocket motor escape system activated and initiated the spacecraft's recovery sequence. As the men on the ground took cover, the spacecraft, then the third stage of the launch vehicle, and then the whole rocket, exploded on the pad. Miraculously there were no injuries or fatalities. It did, at least, provide the Soyuz launch escape system with a chance to demonstrate its recovery of the unmanned Soyuz crew module, although the ground crews would not have wished for such a close view.

The next test, Cosmos 140, was launched on 7 February 1967, and also met with problems in its landing phase. A ground maintenance plug located in the forward heat shield of the Descent Module actually burned through to the module's metal skin, which caused considerable damage to the equipment inside, as the vehicle plummeted back to Earth. Despite surviving re-entry, there was more in store for the unfortunate capsule, because it smashed through an ice floe on the Aral Sea. As we have seen, Soyuz vehicles can be recovered from water, but this one was already damaged, and after only a few minutes floating on the surface it promptly sank in 30 feet of water. A team of divers later recovered it for post-flight inspection.

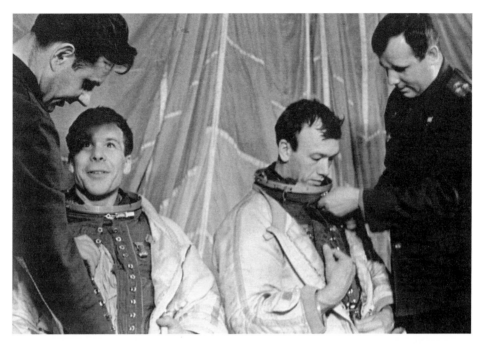

Komarov (*standing left*) with Soyuz 2 EVA cosmonauts Khrunov (*seated left*) and Yeliseyev (*seated right*) during training for their subsequently cancelled mission. At far right is Soyuz 1 back-up pilot Yuri Gagarin. (Courtesy Novosti Press Agency via Astro Info Service Collection.)

Three major safety incidents on three missions obviously indicated that there were still serious flaws in the Soyuz design. The next flight was to be manned, but Mishin tried to cancel the launch, requesting that more unmanned tests first be carried out. He went as far as to refuse to sign the flight endorsement papers, as he strongly believed that the Descent Module was not yet ready for a human crew. He was overruled, and the Communist Party authorised the launch of Soyuz 1 in April. It has never been revealed whether anyone else supported Mishin, or whether anybody queried why a cosmonaut's life should be put at risk before the system was man-rated, in light of the failures on the unmanned tests. But this was at the height of Soviet Communist rule, and central authority was not second-guessed. The 50th anniversary of the Russian Revolution and the 10th anniversary of Sputnik were approaching, and the opportunity to mark the event with yet another spectacular space stunt was too great to miss.

So proud were the Soviets of their programme's new direction that leaks began to filter into the West several months before the event. The new spacecraft would be so advanced that it would outshine all the accomplishments of Gemini in one mission. In March 1967, Lieutenant General Nikolai Kamanin, Director of Cosmonaut Training, suggested that crews were in training and were expected to be ready in the spring or summer. He is also reported to have stated that there must be total confidence in the success of the mission, as it would be far more complicated than anything previously attempted. Preparing the crew and the new hardware would take time. The Soviets, he said, did not intend to speed up the programme unnecessarily, and he quoted the fate of the three American Apollo astronauts, just weeks before. According to Kamanin, excessive haste would lead only to tragedy!

The first manned flight of Soyuz, rumour insisted, would not be simply a solo engineering shakedown flight, as had been planned for the maiden flight of the Apollo programme. There were to be two Soyuz launches, one day apart and with four cosmonauts involved.

Although the dual mission was never announced and, after the loss of Soyuz 1, was denied to have ever been intended, group photographs showed Kamanin with cosmonauts Yuri Gagarin (Komarov's back-up), Valeri Bykovsky, Yevgeni Khrunov and Alexei Yeliseyev. Bykovsky was already rumoured to be the Command Pilot of a second craft, but the significance of the photographs was that Khrunov and Yeliseyev were in EVA spacesuits – and Komarov had also been seen in an EVA suit, although he was probably in training for a contingency EVA rather than for a planned exit. In 1969, Khrunov and Yeliseyev both flew in space for the first time. During the flight they spacewalked from one spacecraft to another (it was stated at the time) in 'a demonstration of space rescue techniques'; but the photographs of them with Komarov had to have been taken before Soyuz 1 launch. Could they have been training for a spacewalk between the second Soyuz and Komarov in Soyuz 1?

In addition, initial launches of new Soviet spacecraft had never had a numerical sequence assigned to them (Sputnik, Luna, Vostok, Venera, Voskhod); but from the very beginning, this mission was designated Soyuz 1. Obviously a second spacecraft (Soyuz 2) would be launched soon and would require separate identification. There were also the stated flight objectives for Soyuz 1 at the time of launch: to test the new spaceship, checking systems and elements (prior to a possible second launch?); to conduct extended scientific, physical (docking?) and technical experiments (hardly the agenda for a short solo flight); continue medical and biological studies of the influence of spaceflight on the human organism (suggesting a flight lasting for more than 24 hours).

This, then, was the flight plan for the first Soyuz mission. Komarov would launch first, alone on Soyuz 1, but with three seats installed. The next day, Soyuz 2, with Bykovsky, Khrunov and Yeliseyev aboard, would launch to begin a rendezvous profile with Soyuz 1, planned for the next day over the Soviet Union. Following the rendezvous, Komarov's Soyuz 1 would then approach slowly, allowing the docking probe of Soyuz 1 to enter the drogue of Soyuz 2. Once the vehicles were mechanically linked, the four cosmonauts would rest.

The spacecraft did not have an internal transfer tunnel. It was more complicated and so was never planned. Instead, Khrunov and Yeliseyev would put on EVA suits and transfer in open space, entering Soyuz 1 to be welcomed by Komarov. The craft would then undock and land separately. Soyuz 1 (now with three cosmonauts onboard) would land first, followed a day later by Soyuz 2 with the lone Bykovsky onboard.

This would be hailed as the first in-space transfer of a crew, in a simulated space

Yuri Gagarin, Soyuz 1 back-up pilot, with Vladimir Komarov, prime pilot for the mission. (Courtesy Novosti Press Agency via Astro Info Service Collection.)

rescue. But it was also the method that Soviet cosmonauts would use to transfer to and from their one-man lunar landing module on the lunar missions. Alexei Leonov had already tested the prototype of the suit that they would use for the EVA.

The training programme for the Soyuz 1 and Soyuz 2 crews finished with a final exam, taken on 30 March. It was reported that all cosmonauts in the training group passed the exam and prepared to move to Baikonur for the final stages of launch preparations. Komarov arrived at the launch site on 8 April.

Pre-launch preparations for the mission would take up to eight days, so the launch was scheduled for no earlier that 24–25 April. With two Soyuz vehicles on two different pads, it would take some time to fuel both rockets. Soyuz 1 fuelling began at 23.00 MT on 17 April, and was concluded 48 hours later. The fuelling for the second launch vehicle began on a second pad on 18 April.

On 17 April, final reviews were held concerning the method of rendezvous and docking between the two spacecraft. One option was a completely automated docking, and the other was an automatic approach to 650–1,000 feet, completed by a manual docking. Mishin preferred the fully automated docking profile, as there were still some doubts about the reliability of the manual docking approach The cosmonauts – fearing they were to lose the ability to demonstrate their piloting skills – argued for the manual docking. After all, that was what they had been training for for more than two years. After much discussion and evaluation, it was agreed to go with the automatic approach by the Igla (Needle) rendezvous system (to within 165–230 feet of the passive target vehicle), and then allow the cosmonauts to manually bring the two Soyuz together.

By 20 April the final launch times were set. Soyuz 1, carrying Komarov, would launch at 03.35 MT on 23 April, followed at 03.10 MT on 24 April by Soyuz 2, carrying Bykovsky, Yeliseyev (Flight Engineer) and Khrunov (Research Engineer). Final meetings between the crews and the designers included discussion of the likely reasons for a postponement or cancellation of the launch of Soyuz 2. This would occur if there were on-orbit problems with the Igla system on Soyuz 1, or if the solar panels were not fully charging the batteries and providing adequate onboard power reserves. Safety was reported to be the most important concern, with the cosmonauts being advised to abandon any attempt at docking if they encountered any malfunction during the approach. At 23.30 on 22 April 1967, while Komarov slept, the launch of Soyuz 1 was finally approved.

Komarov was awakened at 02.00 MT on 23 April, and after attaching biomedical sensors he dressed, not in a pressure suit, but in a light woollen grey jump-suit and a blue jacket. The Soviets relied upon the integrity of the launch vehicle and launch escape system to protect the cosmonaut in the event of a launch malfunction, and on the ability of the Soyuz to sustain the occupant if a serious problem occurred in space. The spacecraft was not equipped with an ejector seat, as in Vostok, but had a parachute recovery system, consisting of a main and a back-up canopy. Although Komarov did not wear a suit, EVA ingress would be possible because Soyuz had two modules with an access hatch between them. The EVA hatch was on the side of the Orbital Module, and Komarov would be safe in the Descent Module. Some photographs of Komarov show him wearing a spacesuit, but this was probably a contingency plan in the event of the EVA hatch not closing or the OM not repressurising. Komarov could have put on his own spacesuit and depressurised the DM, opening the interconnecting hatch and allowing Yeliseyev and Khrunov to enter, shutting the connecting hatch behind them and repressurising the DM. Although not confirmed by the Soviets, if this operation had been attempted it would have been very difficult in such a confined space and with three suited cosmonauts, but it offers an explanation of the unusual EVA-suited Komarov photographs.

Komarov arrived at the launch site at 03.00 MT. By then, 203 faults had been reported in the Soyuz, which required correction before the flight could proceed. These bad omens had a profound effect on Komarov, who was quiet and sombre when preparing for the ride to the launch pad. He discussed his misgivings with fellow cosmonauts, and expressed a sense of foreboding about the mission. But they soon had him singing and smiling by the time they reached the base of the rocket. Komarov ascended to the crew access hatch at the top of the stack, along with members of the launch crew and Gagarin, his back-up. He then climbed into the side hatch in the Orbital Module and dropped gently down through the internal hatch into the Descent Module, taking his place in the form-fitting recliner couch for launch. The countdown and final preparations for launch proceeded smoothly, and soon Gagarin and the launch team closed the hatches and left Komarov alone, awaiting lift-off.

Exactly on time, at 03.35 MT on 23 April, the Soyuz launch vehicle began its maiden flight with a cosmonaut onboard. From that moment, Komarov's fate was

A Soyuz R7 launch vehicle leaves the pad. It was a vehicle of this type that launched Komarov on his ill-fated mission in April 1967.

sealed. He reported light g loads as he left the pad and climbed towards orbit. The powered ascent took 540 seconds, and was carried out with no reported problems. The performance of the first stage went according to plan, with the four strap-on boosters being separated on time and the central core and upper stage finally boosting the lone cosmonaut into orbit. Komarov became the first cosmonaut to enter space twice. It was a feat soon to be overshadowed.

Komarov's initial orbit was given at 125 × 139 miles at 51°, which was a new inclination, as Vostok and Voskhod had been at 64°. Valentina Komarov, the cosmonaut's wife, had been informed of her husband's launch by fellow cosmonaut Pavel Popovich, and said that her husband never told her when he was to go on a 'business trip'. All seemed to be well with the flight, but this was not the case.

Almost as soon as the Soyuz separated from its carrier rocket, it encountered difficulties in preparing for its orbital operations. Initial data received from the spacecraft, at the ground stations in the Crimea and at the launch site, revealed that the left solar array had not unfolded to absorb solar radiation and provide power for onboard systems. A telemetry system back-up antenna had also failed to deploy, and to compound problems the optical surface of the attitude control sensor (used to lock on to the Sun and the horizon, to orientate the spacecraft) had been coated with exhaust emissions from the nearby engines. Of all the problems, it was the sensor contamination that was the most serious, as this would prevent Komarov from orientating his spacecraft in preparation for the docking of Soyuz 2 the following day. Soyuz was already performing a slow rotation as the automatic orientation system had also failed. It was not until the second orbit that direct radio communications were achieved with the cosmonaut via the ultrashort-wave system. The short-wave radio was also out of action.

Komarov initially reported that he felt good. The same could not be said for his spacecraft, however, and in addition to the earlier problems the cosmonaut also informed the ground that the internal pressure of the spacecraft orientation engines had dropped. Ground informed Komarov that he was to shut down as many non-essential systems as possible in order to conserve battery power, and that he was to try to attempt a manual orientation of his spacecraft so that the fully deployed right-hand panel could face the Sun. He had already tried this, and had failed.

The stuck solar panel left the available electrical power far below what was required to complete the planned mission. The onboard storage batteries could cope only with requirements until the 17th orbit, or 24 hours after launch, and the reserve batteries could only provide for two more orbits.

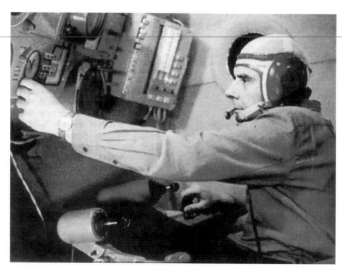

Komarov during training for Soyuz 1. (Courtesy Novosti Press Agency via Astro Info Service Collection.)

The differences between what was reported by the Soviet News Agency Tass, and what was actually unfolding in space, were as follows:

Orbit 3. Tass news report states that Komarov is in good condition. His pulse is 82 beats per minute and his respiration 20 per minute. On Soyuz 1, Komarov actually reported that the solar panel was still jammed and that manual orientation of the spacecraft had failed.

Orbit 5. Tass news report states that the flight programme is being fulfilled and that the cosmonaut is well. Temperature readings show 16° C and pressure is at 750 mm of mercury. On Soyuz 1, Komarov reported that the spinning was getting worse and he was feeling the initial effects of space motion sickness.

Despite all these problems, the State Commission still authorised the launch of Soyuz 2, in the belief that the orientation problem could be solved. During the fifth orbit, Komarov had wrestled with the controls in an attempt to orientate the spacecraft by using Earth's horizon as a point of reference. But the third orientation system on Soyuz consisted of ionic sensors, sensing the spacecraft's orientation with respect to the velocity vector through the ionosphere, and this consistently overrode his manual input and compounded the problem.

By the time Soyuz 1 was on its seventh orbit Komarov was out of the range of ultrashort-wave frequency communication, and as he passed over the Atlantic Ocean and the American continent, he tried to rest. Meanwhile, on the ground high-level communications continued between Moscow, the launch site and the tracking centres to try to resolve the ever-growing problems. Gagarin was sent to Yevpatoriya – the primary communications centre in the Crimea – to help in any way possible.

By the thirteenth orbit Komarov was back in range of the Crimea tracking station, and he immediately informed the ground that further attempts to use the ionic orientation sub-system to position the spacecraft had again failed. As a result of drastic reduction of onboard power usage and the stability of the current at 12–14 amperes, internal temperatures had dropped significantly, adding to the cosmonauts discomfort.

On the ground there were still some who tried to hold out for the launch of the second Soyuz. They proposed using it to try to rescue the cosmonaut, or, as one report suggested, to allow the Soyuz 2 EVA cosmonauts to try to manually pull out the jammed array on Soyuz 1 if they could stop the tumbling. It soon became apparent that the launch of a second Soyuz would not help the situation and, unanimously, the State Commission issued the directive to cancel the launch. The main task then was to try to return Komarov safely to Earth. Everyone, including Komarov, knew this would not be easy, and could well end in tragedy.

By this time, Komarov's wife had been brought to the flight control centre to talk to her husband and provide him with moral support in his obviously difficult 'business trip'. Only hours before she had been unaware that he was to be launched into space for a second time. Now she was fighting back the tears, in what could be an emotional, final, farewell. Meanwhile, US National Security Agents at the USAF base just outside Istanbul were monitoring the progress of Soyuz 1. They reportedly recorded the voice communications between Komarov and ground control, including the anguish of husband and wife when the realisation of his pending fate confronted them. Finally, unable to bear any more, Komarov told his wife to go home. Some of the more sensational reports were later slated as exaggerations and that the ground stations had misheard the comments in the static

Soyuz had three orientation systems: the solar-horizon system (which was inoperable), the ionic system (which Komarov had reported was working) and the manual system (which he had not been able to use effectively due to the ionic system override). One of these systems had to orientate the spacecraft for entry into the Earth's atmosphere. If all three failed, Komarov would be unable to return and would be doomed to die in his spacecraft when his air was depleted.

Komarov's fruitless attempt to try aiming the one open solar panel at the Sun burned valuable propellant in the orientation sub-system that was needed for re-entry. If the fuel reserves dropped too low as Komarov ignited retro-fire, he would not have sufficient fuel to compensate for any deviations in the entry corridor resulting from the change in the centre of gravity caused by the folded solar panel.

Engineers on the ground had only the manual systems, which Komarov had already reported were difficult to use. It was a constant battle for him to keep the Earth's horizon in sight, especially in darkness. The system was normally used during periods when the spacecraft crossed the terminator of the Earth into sunlight. It was also thought that in using the ionic system, sensor disruption from ion pockets could be expected when the spacecraft began its re-entry phase.

By now the spacecraft was on its fifteenth orbit, and the re-entry commands had to be read up to Komarov by the sixteenth orbit to effect a landing on the seventeenth orbit, the earliest opportunity. After much discussion between the centres and with Komarov, it was decided that he would attempt to land the spacecraft by using the automatic ionic sub-system and the back-up set of orientation engines. Gagarin acted as CapCom, reading up the instructions to his friend and relaying best wishes from the ground teams. There was nothing more anyone could do. It was down to Komarov and the questionable reliability of a spacecraft that had already shown too many operational faults. It was certainly not qualified for flight, and should never have been launched. But by now it was too late.

The plan was to initiate entry retro-fire, against the direction of flight, at 02.56 MT on 24 April, on the seventeenth orbit, while out of range of ground control. This would slow down the spacecraft and allow gravity to pull the vehicle towards Earth and begin the recovery sequence.

An initial report, moments later, confirmed that Soyuz was, in fact, still in orbit. Nothing had apparently happened. When communications had been re-established with the cosmonaut, he calmly reported that the deorbit engine had not fired. The ionic system worked as programmed, but as the Soyuz flew over the equator it travelled through one of the predicted pockets of ion particles in the Earth's shadow. This was apparently much less than the onboard sensors could detect, and for once the onboard systems worked as designed, promptly issuing a command to inhibit engine ignition.

A plan was formulated to attempt a second landing on the eighteenth orbit, but Komarov had already entered this orbit, and there was no time to relay instructions up to him. The landing was instead to be tried on the nineteenth orbit. With the two onboard automatic systems not functioning, the only option was for the cosmonaut to attempt manual orientation. Premier Alexei Kosygin spoke directly to Komarov at this time, telling him that his country was proud of him and would always remember him. Komarov's reply was apparently inaudible, but it was evident that neither he nor the ground expected him to survive re-entry.

This entry approach was not only very difficult and complex; it was a technique that none of the cosmonauts had actually trained for, believing that it was an in-flight situation that would never occur on a real mission. To achieve it, Komarov was required to orientate Soyuz on the Sun side during the eighteenth orbit, using the Earth's horizon for reference. Then, prior to entering the shadow he would have to quickly transfer control of the attitude to the onboard gyros, and immediately after exiting the shadow, re-check that the vehicle remained correctly orientated prior to retro-fire on the nineteenth orbit. If he found this was not the case, he would need to perform a fully manual re-entry burn – an extremely difficult operation.

As Komarov prepared for this demanding task he was informed that onboard power levels indicated that he had only sufficient power for up to the twenty-first orbit. This second attempt at re-entry would probably be his final chance. Concentrating on recording the correct information for the operation, he seemed to calm down. Gagarin again relayed the list of instructions up to Soyuz 1.

Under immense strain, Komarov completed his tasks faultlessly. He orientated Soyuz for entry at the programmed time, and the engines burst into life, firing for 146 seconds. Whatever the consequences, Soyuz 1 was coming down.

Komarov reported his success over the radio, but also mentioned that the spacecraft had initiated an 'Accident 2' command. Ground interpreted this as a deviance beyond the designed 8° entry angle as a result of capsule movement caused by the instability of the spacecraft at the beginning of the entry corridor. This meant that Komarov would experience a direct ballistic (high g load) entry, instead of a guided entry. Everything else looked normal for a safe return.

Komarov then issued the command to separate the Descent Module from both the Orbital and Service Modules, exposing the re-entry heat shield. Soyuz 1 entered the period of radio blackout, as ionised air engulfed the spacecraft on its descent through the atmosphere.

As the spacecraft exited the blackout zone, Komarov's voice could be heard relaying information to the ground in a calm and unhurried manner. Observers at the flight control centre began to believe that they had achieved the impossible and brought the cosmonaut safely home.

Several high-ranking officials had already departed by military transport to the reserve recovery site as Komarov completed the re-entry sequence. The recovery site would be west of the usual zone, due to delays caused by the extra orbits. A ballistic re-entry is slower than the gliding re-entry, so a ballistic entry which would be known only half an hour before landing would result in the capsule landing several hundred miles short. The recovery helicopters would have to move to find him, and to allow for a larger coverage, military air bases were probably put on standby to support the search and rescue activities. They were confident that the remaining sequence of events leading up to landing would soon see Komarov home to report to the State Commission on the successful conclusion to another pioneering (if difficult) Soviet spaceflight. But this was not to be.

Recovery of a Soyuz spacecraft is enacted by parachute descent after emerging from the blackout period. The parachute canopy is stored in a container on the side of the Descent Module, behind the crewman's head. At 4.5 miles altitude, the protective hatch near the top, covering the parachute container, separates and pulls out a drogue parachute, which acts as a brake for the descending vehicle. By using the smaller parachute, the vehicle slows to a more controlled 130 fps descent – a safe rate for deployment of the larger, recovery parachute. At a higher speed, the larger canopy would rip to shreds.

On Soyuz 1 the drogue parachute could not pull out the main canopy, which remained stuck in the storage container. Normally onboard safety systems would have detected a failure in the primary system (due to an increasing descent speed), triggered separation of the primary system and activated the back-up system; but unfortunately for Komarov, this did not happen as planned. The system *did* detect a parachute failure, but was unable to separate the failed primary system, as the main parachute was still in the storage container and was not trailing behind the capsule. The single back-up parachute deployed directly into the still-attached drogue parachute, and became entangled as it unfurled. It, too, failed to deploy.

Onboard the Soyuz there was nothing Komarov could do. Although subsequent rumours spoke of anguished screams as he headed towards his doom, he apparently did not cry out when he realised that he would not, after all, survive the landing.

With no means of deceleration, the Descent Module plummeted into the ground, hitting it at a velocity of 90 mph (130 fps). Soyuz 1 crashed into the steppes near Orenberg at 06.24 am MT. At impact, the capsule split open. The solid propellant deceleration motors – intended to be used to slow the spacecraft to a soft landing – exploded, creating a fire. The flight had lasted 26 hrs 47 min 52 sec from lift-off to destruction.

The smouldering wreckage of Soyuz 1. (Courtesy R. v. Beest/British Interplanetary Society.)

Recovery team members search for the remains of Komarov in the wreckage of Soyuz 1. (Courtesy R. v. Beest/British Interplanetary Society.)

Once the actual landing site had been identified following interpretation of the re-entry data search, rescue service helicopters headed towards the area. The first helicopters to arrive at the actual landing site found only the pile of burning remains of what was once Soyuz 1. The whole area was surrounded in a cloud of black smoke, and the fire inside the capsule wreckage was still intense, with molten metal still dripping onto the scorched ground. The capsule was completely destroyed – unrecognisable as the spacecraft it once was, save for the upper circular hatch ring on the top. The recovery teams tried to extinguish the flames with small fire extinguishers and handfuls of soil.

Communications from the site were sparse and garbled, and were censored once it was realised that there had been a tragic conclusion to the troubled flight. The first report from the site to the controllers was a cryptic, coded message that the rescue teams had seen the spacecraft on the ground. It was on fire, and the cosmonaut needed urgent medical attention.

For several hours, no new reports came from the landing area. Kamanin landed at Orsk airport, 40 miles away, two hours after the crash. He received conflicting reports that the spacecraft had landed and was burning but there was no sign of Komarov, or that the badly burned Komarov had been taken to a hospital three kilometres from the landing site. Such was the disbelief that anything could happen to the brave cosmonauts that people even believed rumours that Komarov had landed well off target in Bulgaria or West Germany.

The control centre received no new information for 3½ hours. Kamanin decided to go to the site himself, and was greeted by the burning wreckage. Eyewitnesses said that they saw the capsule descend at great speed, with a trailing, twisting parachute. There were also some explosions after the impact.

The search for Komarov continued, and the rumour that he had been taken to hospital was soon proven false. The ground teams searched the debris, clearing away the wreckage to finally uncover the remains of the cosmonaut. As they shovelled away the debris and moved parts of the instrument panel, they found Komarov's remains, still in his centre couch. The on-site medical team performed an initial examination of the body. They announced that death had been caused by multiple injuries to the head, spinal cord and skeleton.

Once the grim reality of the end of the Soyuz 1 flight had been established, Kamanin flew back to Orsk to report to Central Committee Secretary Dmitri Ustinov, who contacted Soviet General Secretary Leonid Brezhnev. He was in Czechoslovakia, attending an international conference of communist parties. The Soviet news agency was also provided with a scripted report, which was issued twelve hours after the event. Later that day, other members of the State Commission and the design bureaux visited the site. Engineers came to record the scene before the remains were removed and the wreckage finally cleared.

In the early hours of 25 April, Komarov's remains arrived in Moscow, in a coffin accompanied by members of the planned Soyuz 2 crew, their back-ups, and Gagarin. In Moscow, Komarov's widow met the aircraft. His remains were cremated soon after arrival, and were then placed in an urn in the Red Banner Hall of the Central House of the Soviet Army. More than 150,000 Soviet citizens filed passed the urn to pay their last respects.

On 26 April, Komarov's urn was placed in the Kremlin Wall, with full State honours, following a journey across Red Square on an army caisson. A portrait of the deceased cosmonaut was placed against the wall. Supported by family and friends, Komarov's widow suddenly broke away and fell to her knees to kiss the likeness of her husband in a last farewell. A joint telegram was sent by the 47 active NASA astronauts, who expressed their sympathy and condolences to the cosmonaut team and to Komarov's family.

The casket of Komarov is received by fellow cosmonauts in Moscow. (R. Hall Collection.)

Komarov's widow breaks down at the picture of her husband at his memorial stone after his interment in the Kremlin Wall. (Courtesy Novosti Press Agency via Astro Info Service Collection.)

Several days after the internment, it was discovered that further remains of Komarov had been unearthed by Young Pioneers (the Soviet equivalent of the Boy Scout movement) and buried at the site, providing Komarov with two burial places.

Dmitri Ustinov headed the committee investigating the cause of the accident. It was established on 27 April, and included seven subcommittees. The inquiry soon established that the accident was caused by the failure of the parachute descent container to activate correctly or to separate. It was also determined that it was the impact with the ground, and not the subsequent fire, that killed Komarov. As a result of the rapid descent, the front heat shield was not discarded as planned at 3 km altitude, and the soft-landing engines did not fire prior to impact. They actually ignited after landing, and the fire was fuelled from residual concentrated hydrogen peroxide from the attitude control engine subsystem.

Further investigation into the failure of the parachute system found that the parachute simply blocked its own exit due to a difference in pressure between inside and outside of the container. The friction caused by the parachute mass trying to eject caused it to jam in the container. In four drop-tests of the system prior to Soyuz 1, this potential problem was not detected. Chief Designer Mishin also thought that the parachute system had been incorrectly folded during pre-launch packing.

To prevent the same happening again, the Commission recommended a redesign of the parachute container by changing its shape from cylindrical to conical, polishing its interior to make ejection of the parachute smoother, and enlarging the inner volume. An automatic method would be installed to separate the drogue parachute; and to ensure correct packing, photo-documentation would record all stages of folding and stowing.

The earlier in-flight failures could not be investigated as easily, as the hardware had been separated prior to re-entry or were destroyed in the impact. However, the commission determined that the solar panel failure was a result of its becoming caught on the launch shroud. The attitude control sensor had suffered from its optical surface steaming up with water vapour.

Aside from the official commission's report into the loss of Soyuz 1, unofficial claims continued concerning the events in space and the cause of the accident. Over the years, reports in the West continued to evaluate truth and rumours surrounding Soyuz 1. One such unofficial report was included in the history of Mishin's design bureau, published in 1996. It revealed that during the pre-flight preparations for launch both the Soyuz 1 and Soyuz 2 spacecraft had been placed in a high-temperature test chamber to seal the synthetic resin thermal protection system. It was rumoured that during these chamber tests, the spacecraft were 'baked' without the covers over the parachute containers. They became coated with the resin, preventing the eventual deployment of the parachutes on Soyuz 1. The official report does not mention this account, and it was probably kept quiet at the time to avoid incriminating those concerned. (See A.A. Siddiqi, 'Soyuz 1 revisited', *Quest*, Vol. 6 No. 3.)

Mishin had already stated on several occasions that he thought that the parachutes had been incorrectly packaged. If this were true, then the disaster would have been much worse if Soyuz 2 had also launched. Four cosmonauts, rather than one, would have perished.

As a result of the inquiry and two subsequent failures in tests of parachute recovery systems, Chief Designer Pavel Tkachev – the leading designer of parachute recovery systems for both human and automated spacecraft – was dismissed. He headed the institute for research and experiments in parachute-landing systems, which was responsible for designing the Soyuz parachute deployment and recovery system.

It is also evident that the pre-flight checks and tests missed the incorrect packaging of the parachutes, and that quality control for early Soviet missions was lacking in key safety areas. The desire to achieve the first manned flight probably put the same pressure on the Soviets as it had on the Americans. Political pressure to achieve further space 'firsts' (and the loss of Korolyov) added to the pressure on the key figures in the Soviet programme to deliver 'victories' in the Space Race. In America, the target was to reach the Moon. Safety and common sense, to some extent, took a back seat in the achievement of this goal – a philosophy shown by the loss of Apollo 1 to be badly in error. For the Soviets, this same philosophy had cost them Komarov and Soyuz 1. Equally tragic for both programmes was the fact that the lessons were not fully learned at the time, as later disasters would prove.

Many rumours were started during the year following the accident – rumours that Komarov had lost consciousness in orbit, had been frozen to death, or had even burned up during re-entry. Leading the efforts to dispel these rumours were several cosmonauts, including Yuri Gagarin, in training to fly the next Soyuz spacecraft in 1968. He stated that Komarov never reported any problems with the life support system, and that the stages of the recovery sequence were completed normally. During the whole flight, radio communications were stable and excellent, and there was never even a 'shadow of alarm', until the parachute system malfunctioned. For many years this type of guarded reply fuelled additional rumours as to the real truth behind the mission of Soyuz 1.

Initially a triumphal stage in Soviet space exploration on the road to the Moon, Soyuz 1 ended in total failure and ensured that the cosmonauts remained grounded well into 1968. The subsequent loss of Gagarin, in an aircraft accident in March, pushed the return of cosmonauts to space into October of that year.

Following the loss of Soyuz 1, a series of automated missions were flown, pending the return to orbit by cosmonauts in the improved Soyuz. Cosmos 186 performed an automated docking with Cosmos 188 in October 1967, and was followed by a repeat of the achievement by Cosmos 212 and 213 early in 1968. Soyuz 2 finally launched (without a crew) in October 1968, to be joined in orbit by Georgy Beregovoi aboard Soyuz 3 a day later. Although a docking was planned, none was achieved, due to Beregovoi's misalignment of the spacecraft and his attempt to dock with Soyuz 2 'upside-down' – directly to opposite the way it should have been, although the cosmonaut was recovered safely. In January 1969, the crews of Soyuz 4 and Soyuz 5 finally accomplished the original mission of Soyuz 1 and 2 (including the proposed EVA transfer by Yeliseyev and Khrunov). By then, Apollo had almost won the Moon, and Soviet space planners were looking towards a new goal, the creation of a scientific research base in orbit – a space station. Although they still hoped to send cosmonauts to the Moon, attention would soon be fully diverted to the development of permanent space complexes supplied by Soyuz and its variants.

More than 30 years after the loss of Komarov, and long after the grounding of Apollo, Soyuz spacecraft are still flying in space. And they will continue to do so in supporting the construction of the International Space Station. The memory of Komarov, the Soyuz pioneer, endures with every launch and landing.

# Soyuz 11 decompression, 1971

By 1971, humans had been travelling into space for a decade. Five American crews had travelled to the Moon and had returned safely. Further Apollo lunar flights were planned, but there was a shift in the structure of the programme. The Apollo programme would end in 1972, to be replaced in 1973 by America's first space station, Skylab, which had been developed from surplus Apollo lunar hardware. Two Skylab orbital workshops were planned, after which the Space Shuttle would support a permanent space station capable of supporting twelve astronauts. This orbital base was to serve as a jumping-off point for a return to the Moon and to Mars. For the Soviets, having lost the race to the Moon to the Americans, the race to create the first space station was their new goal.

In the spring of 1971 the Soviet Union celebrated the anniversary of Yuri Gagarin's flight by launching the first space station, Salyut 1, marking a new era in manned space exploration. For some time the Soviets had dismissed Western reports that their manned lunar programme had been cancelled because Apollo 11 had landed on the Moon in the summer of 1969, and they even denied that manned lunar flights had ever been their goal. It is now known, however, that this was not the case, and that testing for the manned lunar effort was continued into 1972. According to the Soviets, the most logical way of performing useful research in space was from a permanent manned space platform in Earth orbit, from which trips to the Moon and Mars could be launched. The Moon, it appeared at the time, was not the major goal for Soviet cosmonautics, and by the early 1970s, both in America and in the Soviet Union, the general agreement was that the development of a space station was the next logical step in space exploration.

The launch of Salyut (Salute) 1 on 19 April 1971 was viewed by many as the start of a major programme designed to create large space stations and factories in Earth orbit. But there were problems to be solved before such structures could be made – primarily the unknown effects of long-duration spaceflight on the human body. The longest flights at that time had lasted less than 20 days. The Soviets held the record, set the previous year at 18 days by Soyuz 9. At that time the Americans' longest flight was the 14-day Gemini 7 mission in 1965. Other problems to be addressed included effective work–exercise–rest cycles for large crews, resupply systems, the planning of emergency procedures and the psychological effects such flights might have on the crew. Salyut 1 – the first of a long series of such stations – would take the first steps on that learning curve, but it was a process that was not without cost. As the Soviets celebrated success, placing the Soyuz 11 crew on the Salyut in the summer of 1971, no one could have foreseen the tragic conclusion of the mission – a cruel blow to the mission of the world's first space station.

The Soviet space station programme had evolved from various design studies drawn up during the 1960s. At the time, the Soviets were in a race to beat the Americans to the Moon, initially with a circumlunar mission and then with a manned landing. By December 1968, Apollo 8 had won the first lap, and by the summer of 1969 Apollo 11 had stolen the whole race. No Soviet cosmonaut had gone beyond Earth orbit and, as events transpired, none would do so before the end of the century.

On 9 February 1970, the Soviet Central Committee of the Communist Party and the USSR Council of Ministers directed the space industry to create and operate a Long Duration Station (in Russian, DOS) in order to focus the national effort in a new direction away from the Moon. At that time, two different design bureaux were already working on the first elements of a Russian space station, their Moon-race lost before it had really started.

The design bureau of Vladimir N. Chelomey completed most of the pioneering work on a Soviet military space station. While the Sergei Korolyov bureau was involved in preparing the lunar hardware for trips to the Moon, Chelomey's bureau evolved plans for a 20-ton military space observation platform called Almaz (Diamond), designed to compete with the USAF Manned Orbiting Laboratory (MOL – soon to be cancelled) for the much cheaper and improved unmanned reconnaissance satellites under development. The plans involved spying on American fleet movements, the collection of high-resolution images, and the provision for automatic data return by small return capsules. The station would even carry a rapid-fire Nudelman reactionless cannon, have capacity for spacewalks, and feature experimental apparatus designed for military applications.

By 1970 the Khrunichev construction plant in Moscow had more than ten Almaz stations in various stages of construction, and was experiencing some minor problems setting up the new spacecraft control system. When the Central Committee decided to create a national space station programme, it directed that the first station be orbited within a year. Chelomey's design bureau would not be ready in time, and so the project was transferred to the most experienced design bureau.

The Korolyov bureau – now headed by Vasily Mishin and designated the Central Design Bureau of Experimental Machine Building (TsKBEM) Branch 1 – had the experience needed to complete the space station project. At first reluctant, Mishin soon turned his bureau's attention to the task. To put the station into orbit in 1971, they adapted the control systems and solar panels from the Soyuz spacecraft to integrate with the shell of Chelomey's Almaz station. Instead of a new heavy transport craft, a variant of the Soyuz vehicle was to be employed as a crew 'ferry' craft. The new hybrid station became known as Zarya (Dawn), the name proudly displayed on the hull of the station as it stood on the launch pad in April 1971.

When Almaz became Zarya, no one noticed that the radio call-sign for the ground

Soyuz 11 cosmonauts during training: (*left–right*) Dobrovolsky, Patsayev and Volkov. (Courtesy Novosti Press Agency via Astro Info Service Collection.)

tracking network was also Zarya, and that this would undoubtedly cause confusion with communications to and from the orbiting cosmonauts. In the hours before launch, Zarya became Salyut in a salute to the tenth anniversary of Gagarin's pioneering flight in Vostok 1. It was too late, however, to repaint the name on the station, and Salyut 1 flew with the name Zarya on its hull. (See A.A. Siddiqi, 'Triumph and Tragedy of Salyut 1', *Quest*, Vol. 5 No.3).

Two missions to the station were planned – the first lasting for 30 days and the second for 45 days. To accommodate this, a training group of cosmonauts was formed in the spring of 1970. From this group, four crews were selected, the first two to fly the missions and the others to serve as back-up and support teams. The make-up of the crews was to change over the next few months as a result of operational and personal conflicts, both within the cosmonaut group and within their leadership. The original DOS 1 assignments (with cosmonauts who eventually flew a mission to DOS 1 in italics) were:

Crew 1: *V.A. Shatalov/A.S. Yeliseyev/N.N. Rukavishnikov*
Crew 2: G.S. Shonin/V.N. Kubasov/P.I. Kolodin
Crew 3: B.V. Volynov/K.P. Feoktistov/*V.I. Patsayev*
Crew 4: Y.Y. Khrunov/*V.N. Volkov*/V.I. Sevastyanov

An early change in the crew assignments – for several unrelated reasons – resulted in the removal of Volynov, Feoktistov and Khrunov from the group under the direction of General Nikolai Kamanin, Director of Cosmonaut Training. Volynov was of Jewish origin, and on the orders of the Soviet Central Committee he had been disbarred from future space training (although he would later be restored to flight status and fly a military space station mission in 1976!). Feoktistov had recently been divorced, and would not reflect the perfect example of an heroic Soviet cosmonaut figure; and in 1969 Khrunov had been involved in a hit-and-run accident in which he failed to assist the victim, and was still under 'punishment'. Shatalov was reassigned to Command crew 3, Shonin was promoted to the command of Crew 1, and A.A. Leonov was assigned to crew 2:

Crew 1: G.S. Shonin/*A.S. Yeliseyev/N.N. Rukavishnikov*
Crew 2: A.A. Leonov/V.N. Kubasov/P.I. Kolodin
Crew 3: *V.A. Shatalov/V.N. Volkov/V.I. Patsayev*
Crew 4: *G.T. Dobrovolsky*/V.I. Sevastyanov/A.F. Voronov

Crew training began in September 1970, with the fourth crew joining in January 1971, just weeks away from the planned first launch. By February 1971 Shonin was removed from the first crew as a disciplinary measure for his persistent poor attendance of training. Despite Mishin's push for an all-civilian first crew, Shatalov was reinstated as Commander of the first crew because he was the only cosmonaut who actually had docking experience. The revised teams became:

Crew 1: (Soyuz 10): *V.A. Shatalov/A.S. Yeliseyev/N.N. Rukavishnikov*
Crew 2: (Soyuz 11) A.A. Leonov/V.N. Kubasov/P.I. Kolodin
Crew 3: *G.T. Dobrovolsky/V.N. Volkov/V.I. Patsayev*
Crew 4: A.A. Gubarev/V.I. Sevastyanov/A.F. Voronov

Soyuz 10 was originally scheduled for launch at 03.20 Moscow Time on 21 April 1971, but the launch was abandoned at T–1 min, when one of the service masts on the launch tower failed to retract. This would have initiated a launch abort at T–0, ejecting the crew in their capsule by means of the launch escape tower. The mission was postponed for 24 hours to evaluate the problem. The fault occurred again the following day, but Mishin had learned the reasons for the failure, and as it did not pose a threat to the launch or the crew he personally commanded the launch to proceed.

Soyuz 10 was launched on 22 April 1971, for a mission on Salyut 1 of between 22 and 29 days. However, despite a successful docking with the Salyut 1 station, crew transfer did not take place, and $5\frac{1}{2}$ hours later the two vehicles undocked. The recovery of the crew followed a flight of only two days in space, and the spacecraft soft-landed just 50 yards from a lake. The crew reported on their failure to enter the Salyut, and new plans to man the station were put into immediate effect, amid the disappointment of the failure of Soyuz 10.

Rumours surrounding the early return of Soyuz 10 were not quelled when the official Soviet announcement reported that the crew had completed their planned programme, that entry into Salyut was not part of the agenda, and that they had planned to return to Earth after completing their two-day programme. This was hard to believe, especially after pre-flight reports stated that the crew had completed a month's driving holiday together to assess whether they were compatible as a team.

Cosmonaut Pyotr Kolodin was a member of the original Soyuz 11 prime crew with Alexei Leonov and Valeri Kubasov, but all three were stood down following Kubasov's illness, and were replaced by the back-up crew led by Dobrovolsky. Although disappointed in not making the flight, this reassignment saved their lives, as the tragic mission was flown by their back-up crew. (Courtesy Novosti Press Agency via Astro Info Service Collection.)

Surely this would not be needed for a two-day flight! Additionally, Rukavishnikov was an expert on Salyut systems, and Shatalov and Yeliseyev were flying their third mission together, Shatalov being at that time the leading cosmonaut in rendezvous and docking procedures.

It was many years before the true sequence of events on Soyuz 10 became apparent and before the difficulties they faced in docking with and undocking from the Salyut were explained (see p. 231). The report into the failure was completed by 10 May. It was determined that the Soyuz 10 docking equipment was at fault, that adequate steps should be taken to prevent recurrence of the same problem, and that it would still be possible for two missions to fly to the Salyut before its expected de-orbit later in the year.

Mishin and Kamanin discussed plans to reduce the crew to a two-man team and to allow them to carry bulky EVA equipment to Salyut in the hope of performing EVAs to inspect the docking node and to remove a stuck experiment bay cover. Kamanin argued against this, as none of the cosmonauts had trained for EVA, and the EVA suit planned for a later Salyut station would not be flight-ready in time. Other plans discussed were for the two 30-day missions, because it was felt that there would not be enough supplies to support three-man crews for both missions. It was decided that both missions would become open-ended, and their duration evaluated in flight, based upon the activities of the crew, the use of consumables and the constant tracking of the Salyut's orbital height, which was gradually decaying.

Meanwhile, a new team of three cosmonauts prepared to occupy Salyut 1. Cosmonauts Leonov, Kubasov and Kolodin were training to activate the Salyut from its mothballed state. During a medical examination just one week before the launch of the planned Soyuz 11 mission, Kubasov was found to be suffering from a spot on his lung, and he was grounded. The Soviets contemplated replacing the grounded cosmonaut with his immediate back-up, Volkov, but just two days before the launch it was decided to replace the Leonov crew with the complete back-up crew, since both teams had become well integrated, and splitting them up was deemed 'inappropriate'. Leonov and his fellow cosmonauts would be cycled to a later mission.

Kolodin was not too pleased with this arrangement (as he lost his chance to fly in space), and he argued to join Dobrovolsky's crew to fly the Soyuz 11 mission. So angry was he about being replaced, that he reportedly told Mishin that history would not forgive him for the decision to send the back-up crew to Salyut 1. It would not, but not for the reason that Kolodin cited. To Kolodin's disappointment he met with no success, and he and Leonov were grounded until Kubasov recovered. It is interesting to note that when Leonov and Kubasov were later assigned as a two-man Soyuz crew, and eventually to the Apollo–Soyuz Test Project in 1973, no mention was made of Kolodin. After possibly expressing too much personal resentment, he was not assigned to another crew, and remained grounded.

A new trio of cosmonauts was to become the latest explorers of the cosmos. Cosmonauts Dobrovolsky, Volkov and Patsayev were to fly in space – two of them for the first time. Leonov and Kolodin would be the new back-up crew, but it was not possible to insert a third back-up crewman to replace Kubasov so close to a launch. The crew had been formed just 16 weeks earlier (Dobrovolsky had joined the training group in January) and here they were about to be launched on a three- to four-week mission. Only Volkov had been in space before – for one week, as a member of the Soyuz 7 crew in 1969. None of them had rendezvous and docking experience. It would be a strong test of their training and character to effect a successful mission with such a small amount of preparation.

The launch of Soyuz 11 took place at 04.55 Moscow Time on 6 June, and the vehicle reached Earth orbit at 05.04. Following a course correction at 10.50, Soyuz 11 reached its required 123 × 180 mile orbit, and preparations for transfer to the Salyut station began. Rendezvous operations, enabling the spacecraft to approach the Salyut, began at 07.24 on 7 June, as a 20-second burn of the onboard engine was used to reduce the distance between the two spacecraft. Through the Soyuz periscope, the crew was able to view the approaching Salyut, and at 492 feet distance they aligned for docking.

When the spacecraft were 328 feet apart and closing in at 3 fps, Commander Dobrovolsky took over manual control of the Soyuz. As they approached, a slight drift to the right was noticed on the Salyut. Dobrovolsky compensated for this by using the Soyuz onboard thrusters. At 228 feet the closing velocity of the spacecraft was reduced to just 1 foot/sec. Finally, at 07:49:15 Moscow Time, the probe of the

Soyuz entered the drogue of Salyut and a soft docking of the two spacecraft was completed, and at 07:55:30 the electrical and mechanical connections between the two spacecraft were confirmed. Soyuz 11 was now firmly attached to Salyut in an orbit of 141 × 166 miles. It was just 24 hours after the launch from the Baikonur Cosmodrome, and a new phase of manned space exploration was about to begin.

Three hours after docking – following a complete check of onboard systems and the integrity of the seals securing the two spacecraft – the hatches were opened. Patsayev became the first person to enter a scientific research station in space as he floated into the work compartments of the station. He was followed by Volkov, and finally by Dobrovolsky. Manned occupation of the world's first space station had been achieved, and the Soviets were delighted. Another space 'first' had been attained, and the first of a long programme of such missions had begun.

What was not initially reported in the West were the initial activities of the crew upon entering the Salyut: the repair of the air regeneration system and the immediate replacement of two of the six failed filters. After nearly two months in space with no crew onboard, the station had built up a strong odour of burning air, forcing the cosmonauts to spend their first night attached to the Salyut, sleeping in the relative safety of their Soyuz. By the following day the odour had dispersed, and they began their work programme.

Initial activities aboard Salyut were devoted to the activation of the station's systems and experiments, and mothballing of the Soyuz ferry vehicle. Small manoeuvres were initiated to raise the orbit of the complex to extend its life, and tests were performed on the integrity of the docked configuration. For the next three weeks the three cosmonauts performed their research programme using an array of experiments and hardware. Their experiments ranged from simple evaluations of the living conditions on the station to observations of the Earth's surface. Astronomical observations, a range of biological and technological experiments and a full programme of tests and evaluations of the Salyut station, kept the crew busy.

Regular TV broadcasts from space depicted the three men happy and productive in their work. At a time when the Soviets were overcoming the embarrassment of failing to reach the Moon, the Salyut 1 crew became the new stars of the space programme, pioneering a major achievement in the exploration of space in the name of Soviet communism heroes a decade after Gagarin had led the way.

The other side to the Salyut story was the reported clashes of personality among the crew. They had not been formed as a crew for long and, against expectation, had not had time to become an effective team. The rapid changes in the programme meant that psychologists would not have been able to prepare them as a close unit in the way that other crews had been before flying the longer-duration missions.

On 16 June, Volkov also reported a strong smell of smoke, and the ground controllers suspected a fire had broken out on board. The cosmonauts evacuated to their Soyuz and gradually evaluated the primary and back-up electrical supply systems on the station. This was a tense period for the crew, who found that the electrical system was apparently safe, but it could have created a small fire in a power cable. The crew asked for an immediate return to Earth, but after the station was deemed safe this was refused. Volkov reacted badly to this event, and confronted

Dobrovolsky, the mission Commander and a rookie on his first flight. Volkov – the space flight veteran – declared that he should be Commander, which led to further tension in the crew.

On 19 June, Patsayev celebrated his 38th birthday, and five days later, another record was broken as the crew passed the world endurance record set (at 18 days) by the Soyuz 9 crew during the previous year. Spirits were running high at mission control as the mission drew to a close. Good results were being obtained from the inexperienced and now touchy crew, who were still able to perform efficiently as the flight progressed.

On 20 June they were to observe the third launch attempt of the unmanned giant N-1 lunar rocket, using the 'Svinets' apparatus developed for the Almaz military stations. However, the launch of the rocket was delayed, and when it eventually launched on 27 June, Salyut's ground track was nowhere near the launch complex. The lunar rocket, however, exploded 57 seconds into the flight, and as it fell to the steppes of Kazakhstan it took with it the final drive for a programme to put Soviet cosmonauts on the Moon. Although unable to watch the N-1 launch, Dobrovolsky was able to use the 'Svinets' to observe smaller ICBMs fired from the launch complex.

On 25 June, reports were issued that the cosmonauts were completing their experiments and had begun to load Soyuz 11's descent capsule with the results and items needed for the return to Earth. Apparently the planned four-week mission would end after only three weeks. Left unsaid was the reluctance of the crew to do any physical exercise to maintain their strength for returning to Earth. This was due to faulty exercise equipment and their preoccupation with repairing systems failures around the station. This faulty equipment could have necessitated the early return, and in addition, crew incompatibility was not helping matters in trying to prepare for the recovery.

After three weeks aboard Salyut, they gradually mothballed the space station over the next couple of days as their Soyuz 11 spacecraft was checked out for return to Earth. On the evening of 29 June, at 18.15, the three cosmonauts transferred to Soyuz 11, and Volkov closed the connecting hatch between the spacecraft and the station. On the hatch seal connecting the descent module (in which the cosmonauts would return) and the orbital module of Soyuz, a status light failed to turn off, indicating an incorrect seal, and on discovering this, a very tired Volkov barked at the ground, asking for instructions on what to do next. This would be their only safety barrier between the pressurised crew compartment and the vacuum of space when the OM was separated prior to re-entry, as they did not have any pressure suits. The sensor and the hatch were rechecked, and the cosmonauts were told to place a piece of paper over the sensor. It detected a hatch closure, and the control panel light went out. It was suspected to be a faulty sensor, and to prevent the problem recurring Dobrovolsky placed a piece of tape over it. The three cosmonauts strapped themselves into their couches amongst the scientific results being brought back from their mission. They were already tired and strained from their mission, and the event had unnerved them.

At 18.25 the undocking command was issued, and three minutes later Soyuz 11

As the Soyuz lands by parachute and recovery helicopters close in, the end of a successful mission looks secure, but there is no word from the cosmonauts inside, and a tragedy awaits the team when they finally open the hatches.

slipped, without incident, from its berthing place on the front of Salyut 1. The historic first occupation of a space station had ended, and the heroes would soon be back on Soviet territory to receive the praise they deserved. Following a photographic survey of the exterior of the Salyut, they performed a separation burn to move away from the station. After three orbits, the crew informed ground control that they were about to begin their descent programme.

MCC: 'Goodbye Yantar ['Amber', their call-sign] until we see you soon on Mother Earth.'
Soyuz 11: 'Thank you, be seeing you. I am starting orientation.' These were the last words received from Soyuz 11.

Retro-fire occurred at 22.35, and the seemingly routine forty-minute descent profile began. Shortly after retro-fire, the separation of the Orbital Module and Service Module was completed, leaving the Descent Module to make the return to Earth on an automatic sequence. After the period of blackout, ground control tried several times to establish radio contact with the crew, but with no response. At 23:02 the parachute was deployed and the spacecraft descended slowly under the canopy, with ground control still trying to contact the crew by radio. Thinking it was an equipment failure – perhaps the loss of the reception antenna – the rescue helicopters saw the descent retros fire and the spacecraft gently come to rest on the ground at 23.17.

Expecting to find the three cosmonauts weakened by their record-breaking three

Dust-down at the end of the Soyuz 11 mission after a record of 23 days in space.

weeks in space, the jubilant rescue team opened the spacecraft hatch to find all three cosmonauts lying dead, having perished at the moment of their triumph. Recovery teams pulled the cosmonauts from the capsule and tried in vain to revive them. A preliminary medical inspection of the bodies revealed that all three men had blood in their lungs and had suffered brain haemorrhages. High levels of nitrogen were found in their blood samples. All this pointed to a rapid death by decompression. Initial reports of the crew positions and the layout of the capsule revealed that all of the shoulder straps had been released. Commander Dobrovolsky had become entangled in his harness, probably as he made a vain last attempt to manually close one of two air-equalisation valves that had been found to be open. The radio transmitter had also been turned off.

The news was flashed around the world, which reacted with shock. Rumours about the cause of the deaths circulated almost immediately, ranging from a leaking hatch, to the effects of three weeks in space and the strain of re-entry, and to the health or effectiveness of the crew. The loss to the Soviets was compared with the affect of the assassination of President John F. Kennedy on the Americans, and it once more demonstrated – just over a year after the near-fatal Apollo 13 mission – that spaceflight was still a dangerous and risky occupation.

On 3 July a special state commission was set up to investigate the tragic accident. Still in orbit, Salyut 1 continued to operate in an automated regime. A Soyuz 12 mission had been intended to follow a successful Soyuz 11 mission, probably in September 1971, with a crew commanded by Alexei Leonov. However, after the loss of the three cosmonauts, all flights were abandoned pending the results of the inquiry and the subsequent implementation of the recommendations of the commission. Indeed, after several course corrections, Salyut was commanded to fire its onboard

As the recovery crews open the hatches, all three cosmonauts inside are found dead.

Rescue crews try to resuscitate the cosmonauts, but to no avail. (Courtesy R. v. Beest/ British Interplanetary Society.)

engine system for the last time on 11 October, and completed a planned entry into the atmosphere to burn up over the Pacific Ocean. This was to be the fate of all subsequent Salyut space stations, preventing an out-of-control descent, a problem that was reflected in the Skylab re-entry in 1979. Mir's deorbit in 2000 would be a further example of how difficult it is to control the demise of space debris at the end of a spacecraft's useful life. The larger the structure, the more difficult it is to accurately predict the outcome of its re-entry flight path. It is an area that is already

The bodies of the Soyuz 11 cosmonauts lie under sheets at the landing site

being addressed in the construction of the ISS, at the very beginning of its orbital lifetime.

As the commission began its investigations, several Western theories were put forward as to what might have caused the sudden deaths of the cosmonauts. As the bodies lay in state, growing concern was voiced around the world about the safety of manned spaceflight, leading to talk of the possible cancellation of further efforts.

Perhaps the foremost contemporary theory surrounding the loss of the cosmonauts was that the human body might not be able to survive prolonged periods in the weightless environment. Up to Soyuz 9 (18 days), every mission flown by the Americans or Russians lasted for only a few days, and the three-week flight of Soyuz 11 was a further step in a new area of exploration. During the American Gemini programme, in which the American spaceflight endurance record was set at 14 days, signs were recorded that heart muscles become lazy during extended flight. This was thought to be associated with the cramped conditions onboard the Gemini spacecraft, and in the Apollo programme a rudimentary exercise system was provided for the astronaut crews during 12-day lunar flights. But Apollo was still not as spacious as a space station such as Salyut or the impending Skylab (set for a 1973 launch). The missions of Skylab and Salyut were due to investigate the area of zero-g effects in greater depth. In July 1969, the theory that extended spaceflight caused undue stress on the heart was reinforced when astro-monkey Bonny died shortly after recovery from the nine-day US Biosatellite 3 mission. The post mortem revealed that Bonny had died from heart failure. If this was the cause of the fatalities on Soyuz 11, then plans for extended spaceflights to the planets would have to be shelved until artificial gravity systems were developed. Man would be bound to Earth for years to come.

NASA's acting Administrator George Low expressed his doubts about the heart failure theory, and Dr Walton Jones, NASA's Deputy Director of Life Sciences at the Office of Manned Space Flight, suggested that the crew may have died as a

The Soyuz 11 cosmonauts lie in state. (R. Hall Collection.)

result of rapid decompression of their cabin. The cosmonauts were not wearing spacesuits, because the Soviets thought these were unnecessary for Soyuz. Manned Spacecraft Center Director Robert Gilruth supported this theory. Photographs of the bodies lying in state showed them as they would have appeared due to a leak in the valve system, or if the pressurised shell had been in some way ruptured.

Dr Charles Berry, also from MSC Houston, suggested that a sudden release of a toxic substance might have poisoned or choked the crew (as actually happened, to some extent, to the returning ASTP American astronauts in 1975), or that the automatic mode may have not worked properly or in the correct sequence. Whatever the cause, both Soviet and American space communities realised the seriousness of the investigation and its implications for the future of manned spaceflight.

On 1 July 1971, thousands of Soviets attended the state funeral of the three cosmonauts in Moscow. Soviet President Nikolai Podgorny, Premier Alexei Kosygin and Party Secretary Leonid Brezhnev successively stood watch over the national heroes as part of the honour guard. The cosmonauts were buried in the Kremlin Wall, joining the three 1934 Osoaviakhim balloonists (Fedeseenko, Vasenko and Usykin), Sergei Korolyov, Vladimir Komarov and Yuri Gagarin.

President Richard Nixon sent astronaut Tom Stafford to Moscow as the official representative of the American astronaut group. It was the first time the Soviets had agreed to this, as similar offers for the funeral of Komarov in 1967 and Gagarin in 1968 had been politely but firmly declined. In 1971, however, international détente was in full swing, and talks concerning a possible joint mission in space with an American Apollo crew and a Soviet Soyuz crew had been proceeding for several

The cosmonauts are taken for burial in the Kremlin Wall during a full State Funeral.
(R. Hall Collection.)

months. Stafford was already tipped for the command of the American crew, and it
was fitting that he was the astronaut selected to represent his colleagues at the
funeral. In addition, President Nixon sent a message to the Soviet leaders: 'The
American people join in expressing to you and the Soviet people our deepest
sympathy on the tragic deaths of the three Soviet cosmonauts. The whole world
followed the exploits of these courageous explorers of the unknown, and shares the
anguish of their tragedy. But the achievements of cosmonauts Dobrovolsky, Volkov
and Patsayev remain. It will, I am sure, prove to have contributed greatly to the
further achievements of the Soviet programme for the exploration of space and thus
to the widening of man's horizons.'

Despite this setback, the Soviets were quick to re-establish their commitment to
further manned spaceflights. Cosmonauts would certainly return to orbit to continue
the work started by the Soyuz 11 crew. The Special State Commission set up to
investigate the Soyuz 11 accident presented part of its findings in a public statement
released on 12 July. The statement reported that the flight had proceeded normally
up to the beginning of re-entry, and continued: 'On the ship's descent trajectory, 30
minutes before landing, there occurred a rapid drop of pressure within the descent
vehicle, which led to the sudden deaths of the cosmonauts. The drop in pressure
resulted from a loss of the ship's sealing. An inspection of the descent vehicle showed
that there were no failures in its structure.' The report went on to state that the
investigation concerning the exact cause of the seal failure was continuing. This
comment eliminated all theories into weightlessness, physical deconditioning and
exhaustion as possible causes of the tragedy. NASA immediately contacted the

Soviets and expressed concern that the failure that affected Soyuz 11 could be a possible cause for concern for Apollo 15, in its final stages for launch on 26 July. The Soviets replied with the reassuring statement to NASA that 'the drop in pressure resulted from a concrete failure of one of the elements of the descent vehicle system. Since it is a matter of specific and particular defect we are sure that it cannot be related to the Apollo spacecraft.'

Plans for launching Apollo 15 were not affected, nor were subsequent launches of that class up to 1975. The Apollo 15 crew, incidentally, named one of the craters at their landing site 'Salyut', in honour of the lost crew. All astronauts wore full pressure suits for launch, docking and re-entry, to protect them against rapid decompression. Had the Soviet cosmonauts been wearing similar suits, they could very well have survived. At the time, however, the Americans were negotiating the joint flight of Apollo and Soyuz hardware to take place in 1975, and real concern was expressed outside the agency about the safety of Soviet hardware.

Following almost two years of persistent requests for detailed information on the exact cause of the Soyuz 11 tragedy, American ASTP officials finally received the full explanation during one of the joint meetings held in the USSR in October 1973. The presentation by the Soviets to the American officials was very thorough, with information on the failure, the post-flight investigation, experimental tests recreating the failure and the steps taken to ensure that such an accident would not happen again. The presentation was formulated by ASTP Soviet Programme Director Professor Konstantin D. Bushuyev, who had to convince many senior officials in the USSR that if the joint project was to proceed, the American team needed to know the truth about Soyuz 11.

According to the Soviet presentation, the fatal depressurisation of the cabin occurred when 'A breathing ventilation valve, located in the interface ring between the Orbital Module and the Descent Module, opened inadvertently during the downward path of the descent vehicle.' Following retro-fire, the separation of the two modules occurred at retro-fire plus 723 seconds. On this vehicle the twelve pyro-charges fired simultaneously instead of sequentially to separate the Descent Module from the spent Orbital Module. The force of this discharge caused the internal mechanism of the pressure equalisation valve (a ball joint) to release a seal that was usually discarded explosively much later to adjust the cabin pressure automatically as the vehicle descended into the denser layers of the atmosphere. The first of these should have opened at 17,400 feet, and the second at 14,275 feet. On Soyuz 11, however, this occurred at 105 miles, and forced a gradual but steady loss of pressure, which reached a fatal level within about 30–50 seconds. By 935 seconds after the retro-fire sequence, cabin pressure had reached zero and stayed there until 1,640 seconds, when the pressure began to increase as the descent capsule entered the upper reaches of the atmosphere. By then it was too late for the crew.

The crew must have been aware of the problem. Without the protection of pressure suits, they could hear and probably feel the escape of air. The leak also resulted in slight changes in the attitude of the spacecraft as it dropped towards Earth, causing a thruster to fire automatically to compensate for the unplanned rotation. A hand-operated pump was available near Patsayev, but this took too long to operate. It has been suggested that at least one of the crew did try to operate the pump for a frantic few seconds, but to no avail. As they lay in state, it was noted that Patsayev's body had a bruise on the head, which was attributed in some reports to his efforts to try to manually operate the pump to retain atmosphere in the capsule. The commission reported that apparently Dobrovolsky had tried to get out of his seat and check the hatch, thinking that the hatch seal was at fault, rather than the sensor reported earlier. The hatch was safe and secure. The other crew-members must have released their harnesses to search for the sound of escaping air, but with the static from the active radio it was hard to locate. It was probably then that they shut off the radio transmitter in order to isolate the escaping air. The source was actually beneath Dobrovolsky's couch, but the crew had run out of time to manually close it, and fell unconscious in their seats, their fate sealed in seconds.

It was determined that the events inside the capsule were rapid. Four seconds after the failure of the valve, their breathing rate was three times normal. After the loss of pressure, they were probably unconscious within 10–15 seconds and were dead within 50 seconds of the valve failure. Total pressure loss in the capsule occurred in only 112 seconds. It has been reported that at that time cosmonauts were not trained in the early, emergency activation of the valve.

Once the spacecraft had entered the atmosphere, the automatic descent system continued to function. It deployed the recovery parachute and effected what appeared to the recovery team to be a normal landing, apart from the problem of communication with the crew. The capsule was taken away for detailed examination following the brief preliminary inspection carried out at the landing site.

The tissue damage resulting from the boiling of the cosmonauts' blood in the 700 seconds that they were exposed to the vacuum was soon attributed to a more catastrophic and instant decompression of the vehicle.

The detailed telemetry records of all firings of the attitude control thrusters were analysed, and revealed that unplanned firings had been made to counteract the shift in attitude caused by the escape of atmosphere from the descent vehicle. Examination of the throat of the pressure equalisation valve also revealed traces of pyrotechnic powder. After all this information was gathered and analysed, the Soviets were able to establish exactly how the accident happened, and that it was the sole cause of the cosmonauts' deaths. This information finally solved the mystery surrounding the loss of the Soyuz 11 cosmonauts.

Initially, American experts had been confused when told in early reports that the problem was associated with the spacecraft's 'germerizatsyia,' which could be translated as either the failure of a seal or the loss of air-tightness. Furthermore, private remarks to Ed Smylie (one of NASA's ASTP representatives from Crew Systems) from I.V. Lavrov (his Soviet ASTP equivalent) in December 1971 implied that the problem with Soyuz 11 lay with the pressure equalisation valve. Other Soviets reported that the problem was with the seals that guaranteed the security of the hatch between the crew modules. With this new data, NASA environmental control experts at Houston verified this against earlier information and data provided, and it was concluded that the problem with Soyuz 11 was not one that would directly affect the American crew. This exchange of information about Soyuz

11 and the American Apollo 13 mishap helped forge a partnership that resulted in the highly successful ASTP in July 1975, and an exchange of biomedical information on space matters which has continued ever since.

Once the problem was recognised and identified, the Soviets modified their hardware to prevent the reoccurrence of such an event. They completed a range of ground tests and flew two unmanned missions (Cosmos 496, flown 26 June–2 July 1972, and Cosmos 573, flown 15–17 June 1973), which confirmed the success of the alterations they had devised. Soyuz 12 flew the first manned mission for more than two years in September 1973. Two cosmonauts, Vasili Lazarev and Oleg Makarov, flew a stripped-down ferry-class Soyuz vehicle with great success. Lightweight space suits were worn for the launch and re-entry and all critical parts of the mission, such as docking manoeuvres, to help protect the cosmonauts in the event of cabin decompression.

With the American programme winding down following the end of the Apollo and Skylab programmes, all that remained for the Americans before the start of the Shuttle flights was ASTP. The Soviets, however, soon recovered from the tragedy of Soyuz 11 to continue their series of space station missions in the Salyut and Mir vehicles. The Soyuz 11 crew will always be remembered as the first to prove that living and working on space platforms was possible. Salyut 1 has often been deemed a failure due to the inability of the Soyuz 10 crew to enter it, and to the loss of the Soyuz 11 crew. True, these were difficult and tragic setbacks, but the *mission* of Salyut – to support a human crew in space for up to a month – was a success. The problems with Soyuz 10 rested with the Soyuz docking equipment, otherwise Soyuz 11 would not have been able to transfer its crew. The Soyuz 11 mission, although dogged by equipment failures and an inexperienced and tired crew, suffered its failure *after* leaving Salyut. In its primary objective, Soyuz 11 was a success, but the hardware failed during the return leg, before the cosmonauts' achievements could be appreciated.

Soyuz 11 also graphically demonstrated that even such apparently routine processes as the return to Earth will always pose a serious threat to every space crew. Even after successful completion of flight objectives, the mission is never over until they are safely back on Earth – alive.

# Summary

It can be seen that returning to Earth from space is sometimes as difficult as leaving it. Our atmosphere, while protecting and sustaining us on the surface, also acts as a barrier against any object returning from space. As the molecules of air become more dense, so friction increases, resulting in a flaming end to any unprotected space vehicle. It is also important that entry into the atmosphere is achieved at the right angle, as too steep an entry results in an increased temperature and burn-up to destruction. Too shallow an angle would result in the vehicle skipping off the atmosphere like a stone across water, and out into space, probably unable of a second attempt at entry.

Not only do the conditions have to be correct at the beginning of the entry sequence but also at the end. Weather and equipment play a vital role in the recovery of any vehicle from space, and as with other aspects of spaceflight, the addition of a human crew increases the complexity of the systems needed to provide a safe return.

There have so far been three methods used for the recovery of crews from space: splashdown in the ocean, as favoured by the early American missions due to the size of the recovery area, the absorbing qualities of the water and the resources of the American naval fleet; dustdown in the Soviet/Russian programme, in which the vast and relatively secure expanse of the steppes of Soviet Central Asia provided an easier recovery zone than ocean recovery, although this was used in automated test-flights of the manned lunar craft and early Shuttle variants; and runway landing, as developed for the rocket research aircraft and lifting bodies, and used as the primary landing mode of Space Shuttle missions.

Although the Americans have favoured the ocean and the Soviets have favoured the land, the two countries have prepared for landing astronauts on land and cosmonauts in water (as experienced by the Soyuz 23 cosmonauts in 1976). In addition, although a runway is the primary landing mode of the Space Shuttle, it has a theoretical capability of landing on water, although, as with the Return to Launch Site abort mode, whether it can actually achieve this safely is doubtful, and there is no way to simulate it before it might be called upon to save a crew.

Recovery techniques have also included the use of parachutes and wheels and an intended use of a paraglider for Gemini to return to land, but although tested it was never used operationally to recover a crew from orbit. The Shuttle uses techniques developed during the rocket plane era of unpowered gliding approach, which has to work first time, as there is no option of a missed approach and a second chance. In approximately 100 landings of the six orbiters – *Enterprise, Columbia, Challenger, Discovery, Atlantis* and *Endeavour* – there has never been a major landing incident.

By using parachutes to return a crew from space – as have all the early American and all the Soviet missions – there is an element of risk over which the crew or ground control have no control once the mission leaves the launch pad. Either the parachute will or will not work. Back-up parachutes can be provided, but deployment is certainly desirable to slow down the vehicle before impact with the ocean or land.

Recovery of human exploration craft from altitude by parachute began with the stratospheric balloon programmes in the 1930s, and in 31 US manned missions between 1961 and 1975 all crews were successfully recovered by parachute into the ocean, with only one of the three Apollo 15 parachutes collapsing. Following the Apollo 12 lightning strike during launch it, was feared that the parachute system inside the apex of the CM under the boost protective cover of the launch escape system may have become damaged. It was reasoned, however, that since there was nothing the crew or ground could do if this was so, then the mission to the Moon might as well proceed, and the answer could wait for ten days. The same concerns were redressed on the very next mission, Apollo 13, when after the efforts to return the crew after surviving the in-flight explosion, and four days in an almost frozen spacecraft, everything depended on the parachute working correctly in allowing the crew to survive their ordeal

For the Soviets, the Vostok cosmonauts chose to eject after exiting the blackout period, to descend by personal parachute, while the empty capsule descend to Earth under its own canopy. The redesigned Vostok capsules that were flown as Voskhod 1 and Voskhod 2 could not provide a personal ejection seat for each crew-member, so they remained inside the capsule and lended by parachute. All Soyuz craft have descended by parachute at the end of their missions. Only one of them – Soyuz 23 – landed in water, and although the Soyuz 11 crew died due to a faulty air equalisation valve on their craft, and Volynov had a partial parachute deployment at the end of Soyuz 5, only Komarov on Soyuz 1 encountered a parachute failure that contributed to his death.

The Apollo 7 crew came home pinching their noses to relieve nasal pressure due to catching colds. The crew decision to not wear pressure helmets caused some friction between the flight crew and Mission Control. The lack of pressure garments cost the lives of the Soyuz 11 cosmonauts, and since Soyuz 12 in 1973 all Russian crews have worn suits for launch and landing. Launch and entry suits have been worn on the Shuttle since 1988 as one of the recommendations incorporated following the *Challenger* accident.

All astronauts and cosmonauts receive survival training in the event of an emergency landing in a remote area away from normal landing zones. For cosmonaut Sergei Vozovikov, such training cost him his life when he became entangled in an abandoned fishing net during sea-trials in the Black Sea in 1992, and he drowned.

These techniques were called upon, in part, by Scott Carpenter in 1962, following Aurora 7's overshoot of the landing area, the Voskhod 2 crew in 1965 after coming down in a forest, and the Soyuz 18-A cosmonauts following their launch abort in 1975.

For crews of long-duration missions recovery after landing depended on preparations and regular exercise while in space. After months in microgravity the bones and heart can be weak, and crews need assistance in leaving the spacecraft. Many Soviet cosmonauts have been carried off their capsules at the end of their missions. A crew must maintain fitness in space to withstand the changes when reverting back to 1 g at the end of the mission. Routine exercise may be boring and time-consuming, but it is necessary to ensure a rapid return to pre-flight conditions. The difficulty arises during an event such as the Mir incidents in 1997, when crew time is taken up with surviving, and exercise routines take a lower priority.

Astronaut Shannon Lucid was clearly determined to show that she could walk off the Shuttle at the end of her 188-day tour on Mir in 1996. Soviet Shuttle pilots Igor Volk, in 1984, and Anotaly Levchenko, in 1988, piloted Tu-154 aircraft shortly after landing from their Soyuz mission in a demonstration of their piloting skills after the effects of a week of microgravity followed by a sudden return to gravity, as would be encountered at the end of a shuttle mission. After his Tu-154 flight, Volk suited up, flew a MiG 25 to an altitude of 1.5 miles, and then complete a deadstick landing.

Shuttle pilots had already demonstrated that it was possible to complete a week in space and then, assisted by automated systems, land a winged vehicle on the runway; and with the longest Shuttle flight of up to 17 days, followed by a safe landing, it is known that a pilot still retains limb control. Although the legs can feel heavy and slow to move, there is enough control for returning the vehicle from orbit after three weeks.

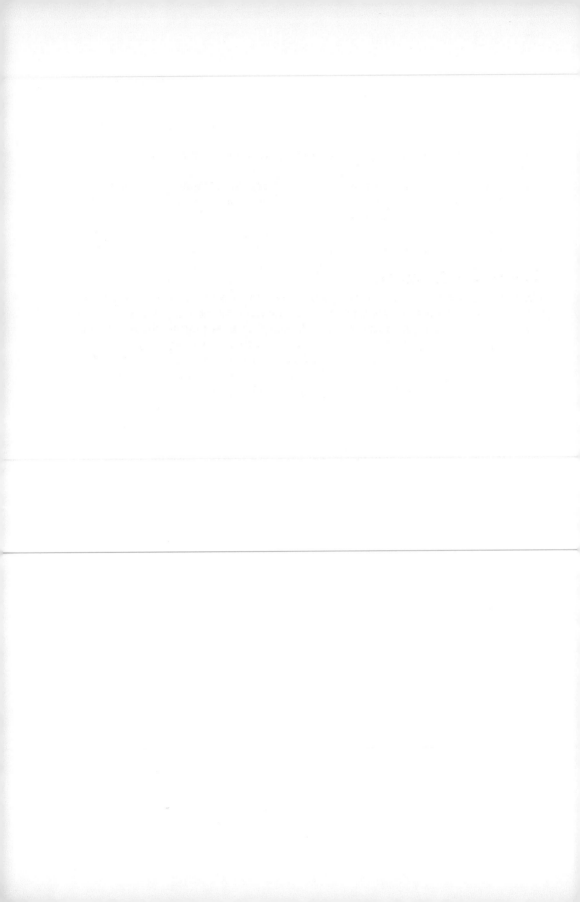

# The future in space

# Overview

The construction of the International Space Station represents the next major development in the human exploration of space. The operational use of the ISS after construction is complete will bring new areas of in flight risk to the human crew, in addition to the risks in training, launch, in-flight and recovery.

After the ISS is established perhaps we shall once more turn our attention to the Moon and out to Mars – the next two closest accessible targets in our Solar System. The hazards that are associated with the exploration of these two celestial bodies will also depend on the method chosen to reach them and on the design of the transportation system. There are, however, certain key elements that feature in any human programme to explore the moons and planets of the Solar System, and these must be addressed before any attempt is made to send humans away from Earth.

Drawing on past experience was a design consideration for the planning of what eventually became the International Space Station. As the ISS design evolved, so NASA established a family of documents, a related database and associated videos under the Man–Systems Integration Standards (MSIS) collection. This was initially a four-volume series that was listed under the NASA STD-3000 designation, which would be expanded upon as needed as new programmes were developed under the NASA Human Spaceflight branch.

These documents would replace all NASA field centre human engineering standards documents, and were designed to provide specific user information aimed at the integration of direct contact between humans and machines in such key areas as launch, entry, on-orbit, and extraterrestrial environments. They were aimed at engineers in the fields of design, systems, and maintainability, in addition to operations analysts and human factor specialists, amongst others, who were involved in the definition and development of all NASA human spaceflight programmes.

Boeing Aerospace Co, for NASA, compiled the document in conjunction with three subcontractors, Lockheed Missile and Space Company, the Essex Corporation and Camus Inc. A review of the final content before publication was completed by a Government/Industry Advisory Group, which consisted of a panel of experts and invited representatives from each of the US prime aerospace companies, support contractors, NASA field centres and Headquarters, as well as other government agencies and non-aerospace companies associated with the ISS program. The initial volumes of the MSIS were published in March 1987, and since then they have been frequently updated.

Drawing upon knowledge and experience from past programmes, and from the latest information on current technology, the work was divided into twelve key areas that influenced the design of human spacecraft. These were: Anthropometry and Biomechanics; Human Performance Capabilities; Natural and Induced Environments; Crew Safety; Health Management; Architecture; Work Stations; Activities Centres; Hardware and Equipment; Design for Maintainability; Facility Management; and Extravehicular Activity. Within these, all design considerations for the health and safety of crew-members were listed under Crew Safety. Specialised subject areas dealt with any aspect of hazards of a mechanical, electrical, thermal or fiery nature; and in addition there were cross-references to other influences on crew safety in the areas of Human Performance, Natural and Induced Environments, Architecture, Workstations, Hardware, Equipment Maintainability and EVA.

By following these guidelines and considerations, many potential hazards and design errors could be omitted from hardware design well before the equipment ever reached the launch processing facility. This in turn would restrict the level of risk and safety infringement for potential crew-members living and working aboard the ISS and follow-on vehicles.

## THE INTERNATIONAL SPACE STATION

In November 1997 the first element of the International Space Station was finally launched by Russian Proton rocket from the Baikonur Cosmodrome in Kazakhstan. The Zarya (Dawn) control module was joined in orbit by the US Unity connecting node in December 1997, delivered by the crew of STS-88.

The construction of the ISS – by an international team of 16 countries supporting 45 launches from launch centres in America, Russia, South America and Asia – would take at least five more years. More than 100 major elements would make up the ISS, and would involve 75 EVAs totalling 1,000 hours outside the complex. With an expected orbital life of 30 years, the ISS would undergo the most demanding international construction project ever attempted.

In the early space station programmes such as the Russian Salyut or American Skylab, the complete station was orbited in one launch, with crews ferried up to an empty station by small single-mission spacecraft. The development of the Mir space complex saw the long-awaited capability of crew exchange, the introduction of further research modules attached to the core Mir module, the further utilisation by the Progress tanker craft and the first major logistics resupply by American Shuttle orbiters.

Although the experiences of past station operations were useful, this was nothing in comparison to the creation of the ISS, in which each launch depended on the total success of the previous mission. The ISS launch manifest was once likened by one NASA employee as 'a row of dominoes stood up in a line'. If just one piece should

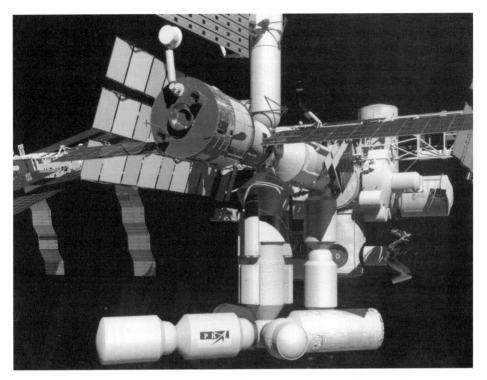

An artist's impression of the International Space Station, due for completion around 2004. The station is expected to be the next major focus of human spaceflight operations for the next 30 years.

topple it would have a devastating effect down the whole line. This 'domino effect' was clearly demonstrated with the repeated delays to the Russian Zvezda (Star) Service Module in the summer of 1999, and difficulties with flight qualification of American hardware. In June 1999 the second and only mission to the ISS in that year was a logistics supply mission that also highlighted additional problems with the quality of the atmosphere within the station modules (see p. 223). Although all of these hurdles could be overcome, it was time-consuming, and pushed the subsequent launches further down the calendar. The launch of the Russian Service Module was necessary before the first resident crew could occupy the station to begin permanent habitation of the complex. By the end of 1999 and the first year of orbital operations of the ISS, the programme was already almost a year behind schedule.

A much more serious delay to the programme would result in the loss or grounding of a Shuttle orbiter, restricting the total lifting capability of the Shuttle fleet by 25%. Indeed, with only three orbiters assigned to ISS operations this would actually mean a loss of one third of the payload lifting capability of the ISS Shuttle fleet. *Columbia* (OV-102) is not equipped to fly ISS missions, and although it could in theory be adapted to perform such missions, the cost is restricting and the chances of a further replacement orbiter (OV-106) like *Endeavour* (OV-105), which replaced the lost *Challenger* (OV-099), is very unlikely. The supplies of flight spares, tooling and a qualified workforce are not readily available to Rockwell/Boeing as they were immediately post-*Challenger*.

During the remaining years of construction of the ISS and throughout its operational lifetime, several areas of concern have been identified which could present serious hazards to the human crew aboard the complex. To address these issues, NASA adopted a four-category level of Caution and Warning for ISS hazardous incidents.

**Crew safety**

As the station grows and supports a rotating resident crew, so the issue of crew safety and rescue becomes a primary and continual concern. The events on Mir in 1997 indicated how even normal, routine space station housekeeping chores or operational activities can suddenly develop into a life-threatening situation without warning.

On board the ISS, several methods of alert and information systems on the current condition of the station modules will be available to the crew at all times. These include illuminated lights on instrument and display panels, and variable tones from onboard audio equipment, as well as text and graphic messages on portable computer systems.

There are four categories of Caution and Warning:

- Class 1, Emergency: in which an event causes a life-threatening condition for the crew, and requires immediate action. There are only three defined emergencies onboard the ISS: fire or smoke in a pressurised compartment; rapid changes in cabin pressure levels; and toxic atmospheric conditions.
- Class 2, Warning: in which hardware and systems failures are detected which require the crew to complete immediate corrective action to avert major impact on the mission, or potential loss of the station or the crew. Such failures could be the total loss of the primary guidance, navigation and control (GNC) computer, the detection of a high cabin pressure or the loss of the control moment gyros (CMG) sub-system which provide attitude control of the complex.
- Class 3, Caution: out-of-tolerance conditions which are not time-critical in nature, but which if left uncorrected indicate that they may develop into a Warning status. These could include the loss of the backup GNC computer, a critical failure of the S-band communications baseband signal processor (which processes the telemetry from the ISS command and control system through the Tracking and Data Relay Satellite System (TDRSS) and then to Mission Control, Houston) or the failure of one of the four GMCs.
- Class 4, Advisory: a non-caution and warning message that provides information to the crew about the status of systems and processes. This could include the non-critical failure of the S-band baseboard signal processor or the trip of the (electrical switch) remote power control module.

In addition, the ISS personnel Crew Health Care System (CHeCS) is a combination of three sub-systems, each of which will meet one of the three major health concerns associated with extended duration human spaceflight. The purpose of the CHeCS is, according to ISS training and familiarisation documents, 'to enable an extended human presence in space by assuring the health, safety, well-being and optimal performance of the ISS crew'.

The three-crew health-related concerns were an early definition during ISS development, and were to include physiological countermeasures to spaceflight, environmental monitoring and medical care. Each of the CHeCS sub-systems addresses one of these concerns:

*Countermeasures Systems (CMS)* evaluates crew fitness and provides counter-measures and an evaluation of the crew members during these countermeasures.
*Environmental Health System (EHS)* monitors air and water quality for chemical and microbial contaminates. It also monitors radiation levels and surface microbial contaminates, and four areas of water microbiology radiation and toxicology.
*Health Maintenance System (HMS)* monitors crew health, an individual response to illness or injury, and provides preventive health care and stabilisation. It also supports emergency patient transport between vehicles such as restraint and life-support measures between ISS and crew return vehicles.

**Crew rescue**
The longevity of the ISS affords the opportunity to provide an on-orbit crew rescue and recovery system when the Space Shuttle is not docked to the complex. Docked Shuttle orbiters supported initial visits to the ISS. The first resident crews will be supported by a docked Soyuz spacecraft which, in addition to providing a transportation role to and from the station in crew rotation scenarios, will provide a capability, at least throughout the completion of assembly (2005) for crew evacuation, as there will always be one or two three-person Soyuz spacecraft docked to the ISS when a resident crew is onboard. The Soyuz spacecraft themselves will be changed out for a fresh vehicle in a similar way that occurred on the Salyut and Mir stations.

The Crew Rescue Vehicle (CRV), or Crew Transfer Vehicle (CTV), will provide emergency crews with return capability beyond 2005 for the whole six-person resident crew. The final design is still under evaluation, but current designs resemble the lifting bodies operated in the 1960s and 1970s. It is planned for the CRV to be available for operational use at the end of the construction phase. The CRV/CTV will feature fully automated functions to deorbit and land, allowing the six-person crew to fully evacuate the station if they are able to reach the CRV location in less than one minute. The crew will have override capability for landing site selection.

**Orbital debris**
Shortly after the return of the STS-96 crew from the ISS in June 1999, the USAF Space Command notified the ISS flight controllers of the possibility of a close approach of a spent Russian rocket stage with the unoccupied space station. Early

For the first few years of assembly, crew rescue when a Shuttle is not docked will be carried out by the Russian Soyuz spacecraft similar to the one depicted here docked to Mir.

predictions indicated a close approach of only $1\frac{1}{2}$ miles, but the actual closest approach, on 13 June, was just over 4 miles. Following this incident, a review of policy and procedures relating to the moving of the whole complex was necessary whenever the subject of a close approach to the complex by orbital debris arose. As a result of this meeting, at the next close approach prediction of orbital debris on 25 October 1999 the ISS was manoeuvred 1 statute mile higher by the Zarya orbital adjustment engines to avoid a Pegasus spent rocket stage. If no burn had been implemented the stage would have passed by the ISS at a distance of at least than 0.8 miles. The 5 second burn increased orbital velocity by a little more than 3 feet per second, raising the orbital altitude. As a result, the Pegasus stage missed the station by 15 miles. The following morning a small adjustment burn moved the station back to its original orbital height.

The risk of orbital impact by an item of debris from an old piece of space hardware not of natural origin (man-made) has long been a concern as the amount of space debris increases and the size of space structures grows. On 16 December 1997 a US National Research Council report on orbital debris indicated that risks to Shuttle astronauts had been inadequately addressed by NASA – a claim that the space agency disputed. The report stated that approximately 95% of material that poses a threat (less than 2.5 inches in size) is undetectable from ground-based tracking devices such as radar and other DoD-operated systems. The council report also stated that meteoroids and orbital space debris pose the single most critical threat to the failure and early termination of a Shuttle mission, and that the chance of a catastrophic collision during a Shuttle mission had been reduced in the past few years from 1 in every 134 missions to 1 in 84 missions. At the time of the report there had been 88 Shuttle missions flown. If the *Challenger* accident was taken out of the equation, as it was classified as a launch accident, then there had been three missions over the odds. A major collision with space debris, according to the NRC report was just waiting to happen.

In reply, NASA Chief Scientist for Orbital Debris and Program Manager for Risk Prevention Nicholas Johnson stated that this had never been true, and that the report was a misinterpretation of the facts. The more the Shuttle was flown, Johnson replied, the more NASA had an understanding of exactly what would cause serious damage. At that time the Shuttle had proven much more resilient than had first been assumed.

The potential of an item of space debris hitting the Shuttle while in orbit has been considered in the NASA planning schedule of every Shuttle mission. In order to identify such potential risks, the space agency calls upon the US Space Command Alert System. Mission Control Center, Houston, is informed when any one of 8,000 items of tracked orbital debris passes within an alert zone that extends 13 miles in front and behind the orbiter and 2½ either side, above or below, the orbiter while in space.

Early in the flight operations, programme engineers assumed that any penetration of the carbon fibre panels that protected the wing surfaces would cause a burn up during re-entry. In addition, penetration of radiator line plumbing or the aerodynamic control surfaces was expected to cause a major systems failure. After almost 100 flights it is believed that the orbiter can withstand small penetrations, depending on the location. Even a large hole in a non-critical area could be sustained and still allow the orbiter to land safely. During orbiter upgrades additional ceramic fibre material – Nextel – inside wing panels and radiator lines covered with aluminium tape and combined with new isolation valves to increase the protection and isolation of a coolant leak, would all contribute to added Shuttle system safety during ISS construction, probably raising the risk of major systems failure due to orbital debris impact to 1 in 700.

As the ISS grows and its orbital life increases, the chance of a serious impact by even the smallest item of debris will increase. Unlike the Shuttle, which has an orbital life of only a few days, the orbital life of the ISS will be several years. Studies from the recovered Long Duration Exposure Facility (deployed by the Shuttle in 1984 and retried in 1990) and experiences from the Mir complex have provided additional information on the seriousness of debris impact on space structures over a long period of time.

## Radiation

A further concern for the safety of ISS crews is the exposure to radiation, especially during the numerous EVAs to be completed during construction of the complex. The total number of EVA hours during the 75 EVAs planned is more than 1,500. With the construction commencing in 1998 and expected to be completed by 2004 at the earliest, this also coincides with the peak of activity of the 11-year solar cycle. This can include severe solar storms that emit streams of high-energy electrically charged particles that intersect the station's orbital path.

The early planning of the Freedom space station, at 28° to the equator, allowed any launch from KSC in Florida to use the rotation of the Earth to assist in placing heavy payloads in orbit, and allowed the Shuttle to carry a maximum payload on each launch. In 1993 the inclusion of Russia in the ISS programme also saw the shift of inclination of the station to 51° to allow Russian launches from the Baikonur Cosmodrome. Both Salyut and Mir stations had previously flown at this inclination.

Exposure levels at this inclination are not life-threatening due to the protection of the Earth's natural radiation belts, but if the space station passes through the South Atlantic Anomaly – in which the innermost level of Earth's torus equilateral radiation belt at 600–3,000 miles dips to only several hundred miles towards the surfaces – poses a serious threat to the station crew. However, data on the size and shape of solar flares could not be acted upon by mission control quickly enough to allow the crew to take immediate action. The ground-based flight directors are restricted by receiving information from secondary sources (such as the National Oceanographic and Atmospheric Administration). NASA was urged to use information from NOAA and other space-based weather data sources to provide models allowing a prediction of 'space weather' immediately around the ISS.

The radiation risk, although not immediately life threatening, does however present a serious situation, which can vary from no more than the equivalent of several hundred chest X-rays to, in extreme cases, serious illness; and in exceeding the prescribed lifetime, doses could be life-shortening (see p. 13).

A dose of radiation absorbed by an ISS crew will originate from three sources: Galactic cosmic rays, trapped radiation, and solar particle events (solar flares). Data measured during Mir missions provide a good baseline for what can be expected on the ISS. On 30 September 1989, cosmonauts Alexandr Viktorenko and Alexandr Serebrov – the then resident crew onboard Mir – were advised of a major solar flare. Normally, the effects of a flare were recorded inside the station for only a couple of days, but the results from this flare lasted for ten days. Both cosmonauts slept in the Kvant 1 module, which was heavily shielded. It was estimated that their exposure to this radiation was, in effect, no more than adding 14 days to their 168-day mission.

Adequate crew shielding will vary throughout the station as it evolves during the construction phase. For the lifetime of the ISS, radiation monitoring and protection will be a constant aspect of crew time management and safety. The provision of personal radiation dose meters for each crew-member for inside the station and outside on EVA will assist in the prediction and documentation of the amount of exposure of each crew-member during orbital stay on the ISS.

The construction and operation of the ISS will also be an important development in understanding how large structures deteriorate when exposed to the environment of space over a period of many years. Data on environmental effects on spacecraft structures and systems have been collected from the recovered LDEF after six years in orbit, from exterior examination of the Mir space complex over its 13-year operational life, and from almost 100 post-flight examinations of the structure of the orbiters after their return from each mission. This information will be used, with data obtained from the ISS, in the design of future space hardware that will be more resilient, and adapted to operation in space. This will result in longer-lived structures, and offer more protection and reliability for future human crews.

## RETURN TO THE MOON AND THE INNER PLANETS

After the creation of the ISS the next logical target for human exploration is a return to the Moon and the initial exploration of Mars. In leaving the relative protection of Earth orbit, new hazards await the crew – whether on the Moon, or exploring the inner planets of our Solar System.

### The Moon

The benefit of exploring the Moon is that we have been there before. The Apollo programme provided us with limited but valuable experience, and knowledge of venturing out to the Moon and living and working on the lunar surface. It also demonstrated the difficulties in safely returning a crew to Earth if systems fail beyond near-Earth orbit. The Apollo 13 incident was a consequence of faulty equipment and pre-launch procedures, and was not a direct result of flying to the Moon. In any future explorations at lunar distance, an adequate system of crew rescue and recovery will be required in case such a situation reoccurs. What saved Jim Lovell and his crew were their training experiences, the availability of the still-functioning Lunar Module and the skills of the ground controllers in adapting the flight plan to ensure a quick return of the stricken craft. For the next generation of lunar explorers a vehicle, or system, capable of supporting them until rescued, or allowing an emergency return to Earth, should be incorporated in the flight schedule.

On the lunar surface the six two-man teams of Apollo astronauts were fortunate that the LM ascent engine fired first time, every time, to begin their return journey. Although simple in design and operation, had this engine not worked then the two unfortunate lunar explorers would have died with no hope of rescue. Part of the original post-Apollo lunar exploration plans, subsequently cancelled, included variants of the LM which were to have been landed without a crew but with habitation modules. This would allow the astronauts, landing in their separate LM, to stay on the Moon for periods of up to two weeks. This could also have doubled as a shelter/lifeboat in the event of a systems failure on the main lander preventing lift-off, and allowing the crew to remain safe until a rescue craft could reach them. It is likely that future human lunar landing vehicles will incorporate elements of habitation modules evaluated on the International Space Station, and in turn these will be used to construct the first research base on the surface.

With variations in temperature variations at the lunar equator between 261° F at lunar midday to –279° F shortly before lunar dawn, ionising radiation reaching the surface from the Sun and other sources, and micrometeorite bombardment on the surface, adequate protection for the lunar explorers will be of immediate concern.

Apollo LM and pressure garments offered adequate short-term protection from the micrometeorites and temperature variations, but they could not protect against the radiation levels. By using the lunar soil as a shield, any habitation module can provide future lunar explorers with adequate protection against extreme temperatures and micrometeorite impacts, and it reduces the radiation penetration to a manageable level. By using radiation counters on EVA suits, the astronauts' personal exposure during long-term activity on the surface can also be monitored to bring it within the guidelines set down by NASA.

Lunar dust contamination was a problem identified early during the Apollo missions. In the lunar $\frac{1}{6}$ gravity, the dust not only covered and clogged surface equipment and experiments, but became a problem inside the LM after EVA. The lunar dust covered the EVA suits and this was brought in at the end of each EVA. Inside the Lunar Module there were no facilities to effectively clean the EVA suits or remove the dust particles from the atmosphere of the LM cabin. Apollo astronauts found that the dust ground into the skin of their hands and fingers and under their nails, and took some time to remove after their return. Fortunately, Apollo landing crews remained on the Moon only a few days and avoided the more serious problem of inhaling lunar dust particles over an extended period of time. Allowing this dust to enter the smallest reaches of the lungs could result in a disease similar to asbestosis or that suffered by coal workers after inhaling coal dust from mine shafts. In a future lunar research station, the problem will be reduced by the provision of an EVA airlock with cleaning facilities and a difference in air flow, restricting the movement of dust into the main habitation.

### Mars

Apollo flights reached the Moon in approximately three days – but not because the Saturn V was unable to propel them there faster, as it could have done so in about 24 hours. The problem lay in the capability of the Apollo Service Module engine to slow the spacecraft enough to allow capture by the weak lunar gravity before flying past into deep space. A human flight to Mars, using the increased gravitational forces of a larger planet, allows for a faster approach than was possible when aiming for the Moon. The added problem in reaching Mars is its distance from Earth.

Mars, the fourth planet from the Sun, is approximately 50% farther from the Sun than is the Earth. The Red Planet orbits the Sun in 686 Earth days, and during periods when its orbit brings it closest to our planet it is a mere 38 million miles away. When the two planets are at their farthest apart, this distance increases to 248 million miles. Unfortunately there is no propulsion system yet designed that can push directly against the gravitational forces of the Sun to attempt a direct flight to Mars when the two planets are at their the closest.

To achieve flight from Earth to other regions of the Solar System, all planetary launches from Earth enter a solar orbit that allows an elliptical transfer orbit to intersect between the Earth and the target planet, which can be the final destination, or can be used in a slingshot manoeuvre to move to the next target. The slingshot method also uses minimal propellant and course corrections en route.

The best launch opportunity for a flight to Mars occurs when the two planets are

Mars, the Red Planet – the next major goal of human exploration. (Courtesy NASA/
STScI.)

at their farthest apart – the Hohmann transfer trajectory. Flight time to Mars can be
as little as 180 days out and 180 days back, with a stay time of approximately 550
days – a mission duration of 910 days, or 2.5 years. However, the longest a human
has spent in space on one mission (at the time of writing is 438 days 17 hours 58
minutes (14 months) by Russian cosmonaut Dr Valery Polyakov, who also has a
career total time in space of more than 679 days. According to the Polyakov, the
greatest barriers to long-duration spaceflight are psychological. The International
Space Station, with its multinational, mixed-gender crew, will help to address these
barriers prior to leaving Earth for more remote targets in space.

With no firm commitment for a flight to Mars, there are currently numerous
proposals on methods and the design of vehicles to achieve the feat. Current opinion
is that the most favoured method would be the landing of the Earth return vehicle on
Mars prior to sending the human crew. This allows for a return/rescue vehicle to be
available fully checked out and fuelled, before the commitment to land a crew. Until
the exact method and hardware is decided upon, exact risk calculation for the human
crew is uncertain.

However, no matter what method of selection or which design of hardware is
chosen, the unavoidable duration of the mission increases the exposure time to space
radiation sources and levels. By flying during a minimum solar cycle period and by

providing a crew radiation-shield facility (lead, or even the water supply) around the outside of any habitation quarters, the levels of radiation would be reduced to acceptable and controlled limits.

An added risk on human flights to Mars is the planet's severe dust storms. Allowing the spacecraft to enter martian orbit to survey the surface prior to landing will provide an opportunity for the crew to wait out a storm that may be taking place when they arrive.

In addition, the capabilities of the landing crew to perform a safe landing in a gravity environment after several months of weightlessness is another factor under evaluation. A small inducement of g forces (similar to the bar-b-que roll on Apollo lunar missions) during the flight out to and back from Mars, built into the vehicle to allow even thermal control of the spacecraft, will affect the design of the spacecraft and operations during the coast phases. If a small gravity-induced system is adapted for the parent vehicles, then the onboard systems that support the human crew (life support, waste management, guidance, navigation and so on) must also be able to operate in microgravity if the gravity system fails in flight.

On the lander, an adequate abort system must be provided during landing, in the event of a termination of the landing attempt. This should also allow for a return to the parent craft or to a safe orbit around Mars, allowing for a rescue attempt. On the surface the crew must have provision for storm protection and a suitable airlock system to prevent contamination of the habitation environment by Martian dust, and capability for an extended stay time if the ascent engine does not fire, allowing time for a second landing craft from orbit, or a rescue mission from Earth.

Self-sufficiency, system redundancy and contingency procedures should be a major feature of an initial Mars exploration mission, until the creation of a Mars research base or orbital research station allows more flexibility in an Earth–Mars–Earth transportation system.

### Venus and Mercury

The two inner planets offer little chance of human exploration. Venus can be used for a gravity slingshot manoeuvre to reach other targets in the Solar System, and will probably be a target for a flyby of a returning Mars crew. Results from automated spacecraft have recorded extremely high pressure and temperature levels, as well as acid rain – none of which are not healthy for a human explorer.

Mercury does not have an atmosphere, and temperature extremes of $\pm 360°$ F and extreme radiation levels render human exploration highly unlikely. The planet is only half the Earth's distance from the Sun, and it therefore receives four times as much heat. The spacecraft would require extensive thermal protection, the pointing of delicate instrumentation or devices away from the Sun, and the placing of the crew behind permanent sunshades. Radiation levels during any human EVA activity inside the orbit of Venus would be highly dangerous, and would probably be conducted by robotic EVA systems.

### Asteroids and comets

More probable targets for human exploration are the numerous asteroids, comets and near-Earth objects (NEOs – migrated asteroids or burnt-out comets) that occupy the inner Solar System. Asteroid missions will probably be rendezvous and proximity operations (prox-ops) rather than landing missions, with the added challenge of achieving a safe orbit without the risk of collision. Sample retrieval missions would probably lead to the mining of resources from the hundreds of possible target asteroids in the asteroid belt beyond Mars.

Comets are much more difficult to reach with a human crew. The approach to a comet is 'safer' when it is well away from the Sun, beyond the orbit of Mars. As a comet approaches the Sun it is heated, and gas, ice and dust erupts from beneath the crust. This material is driven back by the solar wind, to from the comet's tail. To work in such an environment would be like working in a minefield. Even when far away from the Sun, comets can suddenly outburst and throw out material, due to gravitational forces. There is also no firm evidence of the consistency of a comet nucleus. It could resemble a thick and frothy milk-shake, or may be a 'snowy dirtball' or a 'dirty snowball', and anything attempting to land on its 'surface' might just sink right in!

## THE OUTER SOLAR SYSTEM

Beyond the asteroid belt are the gaseous giant planets, Jupiter, Saturn, Uranus and Neptune, and the small rocky planet Pluto. Human exploration of these far locations lies far in the distance future, and will depend on new forms of spacecraft propulsion. Of all the outer planets, perhaps Jupiter and Saturn offer the most tempting targets for future space expeditions.

For human exploration of the jovian system the largest hazard encountered is the radiation belts, which are several thousand times stronger than our own Van Allen belts. Unfortunately the inner three moons of Jupiter – Io, Europa and Ganymede – which are the most interesting, are also well within these radiation areas. It would be near instant death for any EVA astronauts to enter these regions, and even robotic systems would have to have highly developed systems to prevent contamination. The radiation at these levels darkens glass, electric clocks lose their timing, computer components can suffer random flips, and CCD chips can suffer interference.

It might be possible to explore Callisto, the outermost of the largest moons, or support unmanned robotic probes from outside the severe radiation regions. Jupiter also offers a perfect gravity slingshot for reaching other destinations around the Solar System: for example, to the poles of the Sun or out to Saturn, as employed on the automated probes Pioneer 10 and 11, Voyager 1 and 2, and Ulysses.

For human flights to Saturn the most interesting object is its largest moon, Titan, which has a methane atmosphere and temperatures around –302° F. An automated mining facility could be supported by a human research platform, but as with any mission beyond the Moon, the complexity of supporting and protecting a human crew to ensure their survival, adds considerably to the complexity of the outward and return voyages.

Beyond Saturn lie Uranus and Neptune, which were briefly visited by Voyager 2

Journeys to the outer reaches of the Solar System – to Jupiter and beyond – will bring new risks and dangers. (Courtesy NASA/STScI.)

in the 1980s. These worlds occupy a very distant and cold region of the Solar System, and it will be probably be well into the twenty-second century before humans are able to endure the journey to these planets and to Pluto, and return to Earth within a lifetime.

## HUMAN ELEMENTS

All human crewed missions beyond the Moon require a highly reliable and self-supporting Closed Environment Life Support System (CELSS) which the human crew would depend upon not only for mission success but also for survival. If the garden dies, or the atmosphere scrubber clogs up, or the oxygen recycler fails, then without adequate back-up or spares the crew is in real trouble. There will be no opportunity for waiting for spares to be sent from Earth. If the crew cannot use a back-up system or repair the fault, then they will die. The provision of a fully self-contained and totally failsafe CELSS is one of the most challenging objectives for long-range space planners. Not only interplanetary flight, but also colonisation, depends on the capability to provide a survivable environment for the human population. This is the most important element of any space mission away from the Earth–Moon system.

On the positive side, increasing distance away from the Sun reduces the impact of solar events such as flares. The solar wind sweeps away many cosmic rays generated outside our solar system, but the further away from the Sun, the more the background radiation, due to cosmic rays, increases. Radiation in a different form is still of concern for human exploratory missions in space.

Propulsion is another consideration for to the health and safety of a crew. All deep space missions are likely to employ non-chemical propulsion methods which bring their own failure modes. Solar sails are one of the safest forms of interplanetary transport, but there is the added problem – however small the sail rigging, or whatever its construction – of its becoming tangled. In such cases, perhaps one of the new skills to be learnt by future EVA astronauts will be the art of climbing the rigging in a modern-day version of the seafarers of the galleons and tallships of previous centuries (although the risk from pirate solar sail spaceships is not considered to be a danger!).

Nuclear or ion propulsion could result in a reactor leak, or reaction fluid leaks, or power supply problems. The use of nuclear fuels on spacecraft has been the subject of controversy for many years. The danger of a satellite returning to Earth carrying a nuclear power supply was highlighted in 1978 when a nuclear-powered Soviet satellite re-entered and broke up over Canada. In addition, the radio isotope thermal electric generators (RTGs) carried on planetary spacecraft, launched by the Shuttle in 1989 and 1990, caused some concern about the results of a possible launch accident of the type experienced during the *Challenger* tragedy in 1986. What was forgotten was that every Apollo lunar mission from Apollo 12 carried an RTG to power the Apollo Lunar Surface Experiments Package (ALSEP) left on the Moon, and that the Apollo 13 RTG re-entered the atmosphere aboard the LM Aquarius at the end of the mission and sank into the depths of the Pacific Ocean.

The requirements needed to support human life (atmosphere, pressure, humidity, temperature, food and water, waste management) mentioned in the first section of this book all apply to exploring the far reaches of the Solar System, its planets and its moons. In addition, certain requirements for the well-being and productivity of the human crew depend on additional areas that can (and do) affect safety issues and the success of the mission.

Long-duration scientific missions during the Antarctic winter, or extended journeys in nuclear submarines, have resulted in increased levels of stress that have also been reflected in certain missions in space – some of which have been recounted in this book. Stress can produce adverse effects on human activity, and can be the result of a number of factors: forced confinement; isolation from other team members, family communications or outside contact; a lack of privacy or forced socialising with the rest of the crew; monotony and boredom; the fear of equipment failure; excesses of noise, vibration and temperature, which also leads to additional medical problems; and the fear of personal failure or inability to perform certain tasks.

**'A grand oasis in the vastness of space'**
Without doubt the most impressive images brought home by the Apollo lunar astronauts were those of blue planet Earth suspended in the black void of space.

Even though they were near another celestial body, the vision of Earth was so strong that they never felt totally isolated from home, apart from during the short time that they were beyond the far side of the Moon. Even then, the vision of Earth rising over the lunar horizon was a sight that none of them could ever forget, or totally explain.

For any mission beyond the Earth–Moon system, the focus of home will be lost in the background of stars. Even at the height of the Mir crises in 1997, or during the Apollo 13 mission of 1970, the view of Earth (when it could be seen) was a reminder to the crews of how close to home they were, even if they thought they might never return. On a flight to Mars and beyond, this image will not be available to the crew, and the realisation of their isolation from planet Earth will be one that has yet to be experienced or understood.

On a nominal mission to Mars, the separation of the crew from Earth for at least two years would be psychologically the most difficult part of the mission, and no amount of training could prepare them for it. In the event of an emergency such as the Apollo 13 or Mir systems failures, the separation from home will be a true test of character and endurance. The problem of how humans will survive a flight beyond Mars is something for future generations to address. For the current human population, the thought of regular, safe, sustained flights into space, to the Moon and out to Mars, is still some distance in the future – perhaps 50–75 years – comparable with the Wright brothers attempting to plan flight schedules for Concorde. We still have a very long way to go.

# Summary

In 2000 our immediate future in human spaceflight lies in the creation of the International Space Station, which will undoubtedly rely upon experiences gained from both the Russian Salyut and Mir space station operations and the American Skylab and Space Shuttle programmes. These have all contributed to the development of space station operations, resupply and routine access for both crews and logistic transportation.

Crew health and safety issues encountered during those programmes should not be forgotten, and should be drawn upon to help sustain the ISS during its orbital lifetime. It is important that the pressure to launch each element of the ISS during the construction time-frame does not outweigh the safety guidelines established to protect the crews. In addition, careful monitoring of the expanding station hardware, both inside and outside, will ensure that when permanent human presence is achieved then the safety and health of the resident crew-members will be secured.

The experiences gained from Salyut/Skylab/Mir and Shuttle/Spacelab, applied to the ISS, will be an important yet unseen element during the operational lifetime of the station, and one which must be built upon before we once more venture deeper into space. Only with safety at the forefront of each mission and each task will ISS succeed and support research programmes that not only look back at Earth but out into deep space. From then on, humans can once more move out to the Moon and Mars and beyond. With these advances will come the added dangers first encountered during the Apollo 8 mission in 1968.

Apollo 8 was the first human mission to leave the relative safety of Earth orbit and venture 'out there'. Along the way were several danger points that were critical to the success of the mission and the safety of the crew: the correct firing of the Saturn third stage to send the astronauts on the way to the Moon; the correct firing of the service module engine to allow capture into the orbit of the Moon; and most importantly, the firing of the SM engine to break free from orbit around the Moon, allowing the astronauts to begin their journey home. Of these key danger points, the two firings of the engine in the vicinity of the Moon were the most critical. Failure of either could have resulted in disaster and a lost crew, either in deep space or crashed on the lunar surface.

These key points were supplemented by the first lunar landing and lunar lift-off, by Apollo 11. These danger points for a crew have not been faced since the Apollo 17 mission in 1972. When we return to the Moon, these same points will remain, although an adequate rescue system will probably render them less dramatic than they were during the Apollo missions. However, when humans venture to Mars for the first time then the key danger points of distance and planetary encounter once more become major mission milestones and obstacles to be overcome.

For the next 100 years the inner Solar System in the region of the Earth, the Moon and Mars will become the main focus of human space exploration, and will possibly include flybys of Venus and missions to the asteroids. For these deep space ventures the most critical elements for crew safety and survival are reliable propulsion systems and a closed life support system that has a realistic back-up redundancy system in the event of a major systems failure. Emergency rescue from anywhere beyond the Moon is not a realistic option, and future space rescue systems will probably need to rely on system redundancy, crew escape (lifeboat) survival pods, or the capability of becoming 'space wrecked' on the surface of Mars until help arrives from Earth – after a year or so! Whatever lies in the future for human space exploration, surviving the adventure, as well as experiencing it, will be nothing like we have seen before.

# Conclusions

# The lessons learned

Between April 1961 and December 1999, almost 400 human beings ventured into orbit around our planet – some of them as many as six times. But only 24 have ventured to the Moon; and of these, a unique group of 12 became the first humans to walk on the surface of another celestial body. During this same period, four cosmonauts and seven astronauts were killed while completing their missions, and several dozen others were injured or killed in various flying or training accidents.

During the 70-year period covered by this book – including the first four decades of human spaceflight operations – danger and risk have been constant companions on every trip into the stratosphere and beyond. The fact that accidents will happen – no matter how good the preparations and protection – is a primary key in identifying what can be learned from past incidents to help prevent a reoccurrence in the future.

When there is any major disaster involving technology and the loss or injury of human life, a team of investigators painstakingly conducts interviews with survivors and witnesses, closely examines each salvaged item of hardware, and analyses every minute sound from voice recorders and fragments of electronic data. At the end of the investigation a report is produced, and the conclusions are made available to the industry to allow the recommendations for improvements and preventive measures to be incorporated into future vehicles to help prevent the same accident happening again.

In the field of aviation, military and civilian bodies have the opportunity to study accident and crash reports to help upgrade their own fleet of aircraft and to improve training methods for their personnel. When an accident occurs during a spaceflight, post-flight analysis is not so easily accomplished.

Unless the accident happens on the ground during training (such as Apollo 1) or within the atmosphere (*Challenger*, Soyuz 11) the ability to study damaged hardware is restricted. Even recovery of items from a mission can sometimes be impossible due to the re-entry burn-up of hardware (such as the Gemini 8 adapter module, the Apollo 13 Service Module, and the Progress M-34 freighter), explosive destruction (*Challenger*) or impact velocity (Soyuz 1).

The progress of each mission into space is recorded by telemetry on the ground or onboard the spacecraft, to be dumped or retrieved at a later date. From the very first missions beyond the atmosphere, photographic documentation played a key role in human space exploration. Cameras were invaluable not only recording the activities of the crew and the spectacular views they witnessed, but also in providing visual evidence of faulty or damaged equipment for analysis by teams on Earth who had no opportunity to examine the actual hardware.

It was the analysis of images from ground-based cameras placed a variety of vantage points at the Kennedy Space Center that captured the unfolding sequence of the last, short flight of *Challenger* in 1986. It was also the quick snaps taken by the crew of Apollo 13 that gave fleeting impressions of the scale of the explosion that ripped out the side of the Service Module. With these data from the flight vehicles, crew-members and ground resources, a reasonable picture can be built up of the sequence of events leading to the accident. Evidence from ground simulations, mock-up hardware, or more recently, computer graphics, is used to reproduce, as accurately as possible, the actual accident. These simulations and replays are used to provide supportive evidence for the theories and the conclusions, and allow recommendations to be put forward to ensure a similar accident does not (hopefully), happen again.

## THE QUEST FOR SPACE

The stratospheric balloon programmes conducted between the 1930s and 1960s provided a complementary database for the development of aeronautics research in the upper reaches of the atmosphere. At the time advances were also made in the development of aircraft pressurised crew compartments, pressure garments and high-speed flight using jet and rocket propulsion. In addition, methods of crew escape by parachute, and later with ejection seats, were being developed, and studies of the forces of rapid acceleration and deceleration on the human body were underway. It was a pioneering time that produced rapid advances in aviation history, but also some tragic accidents.

During the ascent of the Soviet balloon Osoaviakhim-1 in 1934, the crew sacrificed ballast to achieve a new record height, and in doing so sealed their fate, due to not having enough ballast to land safely. Whether this decision was that of the flight planners or the crew is not clear, but the rapid descent of the gondola increased the strain on the support cables, and they severed, causing the gondola to plummet to the ground. The buffeting prevented the crew from escaping by parachute, and they were killed. The loss of the three aeronauts prompted a redesign of the balloon canopy so that in a failure it could be reconfigured to act as a parachute to assist in a safe landing or to stabilise the vehicle, allowing the crew to escape,

During the flight of the US balloon Explorer in 1934, experience in selecting the right materials and in pre-flight preparations were encountered when launch delays of several months resulted in the fusing together of the rubberised cotton pleated folds of the balloon envelope, so that when expanded in the atmosphere, the sub-zero temperature ripped the material, causing it to rupture the balloon.

During the rocket research aircraft programmes, considerable information was gleaned from aircraft flying in excess of Mach 6 that was invaluable in the development of follow-on research vehicles and production-line military aircraft. Lessons were also learned from the use of Ulmer leather gaskets in the fuel lines of rocket engines, which caused instant explosions when in contact with the fuel supply. Two X-1 planes were lost in this way.

The loss of the X-2 aircraft, which killed pilot Mel Apt, was a further blow to NACA, who were about to take over flight testing from the next flight. Pilot Apt was making his very first flight in the X-2. He was attempting a speed record, and it was his inexperience in handling the X-2, especially at Mach 3 for the first time, that contributed to his death. The X-2 programme was a poorly organised and badly operated programme, and the accident raised questions about the reasons for placing Apt, with such limited experience, in the cockpit. The conclusion was that the Air Force was determined to attain the speed record and altitude record with its last X-2 flight, and actually accepted a calculated risk in losing the aircraft and the pilot. The attempt to achieve records and programme objectives at the cost of safety and experience cost the life of a pilot.

Eleven years later, in November 1967, Mike Adams was lost in the crash of the third X-15 aircraft. Adams had suffered from vertigo on some of this earlier flights, which was not uncommon, as other X-15 pilots had reported disorientation at the end of their powered flights to peak altitude. On one of his earlier post-flight debriefings, Adams had indicated that after encountering severe vertigo, 'I didn't know what the hell I was doing'. On his fateful last flight he apparently encountered these vertigo symptoms again, and probably thought nothing of them; but this time he was apparently disorientated for a much longer period of time.

It has been stated by former X-15 pilot Milton Thompson that Adams was possibly a victim of ignorance, and that in the 1960s the effects of g forces on the human sensory system were only just becoming known and were clearly not understood. Adams could also have not been aware of this, or even decided not to attract attention to the problem in fear of becoming grounded by flight surgeons. Astronauts Alan Shepard and Deke Slayton were grounded for several years due to medical abnormalities, and no top test pilot would wish to change from flying the top aircraft of the day to flying a desk. The investigation board concluded that although the accident was not exactly pilot error, Adams had encountered extreme vertigo during the flight, and this contributed to his disorientation and flight control actions which resulted in the mid-air break-up of the X-15. At the time, pilots were not routinely tested for susceptibility to vertigo. Further research was required in the field of biomedicine as well as engineering to fully understand the effects of high-altitude high-speed flight on both the pilot and the aircraft. Some of these studies continue to be conducted on returning Shuttle flight crews.

## TRAINING FOR SPACE

A period of training is necessary to prepare a person for a flight into space, and, as discussed earlier, this has led to several accidents and the loss of three astronauts and two cosmonauts, one of whom drowned.

The review board report on the loss of the 1967 Apollo 1 astronauts amounted to 2,375 pages, and indicated that the most probable cause had been an arc on a wire located inside the command module that led to the ignition of the 100% oxygen atmosphere. which pumped carbon monoxide through the suit circulation system. The three astronauts died by asphyxiation.

During the investigation and review it was discovered that the command module had defects in its design, workmanship and quality control. The spacecraft crew compartment included patches of flammable material, as well as equipment that was not certified or tested before installation. Safety tests had not been completed, as there was a general feeling that the launch-pad test would not be hazardous, since the launch vehicle was not fuelled; and it was also discovered that there was a lack of fire-fighting equipment in the White Room.

The review also highlighted the use of flammable material in the capsule, even in a volatile 100% oxygen atmosphere, because it was assumed that there would be no ignition source to trigger a fire. During the ground test the cabin pressure test to be conducted on the pad was at 5.5 psi – as it would if the vehicle had been in orbit. However Apollo 1 was only a couple of hundred feet off the ground and as a result of the outside atmospheric pressure of 14.7 psi the crew had to raise their internal pressure level by 5.5 psi to 20.2 psi. This was a potentially dangerous situation, and required only a small spark to trigger a firebomb. That spark, it was suggested, came from the electrical arc in the wiring below Grissom.

The inquiry also revealed poor work at the prime contractor North American, and inadequate checks from NASA. Nothing was hidden during the investigation, and serious mistakes on both sides were admitted and rectified. The fact that by putting a 100% oxygen atmosphere into a spacecraft pressurised to 20 psi increased fire hazards was completely overlooked. The escape of the crew had been foiled by the design of the plug-type spacecraft door, designed to seal it firmly and to be opened from the outside, and no EVA was planned on the Block 1 spacecraft, so there was no need to open the door in orbit. The designers failed to realise that a crew might need to hurriedly escape from the spacecraft while on the ground.

Improved communications and levels of systems safety and testing were all incorporated after the fire, management levels were reorganised, and flammable material was reduced to a minimum inside the spacecraft. More than 30 years after the loss of the three astronauts, the events of 27 January 1967 are still a painful memory to those in America's space programme.

It seems ironic that in 1961 a young cosmonaut trainee died in a flash fire in an isolation chamber just three weeks before Gagarin flew in orbit and more than five years before the loss of the three Apollo 1 astronauts. The incident had been concealed within the Soviet system for more than 25 years before being released in 1986. Had the incident been known in the West at the time, this might have led to a change in procedures for Apollo – although perhaps not, as pressure chamber tests had been conducted in the US for years, and the dangers of 100% oxygen in higher psi were known. What *was* overlooked was the availability of an obvious ignition source. It was poor workmanship and poor quality control that resulted in the Apollo 1 fire. Had there been no arc, the fire would not have occurred, but such was the state of workmanship on the Block 1 vehicle that sooner or later an accident would have occurred.

NASA learned much about communications and systems management as a result of the Apollo fire, but they were lessons that would soon be forgotten in the development of a new type of launch vehicle – the Space Shuttle.

## LAUNCH TO SPACE

The whole mission cycle from preparation of hardware, training of crew and ground staff, to the launch, operating the vehicle during its flight, and recovering its data and hardware, is costly in time, resources and cash. The strong desire to achieve the mission objectives and thereby qualify the time and effort spent on a spaceflight can influence the decision to undertake a mission when the indications are that it would be wise to wait.

Launch falls into the one of the riskiest categories of spaceflight, and the chance of something going wrong during the ascent has been demonstrated during Apollo, Soyuz and Shuttle launches.

The loss of *Challenger* in 1986 revealed a NASA which was different from the organisation that had recovered from the Apollo 1 fire of nineteen years earlier. The agency had ridden high on the success of Apollo, the recovery of Apollo 13, and the continued success of planetary probes, but it had fallen foul of congressional restrictions on the budget. At the time of its creation, the Shuttle was seen as the do-all vehicle which could finally provide reliable and affordable routine access to space for all customers.

The origins of the Shuttle were only a part of a larger and complex space infrastructure, and not a sole element launch system. After annual budget restrictions in the 1970s, continued promotion and marketing of the Shuttle continued to create the illusion that the vehicle was supposed to assume almost all of the American launch vehicle requirements in the future. Had the programme been described as an 'orbital research vehicle' that could offer only limited access to microgravity environment to support and complement expendable launch vehicles, it would probably never have received funds to be built.

By the fifth mission, the system was declared operational, and commercial customers were accepted to fly on future missions – even providing their own representatives to fly missions as part of the launch package deal. By mission 25, a teacher was part of the flight crew of the tenth *Challenger* mission. After a flight of 73 seconds, the Shuttle, the teacher and the concept of a commercial workhorse of the whole US space programme had been lost.

Unlike Apollo 1, in which NASA conducted its own inquiry, an independent Presidential commission was formed to look into the loss of *Challenger* and its crew of seven. What they found was again not pleasant to those in the space agency. The technical failure of the vehicle was soon discovered, and the highlighted cause was that NASA itself was at fault. NASA was trying to turn itself from a research and development agency – a role in which it was extremely competent – into an international competitive launch service business – in which it was clearly *not* competent.

Again, similar to the Apollo 1 investigation, the Shuttle accident investigation found that structures of management, lines of communications, quality control and the pressures to launch to maintain a flight schedule to create a revenue, had all contributed to the loss of *Challenger*.

It was also revealed that the structural weaknesses of the selection of the solid rocket boosters, which had been selected in preference to liquid propellant boosters earlier in the development of the Shuttle during the early 1970s, had led to the loss of *Challenger*. There was a lack of understanding of how the field joints that connected the segments of the SRBs performed under conditions of flight and under extremes of weather and temperature. To the public, the joint seals were invisible during launch, and it seemed incredible that apparent spaceflight experts had allowed a launch if something was wrong. What was not understood was that neither NASA not Morton Thiokol, the contractors for the SRBs, had fully understood the effects of launching in conditions of extreme cold, as experienced on that fateful January day. Even after following the guidelines and flight rules, they had nevertheless made a mistake.

It was found that the temperature specifications had been misinterpreted by engineers at Morton Thiokol and NASA, many years before the accident. Communications between the agency and the contractors were mostly by teleconference at different centres across America. Even basic launch requirements were not understood, and engineers at Morton Thiokol did not generally know flight rules. On the other hand, NASA launch engineers at the Cape were not aware that the SRBs were not tested down to the 31° F lower limit.

During the 1980s NASA became obfuscated with red tape and paper chasing, and employees were moved away from hands-on engineering to offices and paper shuffling. Had face-to-face meetings taken place, an exchange of information and data may have indicated more clearly to both sides that the SRBs were an accident waiting to happen.

In 1969 Apollo 12 had been launched within mission rules – just – and after it had been struck by lightning, NASA decided never again to launch in adverse weather. Flight rules were consequently amended for the Shuttle. As it was a winged vehicle capable of returning to a nearby landing site, certain constraints on to launch – such as weather fronts and wind levels – played a crucial part in deciding if it was safe, and also allowed for a safe landing if an abort was called. The levels of temperature and wind shear on the vehicle on the pad were overlooked when these decisions were made, because the information on the consequences of launching with these parameters was not made clear to NASA; and equally, the fact that temperatures on the pad can drop sharply to form ice on the launch structure and vehicle was not generally known at Morton Thiokol.

Following the *Challenger* accident, weather protection devices were fitted around the launch service structure on both Pads 39A and 39B, and former astronauts moved into positions of safety and programme management and became more involved in launch decision-making processes.

Since 1988, during the first two minutes of every Shuttle flight, *Challenger* is remembered, and launches are not planned to occur on the anniversary of the tragedy.

There have been other launch incidents during the countdown of Shuttle missions, but safety procedures have either halted or cancelled the launch process. STS 51-F successfully demonstrated an aspect of an abort mode when one engine failed during ascent in 1985, but during the 1999 launch of *Columbia*, when a drop in electric current and a leak of hydrogen almost aborted the ascent, it was found that repeated ground maintenance access across the wiring had worn the protective covering. This was apparently missed by pre-flight inspections. As a result, the Shuttle was grounded for six months while repairs were instigated and adequate procedures were implemented to prevent reoccurrence.

## SURVIVAL IN SPACE

Between 1986 and 1999 the Mir space station programme produced a wealth of data on the ability of the human body to live and work for prolonged periods (up to 14 months) in space, and on sustained orbital flight operations and logistics support of a permanently crewed operational space platform (almost ten years). In 1997 Mir also provided valuable – if unwelcome – experience of difficulties in space that could last weeks or months.

The fire on Mir was a surprise, and its effects were alarming. The procedures to control the fire, by the cosmonauts on board, worked well, although with two Soyuz spacecraft docked it was revealed that only one set of re-entry instructions for engine burn-time was available, and a second had to be printed. It was also afterwards realised that both vehicles used the same print-out, and could have entered the atmosphere and landing side by side. A possible contingency procedure for the Soyuz craft had been overlooked.

The collision of the Progress was due to a combination of the inadequate training of Tsibliyev, who was to perform the operation in space several months after training for the operation on the ground. Pressure to achieve the docking was also due in part to the cosmonauts' flight bonuses – awarded after achieving flight objectives and mission goals – inadequate visual references during the approach and Tsibliyev's professional pride as Commander of the mission. The cosmonaut was already fatigued when the operation was attempted, and consequently his reactions were affected.

What was quickly learned following the collision was that close co-operation between the men aboard Mir and the team on the ground was vital in saving the station. Mike Foale's adaptability and determination to be involved in not only the day to day running of the station but in crucial mission activities demonstrated that, on an international mission, clear understanding of language and customs is essential for crew harmony and safety, as well as productivity.

The Mir accidents were unique in the history of human spaceflight because of the duration of the continued levels of stress under which the crews were placed. A crew escape could have been achieved any time, due to a Soyuz being attached, but maintaining a critical power supply and orientation control of the station to allow a safe undocking and separation added to the stress of the crew.

Once again the lines of communications and an understanding of what was happening were tested during the incident. It also demonstrated that adequate preparation for a flight is essential in adapting to any situation encountered.

The unexpected situation on Mir was in stark contrast to that experienced on Apollo 13, in which there was no hope of immediate return to Earth, and the possibility that the crew might be lost. Apollo 13 was a demonstration of resourcefulness both in space and on the ground. What was also learned was that locating vital equipment all in one area of the spacecraft was open to accident. Therefore, following the flight of Apollo 13, a third oxygen tank was installed in an area of the SM separate from the other two, and with upgraded thermostat switches it was hoped that the Apollo 13 incident would never be repeated. However, the fact that it had happened sent shock waves though NASA and Congress. Two more Apollo flights were scrapped, and Apollo would fly only four more lunar missions with hardware already built and paid for. It was learned from Apollo 13 that fate should not be tempted too many times. There was no adequate rescue system in place to recover astronauts stranded in orbit around the Moon or on the lunar surface, but given the chance, and with luck, NASA proved that it could be done.

Apollo 13 also demonstrated that it was necessary to design life-supporting equipment that could fly in more than one vehicle. The lithium hydroxide canisters on the Command Module were of a shape different from those on the Lunar Module, and could not fit into the receptacle in the other vehicle. Designing equipment that can fit other systems ensures redundancy and added safety. The development of a common docking system, beginning with the Apollo–Soyuz Test Project, also adds to the capability of a crew rescue system – if, of course, there are vehicles to fly the rescue mission, and the time to accomplish it.

The Gemini 8 incident was a clear demonstration of mission training and reaction that saved the lives of the two men. Throughout human space exploration, incidents encountered in space have repeatedly challenged flight crews and ground controllers to overcome the hurdle and press on with the mission.

Of almost 100 Shuttle missions, only three have been terminated early due to faulty equipment, and while eight Soyuz missions have failed to achieve docking when planned, the Soviets/Russians have demonstrated a successful series of automated and manual docking manoeuvres over a period of 30 years.

The results of too much pressure on the workload of a crew in orbit was demonstrated early in 1973 during the mission of the third Skylab crew, but what was not learned was that experience from flying round-the-clock scientific Spacelab missions was nothing in comparison to flying long-duration space station mission on Mir. All of the American astronauts sent to Moscow to train for the Mir resident crew programme found it hard going. The training was not the major hurdle; it was the change in culture, attitude and language that were the hardest to overcome. Of the seven, perhaps Foale demonstrated the most adaptability to a new way of conducting a spaceflight.

The contrast between the American and Russian training systems and in-flight procedures was one of the most important lessons learned from Shuttle–Mir. Since the ISS is to be an international station, the application of that experience has a near-term benefit. Closer co-operation between the Russian and America space agencies during the Mir programme also helped identify areas that needed further work – for example, the provision of information on flight systems and procedures, a sharing of work tasks, further language and cultural training.

## RETURN FROM SPACE

Experience derived from the lifting body and X-plane programmes provided valuable information on the return of a winged vehicle from space, and the Shuttle landing sequence has been seen as one of the most successful and trouble-free aspects of the programme.

The parachute recoveries of the early American and all the Soviet/Russian missions have resulted in only two fatalities in almost forty years of operational flying in space. Soyuz 1 was a clear demonstration of a mission launched before the hardware was ready, and as a result, the problems encountered by cosmonaut Vladimir Komarov in orbit, continued to be compounded through to landing. In addition, the packing of the recovery parachute had been cited as a contributing factor to the loss of the vehicle and the cosmonaut. Since 1967 the parachute recovery system of Soyuz has worked remarkably well in more than 80 normal and two aborted mission profiles – obviously a lesson learned the hard way after Soyuz 1.

Even during the return of Soyuz 11 in 1971, the automated parachute deployment and recovery system worked correctly. It was the faulty pressure equalisation valve between the Orbital Module and the Descent Module that caused the deaths of the three cosmonauts during their return. The lesson learned from this tragedy led to the installation of improved life support systems and the cosmonauts wearing full pressure suits for launch and re-entry, which have been modified over the years in line with the upgrading of the Soyuz spacecraft.

As the training for a flight into space is an integral part of any mission, so the readaptation after the return is just as important to ensure crew health is returned to pre-flight levels. Readapting back to live on the planet can be as disorientating an adapting to live away from it. Russian long-duration missions take their toll on cosmonauts for a short while, but there are no indications of permanent damage from spaceflight. Pilots have demonstrated that the skill of landing the Shuttle after a couple of weeks in space is not impaired by microgravity, and therefore manual emergency landings by ISS crew rescue vehicles is not considered to be a major concern.

## THE FUTURE IN SPACE

As early as the second Shuttle–ISS mission, lessons were being learned about the maintenance of the health and well-being of the crew. The STS-96 crew had encountered poor air quality in the Russian Zarya module, which the Americans did not expect and which was not fully explained by the Russians. Once again, an understanding of associated systems and a line of adequate communication were not forthcoming. These have to be established to enable adequate maintenance of crew health and safety.

As the size of the ISS grows, so the lessons learned from past programmes should feature more prominently. Future space systems involving humans will have to take into account past experiences in materials selection (Ulmer leather gaskets on X-planes, flammable materials on Apollo 1); equipment design and location (oxygen tanks on Apollo 13, pressure valves on Soyuz 11, the fire on Mir, the valves near RCS fuel supplies on Apollo 15 and ASTP); system redundancy (Gemini 8 thruster malfunctions, Soyuz docking failures, recovery parachutes on Soyuz 1); crew training and compatibility (long-duration crew training and international training issues, Mir docking errors); crew safety (Gemini 8, Soyuz 1, Apollo 1, Apollo 13, Soyuz 11, STS 51-F, STS 51-L, Mir) interagency communications (Apollo 1, STS 51-L, Shuttle–Mir); and the knowledge that if something *can* go wrong, it *will* go wrong. And it would perhaps be beneficial to occasionally muse on some words of the House Committee on Science and Technology in its report on the *Challenger* accident:

'Though we grieve at the loss of the *Challenger* crew, they would not want us to stop reaching into the unknown. Instead they would want us to learn from our mistakes, correct any problems that have been identified and then once again reach out to expand the boundaries of our experience in living and working in outer space'.

# Appendix 1

## MILESTONES IN THE HISTORY OF HUMAN SPACEFLIGHT

| | | | |
|---|---|---|---|
| 1961 | Apr 12 | Vostok 1 | Gagarin becomes first human to orbit Earth; one orbit |
| | May 5 | Mercury 3 | Shepard becomes first American in space |
| | Aug 6–7 | Vostok 2 | Titov spends 24 hours in space |
| 1962 | Feb 20 | Mercury 6 | Glenn becomes first American to orbit the Earth |
| | Aug 11–15 | Vostok 3 | Nikolayev completes four days in orbit; joint flight with Vostok 4 |
| 1963 | May 15–16 | Mercury 9 | Cooper is the first American to spend a day in space |
| | Jun 14–19 | Vostok 5 | Bykovsky sets world solo spaceflight endurance record of five days |
| | Jun 16–19 | Vostok 6 | First woman in space (Tereshkova); joint flight with Vostok 5 |
| 1964 | Oct 12–13 | Voskhod 1 | First three-person launch (Komarov, Feoktistov, Yegerov) |
| 1965 | Mar 18–19 | Voskhod 2 | First EVA accomplished by Leonov |
| | Jun 3–7 | Gemini 4 | First US EVA and US endurance record (four days) |
| | Aug 21–29 | Gemini 5 | World endurance record, eight days |
| | Dec 4–18 | Gemini 7 | Borman and Lovell accomplish 14-day space marathon |
| | Dec 15–16 | Gemini 6 | First space rendezvous within a few feet of a second manned spacecraft (Gemini 7) |
| 1966 | Mar 16 | Gemini 8 | First space docking; first aborted mission |
| | Jul 18–21 | Gemini 10 | Altitude record of 468 miles |
| | Sep 12–15 | Gemini 11 | New altitude record of 850 miles |
| 1967 | Apr 23–24 | Soyuz 1 | First inflight fatality (Komarov) |
| 1968 | Dec 21–27 | Apollo 8 | First human flight to the Moon; ten orbits by Borman, Lovell and Anders |

| 1969 | Jan 14–18 | Soyuz 4/5 | First docking of two human crewed spacecraft; EVA transfer |
| | Jul 16–24 | Apollo 11 | First manned lunar landing mission (Jul 20) |
| | Nov 14–24 | Apollo 12 | Second manned lunar landing mission (Nov 19) |
| 1970 | Apr 11–17 | Apollo 13 | Third manned lunar landing mission aborted; explosion in SM; crew recovered |
| | Jun 1–19 | Soyuz 9 | Endurance record, 18 days |
| | Jan 31–1971 Feb 9 | Apollo 14 | Third manned lunar landing mission (Feb 5) |
| 1971 | Apr 23–25 | Soyuz 10 | First space station mission; failed to enter Salyut |
| | Jun 6–30 | Soyuz 11 | First space station crew; died during recovery; endurance record, 24 days |
| | Jul 26–Aug 7 | Apollo 15 | Fourth manned lunar landing mission (Jul 30) |
| 1972 | Apr 16–27 | Apollo 16 | Fifth manned lunar landing mission (Apr 21) |
| | Dec 6–19 | Apollo 17 | Sixth and final Apollo manned lunar landing mission |
| 1973 | May 25–Jun 22 | Skylab 2 | First US space station crew; endurance record, 28 days |
| | Jul 28–Sep 25 | Skylab 3 | Endurance record, 59 days |
| | Nov 15–1974 Feb 8 | Skylab 4 | Endurance record, 84 days |
| 1975 | Apr 5 | Soyuz 18-1 | First launch abort |
| | May 24–Jul 26 | Soyuz 18 | Endurance record, 63 days |
| | Jul 15–24 | ASTP/Soyuz 19 | First international space docking mission |
| 1977 | Dec 10–1978 Mar 16 | Salyut 6 | Endurance record, 96 days |
| 1978 | Jun 15–Nov 2 | Salyut 6 | Endurance record, 140 days |
| 1979 | Feb 25–Aug 19 | Salyut 6 | Endurance record, 175 days |
| 1980 | Apr 8–Oct 11 | Salyut 6 | Endurance record, 185 days |
| 1981 | Apr 12–14 | STS-1 | First orbital flight of Shuttle *Columbia* |
| 1982 | May 13–Dec 10 | Salyut 7 | Endurance record, 211 days |
| 1983 | Apr 4–9 | STS-6 | First flight of *Challenger*; first Shuttle EVA |
| | Jun 18–24 | STS-7 | First US woman in space (Ride) |
| | Sep 26 | Soyuz T-10A | First pad abort |
| | Nov 28–Dec 8 | STS-9 | First shuttle flight to exceed ten days |
| 1984 | Feb 3–11 | STS 41-B | First untethered EVAs using MMUs |
| | Feb 8–Oct 2 | Salyut 7 | Endurance flight, 237 days |
| | Jul 17–29 | Salyut 7 | First female EVA (Savitskaya, Jul 25) |
| | Aug 30–Sep 5 | STS 41-D | First flight of *Discovery* |
| | Oct 5–13 | STS 41-G | First US female EVA (Sullivan, Oct 11) |
| 1985 | Oct 3–7 | STS 51-J | First flight of *Atlantis* |
| 1986 | Jan 28 | STS 51-L | *Challenger* explodes 73 seconds after launch, killing crew of seven |
| 1987 | Feb 6–Dec 29 | Mir | Romanenko sets new endurance record of 326 days |
| | Dec 21–1988 Dec 21 | Mir | Titov and Manarov complete one year (366 days) in space |
| 1988 | Sep 29–Oct 3 | STS-26 | Return to flight mission after *Challenger* accident |

| | | | |
|---|---|---|---|
| 1990 | Jan 9–20 | STS-32 | Longest Shuttle flight to date, 11 days |
| 1992 | May 7–16 | STS-49 | First flight of *Endeavour*; replacement for *Challenger* |
| | Jun 25–Jul 9 | STS-50 | Longest Shuttle flight to date, 14 days; finally surpasses Gemini 7 record; fourth longest US flight |
| 1993 | Oct 18–Nov 1 | STS-58 | New Shuttle endurance record, 14 days 8 hrs |
| 1994 | Jan 8–1995 Mar 22 | Mir | Polyakov sets all-time single mission endurance record of 438 days 17 hrs 58 min |
| | Jul 8–23 | STS-65 | New Shuttle endurance record, 14 days 17 hrs |
| 1994 | Oct 4–1995 Mar 22 | Mir | Kondakova sets female endurance record of 169 days |
| 1995 | Feb 2–11 | STS-63 | First Shuttle–Mir rendezvous mission; E. Collins becomes first female Shuttle pilot |
| | Mar 2–18 | STS-67 | New Shuttle endurance record of 16 days 15 hrs |
| | Mar 14–Jul 7 | Mir | Norm Thagard set new US endurance record of 122 days; first US astronaut launched by Soyuz rocket |
| | Jun 27–Jul 7 | STS-71 | First Shuttle–Mir docking mission (return Thagard) |
| | Nov 12–20 | STS-74 | Second Shuttle–Mir docking mission |
| 1996 | Mar 22–31 | STS-76 | Third Shuttle–Mir docking (deliver Lucid) |
| | Mar 22–Sep 26 | Mir | Lucid sets a female endurance record of 188 days |
| | Jun 20–Jul 7 | STS-78 | Shuttle endurance record of 16 days 21 hrs |
| | Sep 16–26 | STS-79 | Fourth Shuttle–Mir docking mission (deliver Blaha, return Lucid) |
| | Sep 16–1997 Jan 22 | Mir | Blaha logs 128 days on Mir |
| | Nov 19–Dec 7 | STS-80 | Shuttle endurance record of 17 days 15 hrs; Musgrave oldest man in space at 60 years (sixth mission) |
| 1997 | Jan 12–22 | STS-81 | Fifth Shuttle–Mir docking mission. (Linenger up, Blaha down) |
| | Jan 12–May 24 | Mir | Linenger logs 132 days on Mir and endures a fire onboard |
| | May 15–24 | STS-84 | Sixth Shuttle–Mir docking mission (Foale up, Linenger down) |
| | May 15–Oct 6 | Mir | Foale logs 144 days on Mir; endures a collision and several power failures |
| | Sep 25–Oct 6 | STS-86 | Seventh Shuttle–Mir docking mission (Wolf up, Foale down) |
| | Sep 25–1998 Jan 31 | Mir | Wolf spends 127 days on Mir |
| 1998 | Jan 22–31 | STS-89 | Eighth Shuttle–Mir docking mission (Thomas up, Wolf down) |
| | Jan 22–Jun 12 | Mir | Thomas spends 140 days on Mir |
| | Jun 2–12 | STS-91 | Ninth and final Shuttle–Mir docking mission (Thomas down) |

| | | | |
|---|---|---|---|
| | Oct 28–Nov 7 | STS-95 | John Glenn returns to space aged 77, as PS |
| | Dec 4–15 | STS-88 | First manned flight of ISS programme |
| 1999 | Jul 23–27 | STS-93 | Eileen Collins becomes first female Shuttle commander |
| | Dec 19–27 | STS-103 | Third Hubble Space Telescope mission – 'the last human spaceflight of the twentieth century' |

# Appendix 2

## SELECTED CHRONOLOGY OF A DIFFICULT ROAD TO SPACE

| | |
|---|---|
| 1934 Jan 30 | After achieving a world record altitude of 22,000 metres in the stratospheric balloon Osoaviakhim-1, three Soviet aeronauts (Fedeseenko, Vasenko and Usykin) perish when the envelope rips and they are unable to control their rapid descent, hitting the ground near Potish-Ostrog, Insar Raion, USSR. |
| 1934 Jul 28 | US National Geographic Society stratospheric balloon Explorer develops a rip on the balloon fabric at 60,000 feet and falls to Earth. The three American aeronauts – Anderson, Stevens and Kepner – manage to escape by parachute seconds before the gondola smashes onto the ground in Nebraska. |
| 1935 Jun 26 | Crew of USSR-1 Bis stratospheric balloon escape major injury when balloon rips. Aeronauts Prilutski and Varigo bail out, enabling Christopzille to control the lighter balloon to a safe descent. |
| 1936 Oct 12 | Aeronauts Fomin, Krikun and Volkov parachute to safety when the canopy of their stratospheric balloon SP-2 Komsomol (VR 60) catches fire by a spark igniting the remaining hydrogen during descent. |
| 1948 May 3 | D-558-1-2 Skystreak aircraft crashes after take-off, due to compressor disintegration. Pilot Howard Lily is killed. |
| 1948 Jun 3 | X-1-1 pilot Lundquist on the 35th flight of the aircraft encounters the left main gear door opening in flight and a collapsed nosewheel on landing |
| 1949 Mar 16 | X-1-1 pilot Boyd encounters an in-flight engine fire and shut-down on flight 41. |
| 1949 May 5 | An engine explosion on board X-1-1 necessitates an emergency landing of the rocket aircraft by pilot Everest during the 113th free flight. |
| 1949 Aug 25 | Loss of pressurisation during the altitude attempt flight of X-1-1 by Everest forces the pilot to use a partial pressure suit for emergency in flight for first time. The pilot survives. |
| 1951 Aug 22 | Launch aborted, but X-1D suffers a low-order explosion during pressurisation for fuel jettison. Pilot Everest moves the B-50 bomb-bay before the unmanned X-1D plane is jettisoned, and it afterwards explodes upon impact with the desert. No injuries. |
| 1951 Nov 9 | X-1-3 destroyed following a planned captive flight for a propellant jettison test. The aircraft and the B-50 launch plane are destroyed in an explosion and fire on the ground. No fatalities. |

| | |
|---|---|
| 1953 Dec 12 | Yeager sets a new speed record of Mach 2.44 (1,650 mph) flying the [?] aircraft, but encounters inertia coupling phenomena and spins out of control. He successfully recovers at 25,000 feet |
| 1954 Oct 27 | X-3 pilot Joseph Walker experiences violent coupling motions on flight 10 during abrupt rolls at Mach 0.92 and 1.05. He survives the load design envelope of the aircraft about all three axes in only 1 second |
| 1955 Aug 8 | X-1A suffers a low-order explosion – later traced to Ulmer leather gaskets – shortly before launch from a B-29. Pilot Walker exits into the B-29 bomb-bay. The extent of the damage forces the jettison of the unmanned X-1A into desert, where it explodes and burns on impact. |
| 1955 Sep 16 | D-558-2-2 Skyrocket pilot McKay uses the emergency hydraulic system to lower the landing gear. |
| 1956 Mar 22 | D-558-2-2 Skyrocket aircraft, with pilot McKay, is jettisoned in flight due to an emergency (runaway prop) on the B-29 launch plane. McKay jettisons fuel and lands safely. |
| 1956 Sep 27 | X-2 pilot Apt sets a new unofficial speed record of Mach 3.196, but encounters inertia coupling and dies in the subsequent crash. |
| 1958 Jul 26 | America's top test pilot Iven Kincheloe is killed in an attempt to eject from a flame-out of his F-104 jet during take-off. A former X-2 pilot, he was assigned as an X-15 pilot and tipped as a future astronaut. |
| 1959 Nov 5 | X-15-2 breaks its back in a structural failure on landing with a heavier than normal fuel load during an aborted powered flight. The fuel does not explode, and Crossfield survives. |
| 1960 Jun 8 | An explosion in the engine compartment of X-15-3 severely damages the aircraft in a static ground test firing of the new engine, the XLR-99. Crossfield survives a massive explosion and fireball. |
| 1961 Mar 23 | Cosmonaut Bondarenko dies from his injuries sustained in a flash fire in an altitude chamber test, three weeks before Gagarin flies in space. The first person selected for space training to die in training. Identity revealed in 1977, but cause of death not released until 1986. |
| 1961 May 4 | Strato-Lab V stratospheric balloon crew member Victor Prather is drowned after landing in the Gulf of Mexico after completing a 113,740 feet ascent with Malcolm Ross. Prather slipped from the helicopter rescue sling and fell into the water, which flooded his pressure suit through an open helmet. |
| 1961 Jul 21 | Grissom nearly drowns when the hatch of his Mercury 4 capsule 'suddenly blows' whilst he is awaiting recovery, and he has to exit the sinking spacecraft. |
| 1962 Feb 20 | Fears are raised as telemetry indicates that the heat shield of Glenn's Mercury 6 capsule Friendship 7 is loose. The information proves incorrect, and the astronaut is recovered without incident. |
| 1962 May 24 | Carpenter's excessive use of thruster fuel forces an overshoot of the planned landing point for Mercury 7. That are fears that he burnt up as he is found floating in his life raft next to the capsule, 250 miles past his intended landing point. |
| 1962 Nov 1 | High-altitude parachutist Dolgov is killed during a planned jump from the stratospheric balloon Volga in tests of pressure suits and altitude parachute jumps. Andreyev, who jumped before him, survives. Dolgov was instrumental in the altitude flight test development of the Vostok pressure garment. |

| | |
|---|---|
| 1962 Nov 9 | X-15-2 crashes on landing when the main landing gear collapses, flipping the vehicle onto its back. Pilot McKay suffers three crushed vertebrae, and lung damage due to inhaling leaking ammonia gas. X-15-2 is rebuilt as X-15-2A, and McKay resumes flying in less than six months, but the injuries force his early retirement from NASA and contribute to his death in 1975. |
| 1963 Aug | USAF ARPS School pilots Dave Scott and Mike Adams – both of them candidates for NASA Group 3 – escape death in the crash of their F-104. Scott decides to stay, while the crafty Adams elects to eject. Had either opted for the other choice, Scott would have been killed using a subsequently evaluated faulty ejection system, and Adams would had been killed by the aircraft engine ploughing into his cockpit area. |
| 1964 Oct 31 | Astronaut Ted Freeman is killed in the crash of his T-38 when a flock of geese strike the jet during landing approach at Ellington Air Force Base, near MSC Houston. The first NASA astronaut to lose his life in training. |
| 1965 Mar 18 | Voskhod 2 pilot Leonov has to manually partially deflate his EVA suit at the end of the historic first EVA to allow him to re-enter the airlock. The next day the cosmonauts are forced to use their back-up retro-rocket system for landing, and then spend a night in the forest, awaiting recovery. |
| 1965 Dec 12 | Gemini 6 astronauts elect to stay with the vehicle when the engines ignite and then shut down. The vehicle does not explode. |
| 1966 Feb 28 | Gemini 9 astronauts See and Bassett are killed when their T-38 jet strikes the roof of the McDonnell factory in St Louis – which houses the Gemini 9 spacecraft they were to fly – during a landing attempt in fog. The back-up crew of Stafford and Cernan, in a second T-38, land safely and eventually fly the mission |
| 1966 Mar 16 | Astronauts Armstrong and Scott survive violent tumbling of the Gemini 8 spacecraft docked to the Agena target, due to a faulty thruster on their spacecraft. They regain control and affect an immediate emergency recovery. This is the first spaceflight mission to be aborted. |
| 1966 May 1 | Balloonist/parachutist Nick Piantanida suffers a pressure suit failure at 57,600 feet during the ascent of Strato Jump III. He is recovered, but remains in a coma without regaining consciousness, and dies on 29 August 1966. |
| 1966 Jun 8 | Former X-15 pilot Joe Walker is killed in a mid-air collision of his F-104 and the XB-70 aircraft. |
| 1967 Jan 27 | Three Apollo 1 astronauts – Grissom, White and Chaffee – are killed in a flash fire in the Apollo 204 capsule at Pad 34 Cape Kennedy. |
| 1967 Apr 24 | Cosmonaut Komarov wrestles with the controls of Soyuz 1 in a trouble-plagued maiden flight for 24 hours. He is killed in the landing phase when recovery parachutes become entangled. |
| 1967 May 10 | Lifting body pilot Bruce Peterson suffers serious injuries in the landing of the M2-F2 lifting body at Edwards, as it somersaults down the landing strip. He survives, and after major surgery and two years in hospital he returns to flying. A film of the accident is used in the opening credits of the TV series *The Six Million Dollar Man* in the mid-1970s. |
| 1967 Oct 5 | Astronaut 'CC' Williams, in training as LMP on Conrad's crew (who eventually flew Apollo 12) dies in jet crash of his T-38 near Talahassee, Florida. He is replaced on the crew by Al Bean. |
| 1967 Nov 15 | X-15 pilot Mike Adams is killed in the mid-air break-up of the X-15-3, |

| | |
|---|---|
| | following attainment of the 50-mile altitude and qualification for his USAF astronaut pilot wings. |
| 1967 Dec 8 | Manned Orbiting Laboratory astronaut and America's first African-American astronaut Robert Lawrence dies in the crash of his F-104 at Edwards Air Force Base, California. |
| 1968 Mar 27 | Yuri Gagarin, the world's first space pioneer, dies in the crash of his MiG-15 during a training flight. He was in training for a second space flight. |
| 1968 May 6 | Neil Armstrong ejects from the Lunar Landing Training Vehicle (LLTV-A1) during LM training at Ellington Field, and lands safely by parachute. |
| 1969 Jan 18 | Cosmonaut Boris Volynov nearly dies when modules of the Soyuz 5 spacecraft fail to separate correctly, forcing a nose-first entry attitude. He only just manages to correct orientation as modules break free before entering the Earth's atmosphere. |
| 1969 May 22 | Flying at 3,700 mph, nine miles above the Moon, Apollo 10 LM Snoopy gives pilots Stafford and Cernan the shock of their lives as the ascent stage performs wild gyrations due to confused computer switch selection. The tumbling clears as quickly as it appeared, allowing the two astronauts to redock with the command module. |
| 1969 Nov 14 | Apollo 12 is struck by lightning during launch from KSC. Despite the loss of the guidance platform for a short while, the vehicle reaches orbit and performs a successful mission. |
| 1970 Apr 13 | Apollo 13, after losing the centre engine of the S-ll launch vehicle at launch on 11 April, suffers a major in-flight explosion when No. 2 oxygen tank is ruptured *en route* to the Moon. The crew uses the still attached LM as a lifeboat for course corrections, and, with round-the-clock help from MCC Houston and support contractors, are safely recovered. The drama of Apollo 13 becomes a major feature film in 1995. Due to suspected contact with German measles, CMP Mattingly was replaced by Swigert. Mattingly never contracted the measles. |
| 1971 Jan 23 | Astronaut Gene Cernan survives a helicopter crash into the Banana River at Cape Kennedy. |
| 1971 Apr 25 | Soyuz 10 crew land safely after an early return due to a faulty docking hatch, landing only 50 metres from a lake. |
| 1971 Jun 29 | After spending three weeks in Salyut 1 space station, Soyuz 11 cosmonauts die as a result of the loss of cabin pressure during automatic re-entry. They were not wearing pressure garments. |
| 1971 Aug 7 | Apollo 15 splashes down safely at the end of the fourth manned lunar landing mission, despite only two parachutes (of three) being correctly deployed. |
| 1972 Apr 21 | Apollo 16 SPS engine fails to ignite on the first attempt to circularise the orbit following separation of LM for landing. The second attempt works, and landing is achieved. The SPS performs normally for return to Earth. |
| 1973 Jun 7 | Astronaut Conrad, on a tethered EVA, is catapulted into space as he and Kerwin struggle to free a jammed solar array in the crippled Skylab OWS. |
| 1973 Jul 28 | Service module RCS quad of Skylab 3 CSM sustains a leak, and a rescue flight is almost launched. The leak is bypassed, and the record-breaking 59-day mission continues. |
| 1974 Aug 28 | Soyuz 15 crew completes an early landing after the docking with Salyut 3 is abandoned. |

| | |
|---|---|
| 1975 Apr 5 | Soyuz 18-1 crew survive first in-flight launch abort when the third stage of their launch vehicle goes off course following the second stage's failure to separate correctly. The crew experience high g recovery and a rough landing down the side of a mountain. |
| 1975 Jul 24 | Three ASTP astronauts encounter poisonous nitrogen tetroxide emitting from an open vent as a thruster vents its fuel during descent. All three subsequently recover from the experience. |
| 1976 Aug 24 | Soyuz 21 cosmonauts hurriedly vacate Salyut 5, due to an 'acrid smell' in the environmental system. Reports of sensory depression affecting the crew are also rumoured. |
| 1976 Oct 16 | Soyuz 23 crew splash down in a lake during a snow storm after an emergency recovery following a failed docking attempt with Salyut 5. |
| 1977 Oct 11 | Soyuz 25 crew performs an early return due to the inability to dock with Salyut 6. |
| 1979 Apr 12 | After almost being stranded in orbit, the Soyuz 33 cosmonauts return safely to Earth after failing to dock with Salyut 6 due to a faulty engine system. This mission is often referred to as the 'Soviet Apollo 13 incident'. |
| 1979 | While undergoing basic training in an altitude chamber for Soyuz and Salyut flight, cosmonaut candidate Viktorenko is badly burned. His injuries force his removal from the team for some time, but he regains his health, completes his training, and makes three flights to Mir. |
| 1980 Sep 8 | Former Soviet Shuttle test pilot Oleg Kononenko is killed while conducting a flight test of the Yak-38A vertical take-off aircraft from the aircraft carrier *Minsk* in the South China Sea. |
| 1981 Nov 12 | After a launch abort at T–31 seconds on 11 November due to clogged fuel filters in the APUs, STS-2 launches successfully. However loss of a fuel cell results in flying a minimum mission of 54 hours. |
| 1983 Apr 20 | Soyuz T-8 aborted docking attempt with Salyut 7 forces an emergency return by the crew. |
| 1983 Aug 30 | Post-flight examination reveals the near burn-through of the recovered SRB casing from STS-8, and corrective measures are implemented for STS-9. |
| 1983 Sep 27 | Soyuz T-10 cosmonauts experience first launch pad abort. As the capsule escape rockets carry them clear of the pad, their launch vehicle explodes beneath them. |
| 1983 Dec 8 | STS-9 suffers onboard computer system failures and RCS nose thruster failures. The crew struggles with the controls for several hours before landing. Small fires are noticed on *Columbia* after landing. |
| 1984 Jun 26 | STS 41-D launch is aborted 4 seconds before lfit-off as computers detect contamination on the main engines as they ignite seconds before SRB ignition. The crew safely leave the vehicle as the water deluge system covers the LSS and Pad in the event of an explosion. |
| 1985 Jul 29 | STS 51-F – after a series of aborted launch attempts – experiences launch Abort To Orbit mode as one of three main engines shut down. The crew attains orbit, and the mission is successfully completed. |
| 1985 Sep 17–Nov 21 | Cosmonaut Vasyutin, onboard Salyut 7, in October 'develops infection' which resists treatment by antibiotics. His temperature rises to 104°F, and he returns home with his crew, earlier than planned, on 21 November. Some reports indicate that he did not adapt well to zero g. |

| | |
|---|---|
| 1986 Jan 28 | *Challenger* explodes 73 second after launch, killing the crew of seven. |
| 1987 Apr 5 | Docking of the unmanned Kvant 1 astrophysics module to the manned Mir core module almost results in a collision. Docking is eventually achieved safely. |
| 1987 Jul 30 | Cosmonaut Laviekin records irregular heartbeats during his extended stay on Mir. He returns to Earth early with a visiting crew, and is later restored to full fitness, but does not fly in space again. |
| 1988 Aug 18 | Soviet Shuttle test pilot and unflown cosmonaut Alexsandr Shchukin is killed in a crash of an Su-26M single-engine propeller-driven aircraft during off-duty activities. |
| 1988 Sep 5 | Soyuz TM-5 Afghan international crew suffer computer errors during their attempted deorbit after a mission to Mir. A delayed landing overnight prompts banner headlines in the world media: 'Stranded In Space'. The crew is successfully recovered at the next attempt. |
| 1989 May 5 | While flying a NASA T-38 to Washington DC for ceremonies honouring the crew of STS-30, astronaut Dave Walker (STS-30 commander) experiences a near miss with an Airbus airliner. |
| 1990 Feb 21 | The launch of STS-36 is delayed for several days due to several technical problems and Commander Creighton's contraction of influenza. The first time a mission has been delayed due to the illness of a crew member. It is too late to replace Creighton from the pool of astronauts without a long delay in the mission. |
| 1990 Mar 25 | During approach, the unmanned Progress M7 supply ship almost collided with Mir, missing the solar panels by just 5 yards, which could have resulted in serious damage to the station and injury to the crew. Progress docks successfully after the crew use their Soyuz to test the docking systems. |
| 1990 Jul 25 | As cosmonauts on Mir perform an unplanned EVA to repair the TM vehicle, cosmonaut Berezovoi stands by in a rescue Soyuz at the launch complex, in the event of their not being able to use their own vehicle for recovery; but they return in their own spacecraft. |
| 1990 Jul 7 | Astronaut Robert Gibson survives a mid-air collision at an air show in New Braunfels, Texas, but a pilot of a second aircraft is killed. Gibson is grounded for one year, and loses command of STS-46, although he later flies two further missions. Former Mercury astronaut Deke Slayton was also in the race. |
| 1990 Sep 9 | Soviet Shuttle civilian test pilot Stankyavichus is killed in the crash of an Su-27 during the Salveda air show near Treviso, Italy. He was in training for a first manned Buran flight, then planned for 1992. |
| 1991 Apr 5 | Astronaut Sonny Carter is killed in the crash of a commuter aircraft at Brunswick, Georgia, while on NASA business. |
| 1991 Jul 27 | Cosmonaut Artsebarski's EVA suit visor fogs during EVA, so that it is impossible for him to see. Colleague Krikalev rescues him and guides him back to the EVA hatch. |
| 1993 Mar 22 | STS-55 launch is aborted at T–3 seconds, due to an incomplete ignition of SSME No.3. |
| 1993 Jul 11 | Cosmonaut Vozovikov is drowned during a Soyuz survival training exercise. |
| 1993 Aug 12 | STS-51 launch is aborted at T–3 seconds, due to a faulty sensor monitoring fuel flow in the SSME No.2. |

| | |
|---|---|
| 1994 Aug 18 | STS-68 launch is aborted at T–1.9 seconds, due to the detection of a high discharge temperature in the SSME No.3 engine. |
| 1997 Feb 23 | A defective oxygen generation 'candle' explodes on Mir, causing a fire which fills the space station with smoke. |
| 1997 Apr 4 | The flight of STS-83, carrying the Microgravity Science Laboratory, is cut short from the planned 16 days to four days, due to a faulty fuel cell. Less than three months later the mission is reflown as STS-94. |
| 1997 Jun 25 | Progress M-34 collides with Mir's Spektr research module during the docking approach. |
| 1999 Jun 1–3 | STS-96 astronauts encounter a solvent-like odour inside the International Space Station, leading to headaches, nausea and, in one case, vomiting, during FD 6, 7 and 8. |
| 1999 Jul 23 | During the launch of STS-93 a voltage drop is recorded in one of the orbiter electrical buses, causing one of two redundant main engine controllers on two of the three SSME to shut down. In addition, a hydrogen leak in the No.3 engine nozzle results in a 7-mile shortfall in the orbit attained. The mission is completed successfully, but the resulting post-flight investigations last for nearly six months, grounding the Shuttle fleet. |

# Appendix 3

## ASTRONAUT AND COSMONAUT DEATHS

This list includes only those selected for NASA, USAF, Russian CIS (listed as USSR) and international programmes involving a flight of more than 50 miles. Those selected for stratospheric balloon or altitude parachute descent, rocket research plane (excluding X-15/X-20) and lifting body programmes are not included. Those who died of natural causes or off-duty accidents, or after leaving the programme, are also omitted. The age is included in brackets after the name.

| | | | |
|---|---|---|---|
| Adams, M.J. (37) | USA | 1967 Nov 15 | X-15-3 crash |
| Bassett, C.A. (34) | USA | 1966 Feb 28 | T-38 jet crash |
| Bondarenko, V.V. (25) | USSR | 1961 Mar 23 | Flash fire in isolation chamber |
| Carter, M.L. (43) | USA | 1991 Apr 5 | Commuter plane crash on NASA business |
| Chaffee, R.B. (31) | USA | 1967 Jan 27 | Apollo 204 pad fire |
| Dobrovolsky, G.T. (43) | USSR | 1971 Jun 30 | Soyuz 11 entry |
| Freeman, T.C. (34) | USA | 1964 Oct 31 | T-38 jet crash |
| Gagarin, Y.A. (34) | USSR | 1968 Mar 27 | MiG 15 jet crash |
| Grissom, V.I. (40) | USA | 1967 Jan 27 | Apollo 204 pad fire |
| Ivanov, L. (30) | USSR | 1980 Oct 24 | Killed in test flight of MiG 23 |
| Jarvis, G.B. (41) | USA | 1986 Jan 26 | *Challenger* explosion |
| Komarov, V.M. (40) | USSR | 1967 Apr 24 | Soyuz 1 crash |
| Kononenko O.G. (42) | USSR | 1980 Aug 9 | Killed during a flying accident during sea trials aboard carrier *Minsk* in the South China Sea |
| Lawrence, R.H. (32) | USA | 1967 Dec 8 | F-104 jet crash |
| McAuliffe, S.C. (37) | USA | 1986 Jan 28 | *Challenger* explosion |
| McNair, R.E. (35) | USA | 1986 Jan 28 | *Challenger* explosion |
| McKay, J.B. (52) | USA | 1975 Apr 27 | Complication from X-15 crash in 1962 after the leaving the programme |
| Onizuka, E. S. (39) | USA | 1986 Jan 28 | *Challenger* explosion |
| Patsayev, V. (38) | USSR | 1971 Jun 30 | Soyuz 11 re-entry |

| | | | |
|---|---|---|---|
| Resnik, J. A. (36) | USA | 1986 Jan 28 | *Challenger* explosion |
| Scobee, F.R. (46) | USA | 1986 Jan 28 | *Challenger* explosion |
| Shchukin, A. V. (42) | USSR | 1988 Aug 18 | Crash of SU-26M prop-driven aircraft when off duty |
| See, E.M. (38) | USA | 1966 Feb 28 | T-38 jet crash |
| Smith, M. J. (40) | USA | 1986 Jan 28 | *Challenger* explosion |
| Stankyavichus R.A.A. (46) | USSR | 1990 Sep 9 | Crash of Su-27 aircraft crash during an air show in Italy |
| Volkov, V.N. (35) | USSR | 1971 Jun 30 | Soyuz 11 re-entry |
| Vozovikov, S.Y. (35) | USSR | 1993 Jul 11 | Drowned in Black Sea during survival training exercise |
| White, E.H. (36) | USA | 1967 Jan 27 | Apollo 204 pad fire |
| Williams, C.C. (35) | USA | 1967 Jun 5 | T-38 jet crash |

# Bibliography

Works included in the list of further reading are primary references. The hundreds of other books, articles and reports researched over many years are too numerous to be listed here, and are summarised as follows.

Magazines: *Flight International*; *Aviation Week and Space Technology*; *Journal of the British Interplanetary Society* and the BIS magazine *Spaceflight*; *Time*; *Life*; *Newsweek*; *Soviet Weekly*; *Soviet News*; *Quest*; *National Geographic*; and the newspapers *Washington Post*, *Houston Post*, *Houston Chronicle* and *Florida Today*.

Reports: NASA Reference Series SP-4000, *Astronautics and Aeronautics* (1963–1990); *Soviet Space Program Reports*, US Library of Congress (1966–1987); *Astronauts and Cosmonauts Biographical and Statistical Data*, US Library of Congress (1975–1993); NASA *Chronology of KSC Activities and Related Events*, KSC History Office (1980–1997); and *The Soviet Year in Space*, Teledyne Brown Engineering (1982–1990).

In addition, flight documentation covering human stratospheric and space exploration has been consulted at NASA field centre History Archives and Public Affairs Offices at HQ Washington DC; Johnson Space Center and Rice University, Houston, Texas; Marshall Spaceflight Center, Huntsville, Alabama; Kennedy Space Center, Florida; Dryden Flight Research Center, Edwards Air Force Base, California; the major US contractors (North American/Rockwell International, McDonnell Douglas, Grumman, Boeing, Martin Marietta, Morton Thiokol, Bell and Hamilton Standard); the private collections and archives of a number of space researchers and historians; and Astro Info Service's own collection of historical material, built up over 30 years.

## FURTHER READING

1936 *Exploring the Stratosphere*, Gerald Heard (Nelson)
1953 *Sound Barrier*, Neville Duke and Edward Lanchberry (Cassell)
1956 *The Lonely Sky*, William Bridgeman and Jacqueline Hazard (Cassell)

1958   *The Fastest Man Alive*, Frank Everest, as told to John Guenther (Cassell)

1960   *Man High*, David G. Simon with Don A. Schanche (Doubleday)

1962   *We Seven: By the Astronauts Themselves*, Carpenter, Cooper, Glenn, Grissom, Schirra, Shepard and Slayton (Simon & Schuster)

1963   *Project Mercury: A Chronology*, James Grimwood (NASA SP-4001)

1966   *This New Ocean: A History of Project Mercury*, Lloyd Swenson, James Grimwood and Charles Alexander (NASA SP-4201)

1968   *On Course for the Stars: The Roger B. Chaffee Story*, C. Donald Chrysler and Donald L. Chaffee (Kregel Publications)

1969   *Living in Space: The Astronaut and his Environment*, Mitchell R. Sharpe (Aldus)

  *Project Gemini: A Chronology*, James Grimwood and Barton Hacker, with Peter Vorzimmer (NASA SP-4002)

  *The Apollo Spacecraft: A Chronology*, Vol. 1, Ivan Ertel and Mary Morse (NASA SP-4009)

1970   *First on the Moon*, Neil Armstrong, Michael Collins and Edwin Aldrin with Gene Farmer and Dora Jane Hamblin (Michael Joseph)

  *Apollo 13: 'Houston we've got a problem.'* (NASA EP-76)

1972   *Always Another Dawn: The Story of a Rocket Test Pilot*, Scott Crossfield with Clair Blair (Arno Press, originally published 1960)

  *Russians In Space*, Evegenny Riabchikov (UK English translation, Weindenfield & Nicholson)

1973   *Soviets in Space*, Peter Smolders (Revised English translation, Lutterworth Press)

  *13: The Flight That Failed*, Henry S.F. Cooper Jr. (Angus & Robertson)

  *The Apollo Spacecraft: A Chronology*, Vol. II, Mary Morse and Jean Bays (NASA SP-4009)

1974   *Carrying The Fire: An Astronaut's Journeys*, Michael Collins (Farrar, Straus & Giroux)

  *Petersen's Book of Man in Space:* Vol. 1, *The First Small Step;* Vol. 2, *A New Environment;* Vol. 3, *The Power and The Glory;* Vol. 4, *A Giant Leap For Mankind;* Vol. 5, *Beyond the Threshold*, Al Hall (*ed.*) (Petersen)

  *Starfall*, Betty Grissom and Henry Still (Crowell)

1975   *Apollo Expeditions to the Moon*, Edgar M. Cortright (*ed.*) (NASA SP-350)

1976   *The Apollo Spacecraft: A Chronology*, Vol. III, Courtney Brooks and Ivan Ertel (NASA SP-4009)

1977   *On the Shoulders of Titans: A History of Project Gemini*, Barton Hacker and James Grimwood (NASA SP-4203)

  *Skylab: A Chronology*, Roland Newkirk and Ivan Ertel with Courtney Brooks (NASA SP-4011)

  *Skylab: Our First Space Station*, Leland F. Belew (*ed.*) (NASA SP-400)

  *The All American Boys*, Walter Cunningham (Macmillan)

  *The Apollo Spacecraft: A Chronology*, Vol. IV, Ivan Ertel and Roland Newkirk with Courtney Brooks (NASA SP-4009)

1978   *The Partnership: A History of Apollo Soyuz Test Project*, Edward and Linda Ezell (NASA SP-4209)

1979   *Chariots for Apollo: A History of Manned Lunar Spacecraft*, Courtney G. Brooks, James M. Grimwood and Lloyd S. Swenson (NASA SP-4205)

  *The Right Stuff*, Tom Wolfe (Jonathan Cape)

1980   *Handbook of Soviet Manned Space Flight*, Vol. 48, AAS Science and Technology Series, Nicholas L. Johnson (American Astronautical Society Publications)

1981   *The History of Manned Spaceflight*, David Baker (New Cavendish Books)

*The Illustrated Encyclopedia of Space Technology: A Comprehensive History of Space Exploration*, Ken Gatland (principal author) (Salamander Books)

*Test Pilots: The Frontiersmen of Flight – An Illustrated History*, Richard P. Hallion (Doubleday)

*Red Star in Orbit*, James E. Oberg (Harrap)

1983 *Living and Working In Space: A History of Skylab*, W. David Compton and Charles D. Benson (NASA SP-4208)

1984 *On The Frontier; Flight Research at Dryden 1946-1981*, Richard Hallion (NASA SP-4303)

1985 *An Illustrated History of Space Shuttle: US Winged Spacecraft: X-15 to Orbiter*, Melvyn Smith (Haynes)

*Living Aloft: Human Requirements for Extended Spaceflight,* Mary Connors, Albert Harrison and Faren R. Akins (NASA SP-483)

*North American X-15/X-15A-2. Aerofax Datagraph 2*, Ben Guenther, Jay Miller and Terry Panopalis (Aerofax)

*Yeager: An Autobiography*, Chuck Yeager and Leo James (Century)

1986 *Challengers: The Inspiring Life Stories of the Seven Brave Astronauts of Shuttle Mission 51L*, By the Staff of the Washington Post (Pocket Books)

*Report of the Presidential Commission on the Space Shuttle* Challenger *Accident* (five volumes), 6 June 1986, Washington DC

*Heroes of the Challenger*, Daniel and Susan Cohen (Archway Paperbacks)

*Ellison S. Onizuka: A Remembrance* (Second Edition), Bennis Ogawa and Glen Grant (Signature/Mutual)

*'I Touch the Future ...' The Story of Christa McAuliffe*, Robert T. Hohler (Random House)

*Manned Spaceflight Log: New Edition*, Tim Furniss (Jane's)

1987 *Across The High Frontier*, William R. Lundgren (Bantam Paperbacks)

*Challenger: A Major Malfunction*, Malcolm McConnell (Simon & Schuster)

*Heroes In Space: From Gagarin to Challenger*, Peter Bond (Blackwell)

*NASA Space Shuttle: From The Flightdeck 2*, H.R. Siepmann and David J. Shayler (Ian Allen)

*Prescription For Disaster: From the Glory of Apollo to the Betrayal of the Shuttle*, Joseph J. Trento (Crown)

*Shuttle Challenger: Aviation Fact File*, David J. Shayler (Salamander Books)

1988 *Bell X-1 Variants, Aerofax Datagraph 3*, Ben Guenther and Jay Miller (Aerofax)

*Challenger: The Final Voyage*, Richard S. Lewis (Columbia)

*Countdown: An Autobiography*, Frank Borman with Robert J. Sterling (Silver Arrow Books)

*Liftoff: The Story of America's Adventure in Space*, Michael Collins (Grave Press)

*Schirra's Space*, Walter M. Schirra with Richard N. Billing (Quinlan Press)

*The Soviet Manned Space Programme: An Illustrated History of the Men, the Missions and the Spacecraft*, Phillip Clark (Salamander Books)

*The X-Planes: X-1 to X-31*, Jay Miller (Orion Books)

*Uncovering Soviet Disasters*, James E. Oberg (Random House)

1989 *Apollo: The Race to the Moon*, Charles Murray and Catherine Bly Cox (Simon and Schuster)

*Where No Man Has Gone Before: A History of Apollo Lunar Exploration Missions*, William David Compton (NASA SP-4214)

*Survival In Space*, Richard Harding (Routledge)

*Race to the Stratosphere*, David H. DeVorkin (Springer-Verlag)

1990   *Moonwalker*, Charlie and Dotty Duke (Oliver Nelson)
       *Judith Resnik: Challenger Astronaut*, Joanne E. Bernstein and Rose Blue with Alan Jay
          Gerger (Lodestar Books)
       *Press On! Further Adventures in the Good Life*, Chuck Yeager and Charles Leerhsen
          (Bantam Paperbacks)
       *The Soviet Cosmonaut Team:* Vol. 1, *Background Section, The Soviet Cosmonaut Team;*
          Vol. 2, *Cosmonaut Biographies*, Gordon R. Hooper (GRH Publications)
       *Almanac of Soviet Manned Space Flight*, Dennis Newkirk (Gulf Publishing Co.)
1991   *Ronald McNair: Astronaut. Black Astronauts of Achievement Series*, Corinne Naden
          (Chelsea House)
1992   *At The Edge of Space: The X-15 Fight Program*, Milton Thompson (Smithsonian
          Institute Press)
       *Tex Johnston: Jet Age Test Pilot: From the First Jet to the Saturn Booster*, Tex Johnston
          with Charles Barton (Bantam Air & Space Paperback 22)
1993   *A Journal for Christa: Christa McAuliffe, Teacher in Space*, Grace George Corrigan
          (University of Nebraska Press)
       *Suddenly Tomorrow Came ...: A History of the Johnson Space Center*, Henry C.
          Dethloff (NASA SP-4307)
1994   *A Man on the Moon: The Voyages of the Apollo Astronauts*, Andrew Chaikin (Michael
          Joseph)
       *Deke: US Manned Space: From Mercury to the Shuttle*, Donald K. Slayton with
          Michael Cussutt (Forge)
       *Into the Unknown: The X-1 Story*, Louis Rotundo (Airlife)
       *Lost Moon: The Perilous Voyage of Apollo 13*, Jim Lovell and Jeffery Kluger (Houghton
          Miffin)
       *Moonshot: The Inside Story of America's Race to the Moon*, Alan Shepard and Deke
          Slayton with Jay Barbree and Howard Benedict (Virgin)
       *They Had a Dream: The Story of African-American Astronauts*, J. Alfred Phelps
          (Presido)
1995   *Mir Hardware Heritage*, David S.F. Portree (NASA RP-1357)
       *The Pre Astronauts*, Craig Ryan (Naval Institute Press)
1996   *Contest for the Heavens: The Road to the Challenger Disaster*, Claus Jensen (English
          translation, Harvill)
       *Silver Lining: Triumph of the Challenger 7*, June Scobee Rogers (Peake Road)
       *The Challenger Launch Decision: Risk Technology, Culture and Deviance at NASA*,
          Diane Vaughan (Chicago)
       *Space Shuttle*, Dennis R. Jenkins (Motor Books International)
       *The New Russian Space Programme*, Brian Harvey (Wiley–Praxis)
1997   *Korolev. How One Man Masterminded the Soviet Drive to Beat America to the Moon*,
          James Harford (Wiley)
       *The Quest for Mach One: A First Person Account of Breaking the Sound Barrier*, Chuck
          Yeager, Bob Cardenas, Bob Hoover, Jack Russell and James Young (Penguin Studio
          Books)
       *Walking to Olympus: An EVA Chronology. Monographs In Aerospace History Series #7*,
          Davis S.F. Portree and Robert C. Trevino (NASA)
       *Wingless Flight: The Lifting Body Story*, R. Dale Reed with Darleen Lister (NASA SP-
          4220)
       *The Mir Space Station*, David M. Harland (Wiley–Praxis)
       *Living In Space*, G. Harry Stine (M. Evans & Co.)

1998  *Dragonfly: NASA and the Crisis Aboard Mir*, Bryan Burrough (Harper Collins)
      *Genesis: The Story of Apollo 8 – The First Manned Flight to Another World*, Robert
         Zimmerman (Four Wall Eight Windows)
      *Who's Who In Space: International Space Edition*, Michael Cussutt (Macmillan)
      *The Space Shuttle*, David Harland (Wiley–Praxis)
      *Starman*, James Doran and Piers Bizony (Bloomsbury Publishing)
1999  *Flying Without Wings: Before the Space Shuttle – Test NASA's Wingless Aircraft*,
         Milton Thompson and Curtis Peebles (Crecy Publishing)
      *The Last Man on the Moon: Gene Cernan and America's Race in Space*, Eugene Cernan
         with Don Davis (St Martin's Press)
      *The Saga of Bell X-2: First of the Spaceships – The Untold Story. X-Planes, Book 4*,
         Henry Matthews (HPM Publications)
      *The Shuttle Decision: NASA's Search for a Reusable Space Vehicle*, T.A. Heppenheimer
         (NASA SP-4221)
      *Waystation to the Stars: The Story of Mir, Michael and Me*, Colin Foale (Headline
         Books)
      *NASA Historical Data Book, Volume 5: NASA Launch Systems; Space Transportation;
         Human Spaceflight and Space Science, 1979–1988*, Judy A. Rumerman (NASA SP-
         4012)
      *Towards Mach 2: The Douglas D558 Program*, J. Hunley (*ed.*) (NASA SP-4222)

# Index